EINSTEIN ON TRIAL

OR

METAPHYSICAL PRINCIPLES

OF NATURAL PHILOSOPHY

by

Jorge Céspedes-Curé

Editor: Shirley Ramsey

Publisher: et al Organization.

2002

ISBN: 0-9713873-0-3

et al. Organization
P. O. Box 90373,
Caracas 1083 A
Venezuela

First Edition 2002
Second Edition, 2005
Third Edition, 2009
Printed in the UK by Antony Rowe Ltd.
This book is available through **Amazon.com** in the USA
and through **Amazon.co.uk** in the United Kingdom

To Shirley Ramsey and to the future generations.

CONTENTS

Index of Tables **Page**

Index of Figures

PREFACE

I had at least five reasons to write this book. The first was to defend Newton's classical mechanics from false accusations raised by Einstein. The second reason was to deliver Natural Philosophy (physics), from the mathematical bondage of which it was submitted during the 20th century. The third reason was to liberate the minds of scientists in general, and physicists, in particular, from the corrupted 20th century slogan "publish or perish." The fourth reason was to help induce an implosive neo-renaissance by bringing all sciences back to the womb of mother philosophy. The fifth and last reason to write the book, *Einstein on Trial*, was to synthesize or fuse science, philosophy and theology of the 21st century in one solid epistemological doctrine or school of thought.

First Reason.- The first reason shows the following false accusations of Einstein against Newton's *Principia*, or classical mechanics:

1. Classical mechanics is powerless to explain the anomalous motion of planet Mercury.
2. Classical mechanics is incapable of explaining the equality between inertial mass and gravitational mass.
3. Classical mechanics is incompetent to deduce formally (logically) the mathematical structure of forces.
4. Absolute space of classical mechanics does not exist.
5. Absolute time of classical mechanics does not exist.
6. The cosmic ether (*Sensorium Dei* of Newton) does not exit.
7. In classical mechanics absolute rotation does not exist.

With the help of forerunners, I was able to disprove all the previous false accusations against Newton's classical mechanics.

Second Reason.- The second reason shows an invasion into the temple of Natural Philosophy by a herd of modern barbarians who were expelled from the temple of Mathematics. These 20th century "attilas," riding horses with symbolic hooves, annihilated the growth of all ontological grass in the meadows of Natural Philosophy.

Obviously, after Einstein created his GRT, he wrote, "*Our experience hitherto justifies us believing that nature is the realization of the simplest conceivable mathematical ideas*." At least Einstein is only proposing to believe in his statement, like any dogma of faith has to be believed. Mathematics, the queen of the formal sciences, should be treated as an almost sacred dictionary in the hands of natural philosophers, because it allows us to translate our philosophical speculations about nature into a precise language. For this reason, Mathematics is a wonderful medicine to cure mental verbiage.

Third Reason.- Many years ago, when I was teaching physics to engineering students, I gathered many definitions of the concept "engineering." I put aside one of the definitions, and I determined the "average" definition of the rest. The result was this: *Engineering is the application of science for the benefit of human kind.* What about the definition I left aside? In my opinion it is very accurate: *Engineering is the art of making two dollars with one dollar*. In the last century, we introduced the word *technology* to replace the word *engineering*. Thus, technology is still the art of making a profit. By the way, "art" in Greek is "techné." After the second world war, technology corrupted the minds of the lovers of wisdom. The slogan "publish or perish" was a translation of the previous slogan "don't think, do it." Today, at the entrance of the new millennium, we are buried under a mountain of irrelevant papers, though everyone calls them scientific papers. We still do not know that diseases are caused by electrodynamic imbalances at a cellular level. Today, we have to wait minutes to activate terrestrial rovers on planet Mars, and hours to activate instruments on Pioneer 10. Today, we are empirically discovering that "light can travel faster than light." You do not need to be an expert in the ontological and the logical principles of non-contradiction to conclude that the previous proposition about light is an extremely idiotic statement. Confucius said that knowledge is good, but better is its application. If Confucius referred to factual knowledge, then we better listen to the German poet Goethe saying - If Nature does not want to reveal her secrets to your spirit, you will never unveil them with hammers and wedges. - In the present (August, 2000), physicists do not know there are longitudinal electric waves, which travel millions of times faster than the speed of light. However, they are discovering electric signals which they believe are transversal waves. Theoreticians from Cambridge University, in 1837, knew about the probable existence of these new longitudinal signals, which should be the minimum requirement to participate in truly intelligent cosmic dialogues. We must teach the young professors to write, at the end of their careers, treatises on their subjects, and not a farrago of technological papers, which stagnate true theoretical science. The list of unfulfilled applications of science is interminable. We are still using Chinese technology, which is more than three thousands years old, to travel into space, making electromagnetic smoke signals between mother earth and space-capsules (I do not consider them spacecrafts.) After seven thousand years of recorded history, we should propose a toast to our magnificent engineers for having accomplished such extraordinary technological feats with the primitive physics still used on this planet.

Fourth Reason.- The first renaissance in science was explosive. Every natural science broke loose from the womb of mother philosophy. When *Experimental Philosophy* was born, from the mind of Galileo, it was almost immediately expelled from the Mediterranean shores. *Galileo on Trial*, in 1633, by the fanatic Inquisition of those days, was a signal for the young *experimental philosophy* to emigrate from the temperate waters of the Mediterranean Sea to the cold climate of northern Europe. Nordics, if they are idle in the summer, are not prepared for the cold winter months ahead and often die. For this reason they adopted the experimental aspect of *experimental philosophy*. Thus, experimental knowledge (*scientia*) was developed by the Nordics to survive the implacable winters. Philosophy remained exclusively in the Mediterranean hands of the powerful theologians. Galileo was forced to retract the heliocentric system as a philosophical truth. However, Galileo was permitted to maintain the heliocentric concept as a mathematical hypothesis. Hence, to avoid torture and even the penalty of death, experimental or natural philosophy suffered a massive explosive Inquisitorial epistemological "big bang" during the first renaissance.

In the 20th century, the new scientific priests have recreated a more subtle or Scientific Inquisition. This modern Inquisition burns the new Giordano Brunos in the *flames of silence*. The works of these modern rebels are seldom, if ever, published in the journals of the new Scientific Inquisition. Bewildering as it may seem, the most democratic country in the world, where we have a first amendment of freedom of speech, there is no freedom to publish new concepts or theories, which questions the *sacred scientific scripts* of the so-called prophets of the 20th century. But the days of these new inquisitors are counted because of the global electronic consciousness of our new century. It must be noted that no man or institution has the right to silence the thoughts of any single person. The new implosive renaissance is already on its way. All sciences are beginning to return to mother philosophy in order to master the philosophical scaffolding of the new advanced scientific theories.

Fifth Reason.- There is nothing worse than an arrogant-ignorant university professor. Ortega-y-Gasset refers to them as *ignorant sages*. These new scientific barbarians are experts with crumbs of knowledge, but as any expert, they ken (are acquainted with) their restricted gnoseological fields, but truly understand almost nothing. The essence of this last statement came from Einstein when he referred to one physicist, in particular. After this new implosive renaissance, which will be well developed in the next few decades, the next generation will fuse, condense and synthesize the fundamental knowledge of the formal disciplines, with the philosophical, theological and scientific intellectual disciplines. They will fuse this knowledge into one epistemological quantum. As the author of *Einstein on Trial*, I have been honored to develop Einstein's vision of a theology not based on a fearful God, but a theology based on *Un Amor Intelectual por Dios*.

I will let the readers decide for themselves if I was able to reduce all physics to one dynamical principle, to Newton's second principle of motion. It is my impression that my efforts in writing this book were focused on proving that everyone was wrong, because everyone was only partially right. My last thoughts are for those two great men, Newton and Einstein, who were lifted up to the shoulders of giants so that they might look beyond where no man has ever looked before.

Jorge Céspedes-Curé
Maggie Valley, North Carolina, U.S.A.
August 15, 2000

ACKNOWLEDGMENTS

The writing of this book, which began to formulate in my mind 25 years ago, has been a difficult task for me, mainly because I had no close colleague with which to discuss my "strange" concepts about the being of nature. Seven years ago, with the assistance and encouragement of a few people, I seriously decide to put my thoughts on paper. Without these few people, I never would have succeeded in bringing my thoughts to fruition.

From Chile, S.A. my first thoughts go to my parents, who have long since departed, but who gave me a Proverbial discipline, which has lasted a lifetime. Professor Heinrich Hauser, from the Naval Academy of Chile, who taught me geometry when I was 11 years old. My numerous professors from Universidad Católica de Valparaíso, Chile: Professor Luis Cortazar, who taught me advanced mathematics, and instructed me to use it as a dictionary to translate the philosophical adventures of my mind. Dr. Luigi Farezzi, who taught me to study the History and Philosophy of Science, because there I would find the many sources of creativity. Professor Tomás Muzzio who introduced me to Bohr's initial quantum theory, and Professor Vadim Praus, who taught me Analytical Geometry, and advised me not to draw geometrical figures if I could translate them into algebraic expressions.

From the University of Oklahoma, I look back with cherished reflections on Dr. Stanley Babb. He taught me that my hands are the means to empirically give birth to my mental children. From the same university my gratitude goes to Dr. Jack Cohn, who taught me that the existence and uniqueness of solutions of nonlinear integral equations are a serious concern for mathematicians, but not for natural philosophers. My gratitude goes to the late Professor Dr. Pavel Parshin, from St. Petersburg, Russia, founder of the Galilean Electrodynamics East (GED-East), who after 1998 encouraged me to finish writing this book. He was most interested in the ontological foundations of the Newtonian Quantum Mechanics that is presented in this book.

There have been many friends who have encouraged and helped me along the way. To name a few, Jeanne P. Rindge, of Waynesville, NC, President of *The Human Dimensions Institute,* who has influenced me immensely in transrational science, and has always had faith in me when the path looked bleak. When it came to chapter 7, she was an extremely helpful critic. I will forever be grateful to her. My dear departed friend, Harry Eichler from Florida, who was President of *Neo-Dynamics Research Corporation*, was another who always imparted optimism in my work and inspired me to continue laboring on more than one of my "strange" concepts about nature. Another close friend I want to acknowledge is Ray Kaloustian. He was always there for me whenever I needed someone to verbalize my thoughts. Not only that, he forever provided references from his electronic book store www.flinet.com/~rasci, which were hard to find, and were of the utmost value to me. Another friend I want to thank is Charles B. Padgette for his ability to convert drawings into computer graphs to use them in this book. I also thank Dennis Davis, another friend, who made the schematic drawing of the Millikan Apparatus.

There are those who physically helped me with this book by checking my material for accuracy of content and for my many errors in English structure. I can never completely thank my faithful friend and colleague, Dr. Eduardo D. Greaves, from Caracas, Venezuela. Without him, I would still be struggling with the references and other parts of the book, such as the Introduction. He is the only scientist who read the complete manuscript of "Einstein on Trial." He made valuable suggestions and always encouraged me with one descriptive word to qualify each chapter. Finally, Dr. Greaves decided to do the typesetting of this book. He always reminded me of Edmund Halley, the young friend of Isaac Newton, who was the only one who truly understood the Third Book of the *Principia*.

My intellectual gratitude goes to Dr. Thomas Streit who translated the papers of Lense and Thirring from German into Spanish. He did this while he was an undergraduate student in physics, Caracas, Venezuela, in 1978. My deepest appreciation goes to the late Prof. Sergio Luci, from Santiago, Chile, for the long enlightening discussions on epistemology in Valencia, Venezuela. From the same city, I thank Prof. Demetrio Bidirini, professor of chemistry, for his encouragement, since 1976, to publish my new concept of the neutron as a "Lilliput" hydrogen atom. My gratitude goes also to my Florida friend, Janet O'Brien, who kindly checked for typographical errors. As she did not know physics, she either jokingly or gratefully told me this book helped to put her to sleep at night.

I want to thank all my children for the time they gave me to attack the many windmills of ignorance that grow in this planet. I want to express my very sincere spiritual, as well as material gratitude to my devoted wife, Shirley Ramsey-Curé. Without her this book would never have seen the light of day. She never lost faith in me and helped pave the way for me to finish my life's creation. She, herself, worked endlessly at the computer correcting my sentence structure. Not only did she work at the computer, but each morning, at breakfast time, she patiently listened to a lecture on physics, philosophy or theology. Not being a physicist, a philosopher or a theologian this was not an easy exercise for her, but for me, it helped to organize my thoughts for another day of writing.

I also would like to thank different authors for short quotations, which I used from their books for critical analyses. The first is Dr. André K.T. Assis, from Brasil, for his many significant and transcendental papers and books on electrodynamics and gravitation. Then follows Dr. J.P. Wesley, from Germany, for his prolific production of landmark papers and books on electrodynamics, and Weber's electromagnetic field equations, which were created by Wesley himself. I should also mention his unusual book on Quantum Theory, which taught me that I was not alone in these neo-renaissance endeavors. My intellectual gratitude goes also to Prof. Ludwik Kostro, from Poland, for his magnificent research of unknown epistolary correspondence of Einstein with different persons and scientists on fundamental topics in electromagnetic wave propagation before the publication of SRT, in 1905. Kostro's research of 1988, and Einstein's lecture delivered in Leyden, in 1920, and published by Dover Publications, Inc., in 1983, under the title *Sidelight of Relativity*, were two solid rocks with which to build an *energetic-ether* theory presented in chapter 5. Then comes my Aquarian friend, the late Dr. Stefan Marinov, to whom I owe an overwhelming inspiration to attack windmills of ignorance. I also would like to express my gratitude to Stefan for his prolific writings in his thorny, and painful *Tao* of truth. My deep gratitude goes to the late Dr. Bruce DePalma for his constant encouragement about my energetic conception of the cosmic ether, which he called *Primordial Energy Field.* To Dr. Thomas Phipps, from Illinois, for his book *Heretical Veritas*, and the many emails we exchanged on Marinov's electrokinetic force, $q\,(\mathbf{v}\cdot\nabla)\mathbf{A}$. I also would like to thank Prof. F.J. Müller, from Miami, Florida, for the innumerable conversations about the existence (mental or real) of magnetic lines and unipolar induction. Müller is an old fashion physicist who still makes important and interesting new discoveries "a la Faraday," instead of using Newtonian-German's electrodynamics without mentioning the magnetic induction **B**. My gratitude also goes to Dr. Franco Selleri for his enlightening book *Quantum Paradoxes and Physical Realities*, and for the exceptional conversation I had with him about the EPR paradox. This took place last June, 2000, at the University of Connecticut, during the International Meeting of the Natural Philosophy Alliance. Finally, I want to thank Millennium Twain, from New Zealand, for showing us that still we have "intellectual knights" ready to fight any Lord of the new Scientific Inquisition. I have personally met all of the above colleagues mentioned above.

Among the authors I want to thank but was not fortunate enough to meet them personally are the following: Dr. Max Jammer, whose masterly books I have studied and thoroughly enjoyed. Among his books, special mention should be given to *The Conceptual Development of Quantum Mechanics* (1966), and *The Philosophy of Quantum Mechanics* (1974). These two books were absolutely essential in helping me to write chapter 2. Unfortunately, his recent book *Einstein and Religion* (1999), came to my attention only this past August, 2000. However, this last book of Jammer served as a confirmation of my thoughts about Einstein's theological beliefs. An enormous feeling of gratitude goes to the late Professor Richard S. Westfall for his extraordinary book, *Never at Rest*, a biography of Isaac Newton. Furthermore, there was also his other book, *The Construction of Modern Science*, both of which showed me unknown aspects of Newton's life and work. Then there is Mario Speiser with his excellent Master's Thesis, *Mach's Principle,* where I found Speiser's collection of 47 different definitions of the so-called Mach's Principle.

My eternal gratitude goes to all of my silent forerunners, mentioned in the reference section of each chapter of *Einstein on Trial*. Though it is true, as I said in the beginning of my acknowledgments, I never discussed the content of this book with any physicist for a quarter of a century. It is also true, that with so many wonderful books, I was constantly in the company of the best minds of this planet from Thoth to Einstein.

I want to express my thanks to the following publishers for granting me permission to quote from their publications. Initially, I want to thank the University of California Press, Berkeley, CA 94720 (Sir Isaac Newton, *Newton's Principia or Mathematical Principles of Natural Philosophy,* translated into English by Andrew Motte in 1729, translation revised by Florian Cajori, 1934); Dr. Cynthia K. Whitney, Editor and Publisher of Galilean Electrodynamics, **2**, no. 3, pp. 43 - 47 (1991) (J.C. Curé, *The Perihelic Rotation of Mercury by Newton's Original Method*); Open Court Publishing Company, a division of Carus Publishing Company, Peru, IL 61354 (Ernst Mach, *The Science of Mechanics: A Critical and Historical Account of its Development,* translated by Thomas J. McCormick, sixth edition, 1960); Dover Publishing Company, Mineola, NY 11501 (James Clerk Maxwell, *Electricity and Magnetism, 1954 republication*); Dover Publishing Company, Mineola, NY 11501 (James Clerk Maxwell, *Matter and Motion*, reprinted by Sir Joseph Larmor, Lucasian Professor of Mathematics); Dover Publishing Company, Mineola, NY 11501 (Albert Einstein, *Sidelights and Relativity,* republication of the translation by G. B. Jeffery, D. Sc. and W. Perrett, Ph.D, 1983); Crown Publishing, Division of Random House, New York, NY 10171 (Albert Einstein, *Ideas and Opinions,* based on "Mein Weltbild," edited by Carl Seelig, new translation and revision by Sonja Bargmann, 1954); Princeton University Press, Princeton, N. J. 08540, copyright 1956 (Albert Einstein, *The Meaning of Relativity*).

Finally, I want to mention Dr. William A. Whitaker, Lt. Colonel USAF, Military Assistant to the Director of The Defense Advanced Research Projects (ARPA) of The Department of Defense of the United States of America. I thank him for his encouragement and the reviewer's recommendations about the new Newtonian Relativistic Electrodynamics and Gravitodynamics, which after 25 years, I have now published in this book. I waited a long time to publish all these new conceptions about Nature, but I had two reasons to delay my publication. The first one being that I wanted to gather as much pragmatic evidence as possible about the many predictions of my theory. In the beginning, I considered my theory as perhaps pure speculation infested with metaphysical verbosity. The second reason was more important. It was a combination of philosophical, political, and ethical reasons. As I used the Newtonian *materialistic philosophical* conceptions of the universe to develop my theory, I thought that a new technology, based on my theory, could be developed before and outside democratic countries. However, the international political arena has changed in the past decade. I now feel it is time to publish my conceptions for the entire world. Below is a transcript of a letter I received from Dr. Whitaker in 1975.

DEFENSE ADVANCED RESEARCH PROJECTS AGENCY
1400 WILSON BOULEVARD
ARLINGTON, VIRGINIA, 22209

October 17, 1975

Professor Jorge C. Curé
7207 Dartmouth Avenue
College Park, Maryland 20740

Dear Professor Curé:

As you requested, I had your report, "A New Electro-dynamic Law and Its Applications," reviewed by several scientists of the Department of Defense in order to determine whether the best interests of the US Government would be served by its open publication.

Our normal policy is to encourage the widest possible publication and open scientific discussion of new basic theoretical research, provided it does not have short-term weapons applications of an extraordinary impact. Thus, DoD-sponsored basic research in masers, lasers, superconductivity, atomic and molecular physics, explosives, etc., are given the widest possible distribution. On the other hand, we would have been remiss not to have guarded very carefully the initial work on nuclear fission. In general, however, the wide publication allows the greatest possible exploitation which is the strength of our free society. Our review has recommended such a procedure for your new theory.

The reviewers found the approach very interesting, although the briefness of the report allowed only a glimpse of the development. Their main area of concern was the predictions and applications. The unusual effect here discussed seems to be formulated on the basis of reasoning in analogy to the Mach-Thirring effect in general relativity, which itself has not been subject to experimental verification. The take-off from such a controversial background, combined with the apparent violation of conservation of momentum and its implied violation conservation of energy and the laws of thermodynamics, will require the widest possible discussion in the physics community to explore these matters. On the other hand, the proposed experimental test is relatively simple and should provide a straight-forward answer to this problem; and if positive, would therefore have implications towards the general relativistic question. The applications, if true, would be important in the range of very low thrust space flight and did not appear to have such unique and immediate military impact that we would be justified in monopolizing it.

For your information, the reviewers also suggested that one of the main benefits of the theory might be pedagogical and that considerably more development of the rationale would be useful in publication. Further, they would have also appreciated in appendix the detailed calculations of observables and comparison with normal theory, for instance, the perihelion motion of mercury which you said checked with general relativity. The illustrative effects of such calculations were thought to be very powerful.

Another matter which we discussed was the experimental test and the possibility of executing it at some DoD laboratory. A review of such laboratories reveals that the specific equipment of such a test, for instance large current superconductive rings, is not conveniently available in our laboratories, although available in university labs. It would, therefore, be much more expensive for a DoD lab to initiate such a test. The simplicity of the test, to those who have the right equipment, should provide sufficient incentive once open discussion of the theory is available.

Again I thank you for your consideration of the national security and look forward to seeing further development and publication of your theory.

Sincerely,

William A. Whitaker
Lt Colonel USAF
Military Asst. to the Director

INTRODUCTION

It is a great fortune for modern and future science that this book comes into print. In it is contained the scientific foundations for a new understanding of the fundamental forces of nature, which govern phenomena from the nuclear to the cosmic scale. More importantly than a trial to Einstein, Jorge C. Curé brings to modern physics orthodoxy an extension to Newton's principles of his dynamic theory. He sets down the basis for a classical Newtonian relativistic electrodynamics, for a Newtonian quantum mechanics and for a Newtonian relativistic gravitodynamics. He sets a sound philosophical basis for the principles of natural phenomena and then translates the phenomenology and philosophical speculations into the rigorous language of mathematics. The final consequences of these epistemological, ontological and mathematical explorations are overwhelming. They imply a revision of modern cosmology and suggest a glimpse of the path to that most aspired dream of physicists; a unified theory of natural forces.

In chapter 1, *Ontological Principles*, he examines the philosophical knowledge Einstein had about the ontological and epistemological foundations of physics. He finds, in this respect, that Einstein was one of the few creators of 20th century physics who knew precisely what he was doing in the philosophical foundations of physics. In this chapter, Einstein is found not guilty.

At the beginning of the 20th century, it was clear that Maxwell's wave electromagnetic theory had predicted an effect, which was never verified by Michelson-Morley's experiment. Also, the experiments of Trouton and Noble never verified the prediction of Lorentz's electrodynamic theory. At these crossroads in the history of physics, Einstein opened a new path with his 1905 Special Relativity Theory (SRT). With it he killed the concept of the "Luminiferous Ether," chose to uphold Maxwell-Lorentz's electrodynamics and modified Newton's dynamics. This is the road which physical science has followed for a whole century.

In chapter 2, *Newtonian Quantum Mechanics*, the reader is to find some surprises. Curé deduces a generalized Hamilton-Jacobi equation (HJ) from Newton's axiom of motion in a few mathematical steps. This generalized HJ equation contains an extra term, which Curé calls the "quantum collective potential" (QCP). Curé goes on to show the ontological origin of the QCP based on a philosophical consideration when applying Newton's axiom of motion. He shows this QCP to be identical to the so-called mathematical Bohm "quantum hidden potential." Curé calls the generalized HJ equation, Hamilton-Jacobi-Bohm's equation (HJB). He then shows that

Schrödinger's equation is a *particular* case of HJB's equation. In this chapter, a wave equation is deduced for the energy-momentum potential S, or generating function in mathematical physics. As a corollary, Curé deduces Planck-Einstein's and de Broglie's hypotheses. Einstein is completely redeemed from unfair reactions about his concept that "quantum mechanics is an incomplete theory." In this chapter, Einstein is found not guilty, at all, in quantum mechanics.

In chapter 3, *Compendium of Electrokinetics and Electrodynamics*, Curé takes the reader to the crossroads of the beginning of the 20th century in order to question that which Einstein upheld: Maxwell-Lorentz's electrodynamics. A vast collection of Electrokinetics (EK) and Electrodynamics (ED), which were scattered in the history of physics of the 19th and 20th centuries, are examined. They are classified and also translated to a modern vector notation. Then a new Newtonian Relativistic Electrodynamics is advanced using a modern version of the old concepts of Newton when he talks about absolute space and absolute time. It is emphasized that Einstein did not change anything in Maxwell's field equations, nor in Lorentz's electrodynamics. Thus, if Lorentz's ED is incomplete, then SRT is incomplete, and consequently both are wrong. However in this chapter, no judgment is expressed against or in favor of Einstein.

In chapter 4, *Newtonian Relativistic Electrodynamics*, Newton's three principles are extended from three to five. Curé accepts Einstein's challenge to deduce "formally and logically" a Newtonian Gravitodynamics. In the elaboration of the challenge he ends by deducing theoretically a new Newtonian Relativistic Gravitodynamics and a new Newtonian Relativistic Electrodynamics identical to the one established by inspection from empirical and semi-theoretical EK and ED at the end of chapter 3.

Curé presents several other outstanding contributions in this chapter: Experimental proof, which shows there is a new ED field proportional to the square of the electric current. The revival of Eddington's model of the neutron as a miniature hydrogen atom. This model is a consequence of applying the new Newtonian Relativistic Electrodynamics to the hydrogen atom. This application provides an example, which the new Electrodynamics may provide a rational and rigorous explanation to the wealth and complexity of phenomenology in the nuclear and atomic realm. It also provides an indication of the huge magnitude of the outstanding work to be done.

Other topics in chapter 4 are the following: Three hybrid modern EDs in which the variation of mass is maintained as in SRT, the "deduction" of Hertz's "hypothesis" and an interesting mathematical analysis of the commonly used "convective operator," which results in a logical and classical relativity law of velocities. Curé comes to the conclusion that the price we have paid for making space and time relative has been high. Mass variation has caused total stagnation in electrodynamics for almost a century. Curé demonstrates that apparent mass variation is a consequence of electrodynamic interaction of fast moving charges. Einstein, in this chapter is found guilty, and the reader becomes aware that at the crossroads of the beginning of the 21st century an alternative path is to uphold Newton's dynamics and modify Maxwell-Lorentz's electrodynamics.

In chapter 5, *On the Origin and Identity of the Cosmic Ether,* Curé re-examines the ether concept discovering how Einstein himself resuscitated the concept in 1920. However, the physics community never identified or recognized the very nature of the ether, which now is re-appearing with the new name "zero point energy." In this chapter, he postulates the Primordial Energy Field theory, and Einstein, in a final analysis, is found not completely guilty, but a hermeneutic victim of his time. With respect to the luminiferous ether in relation to his General Relativity Theory, Einstein is found not guilty.

In chapter 6, *Newtonian Relativistic Gravitodynamics*, Curé has a further store of surprises. He demonstrates that Newton provided, in his *Principia* in 1687, an original explanation about the perihelic rotation of planet Mercury. This is presented as solid proof of the falsity of Einstein's accusation against Newton's dynamics. In this section, Einstein is definitively found guilty. However, in the next section, Curé defends General Relativity Theory (GRT), in the event that the sun is oblate. Using Lense-Thirring's solution of Einstein's field equations of 1918, he calculates the intrinsic angular momentum of the sun. For this he uses the astronomical measurements of the excess perihelic rotation of planet Mercury. In a further application, Curé uses the same solution of Lense-Thirring to analyze the case of the jovian satellites of planet Jupiter, in order to determine the intrinsic angular momentum of the giant planet.

In the following section, Curé makes a thorough analysis of the concepts of inertial mass and gravitational mass. He demonstrates the serious mistake made in physics for having assumed the existence of two "essentially" different entities, when in reality they were identical. Thus, Einstein's Principle of Equivalence is found to be the result of an unhappy act of philosophical, etymological and historical ignorance of the physicists of the 19th century. Here Einstein is found guilty for relying too strongly on the undeserved authority of Mach.

Curé explains how Einstein assigned to his field equations of his GRT the name "Mach's Principle." A "principle" that has brought much discussion into physics. In this section, he identifies 47 statements of Mach's Principle put forth by the physics community. Only a few coincide with Einstein's original assertion. The rest of the statements are free interpretations of Mach's writings. However, the concept has implications on the gravitational action of the entire universe; action on each one of the material elements, represented by Einstein's energy-momentum tensor. Curé calls this the Collective Cosmic Universal Gravitational Potential, in total analogy with the Quantum Collective Potential. This tensor has far more implications. These implications Curé explores in chapter 7.

In another section, Curé uses the measured starlight deflection by the energy field of the sun to determine the stellar density of energy. He does this by using the Primordial Energy Field theory, which he developed in chapter 2. Curé then speculates about the concept of Ritz where gravitational forces may be statistical residues of electrodynamic fields caused by electric dipoles. At the end of this chapter, Curé speculates on a nonlinear electrodynamic theory based on GRT and on an alternative explanation of the starlight redshift. This work has profound implications on the Big Bang theory and modern cosmology. In this chapter, Einstein is found guilty of false and unmerciful accusations, which he made against Newton. He was critical of Newton in relation to the anomalous motion of planet Mercury, and the incapacity of Newton's dynamics to explain the identity of inertial mass with gravitational mass.

In chapter 7, the last chapter, *Einstein's Theological Beliefs and Scientific Theology or Cosmotheism* begins with the analyses of four essays written by Einstein, between 1930 and 1948, about "Science and Religion." He illustrates how at times, people thought that Einstein was an atheist but, he states, this was never true. Einstein was a pantheist, i.e., a person who believes that God is everything in the universe. Curé explains pantheism as one of the rational theologies: For a pantheist the universe is a part *of* God, while for other theological doctrines (Christianism, for example) the universe is apart *from* God. This last position establishes that God's creation (the entire universe) is separated from God. Pantheism, on the other hand, establishes that the universe is God. For a pantheist the question - what created the universe? - is an absurd question, because it is equivalent to this other question - what created God? Curé points out that a true pantheist cannot ask the question - *who* created the universe? - Because for any pantheist, in particular for Einstein, God is not "personal." Einstein believed the concept of a "personal God" (that God is a universal Person) is the strongest point of bitter disagreement between Science and Religion. After commenting on Einstein's essays on Theology, Curé finds Einstein guilty of lacking knowledge about the concepts of "consciousness" and "person."

However, Curé points out, Einstein, through the four essays, foresaw the advent of a future scientific theology. He believed that through a "cosmic religious experience" man could acquire transrational knowledge by a transcendental re-connection with a Supreme Intelligence. In the rest of this chapter, Curé pursues, to its finality, the consequences of these initial theological intuitions of Einstein. In this way, Curé establishes the foundations of "Cosmotheism" or Scientific Theology.

The first proposition to prove scientifically is God's existence in the real world, outside and totally independent of the human mind. This is accomplished with the acceptance by Einstein of the luminiferous ether as pure energy, and his principle of equivalence of mass and energy. The second proposition to ascertain scientifically is the existence of a Collective Cosmic Universal Consciousness who is God. Curé accomplishes this through a pragmatic definition of consciousness based on concepts advanced by Pierre Teilhard de Chardin. The final conclusion of this new Cosmotheism is, that "God is Universal Person," a step Einstein never made in his writings on theology. Nevertheless, Curé states, Einstein without realizing it, expressed mathematically the Universal Consciousness in his General Relativity Theory. Curé demonstrates this at the end of the last chapter. Also, in the last chapter of *Einstein on Trial*, there is this sense that Einstein motivated, in the mind of the author of this book, the creation of a Cosmotheism. Some people say that a pantheist is a person that has not had time to become an atheist. Had Einstein had that time, he would have created Cosmotheism, and would have become the first cosmotheist.

This book has been many years in the making. The painstaking historical and physical research, the rigorous theoretical and mathematical derivations are the results of a lifetime dedicated to the pursuit of truth in natural philosophy. Why is it that this other alternative path at the crossroads of the beginning of the 20th century has never been presented to the scientific community in the established periodicals? The answer lies in the tight framework of the established scientific orthodoxy.

After the apparition of the noosphere, the developing human species invented language to describe concrete objects, actions, qualities, needs and eventually not so concrete abstract ideas. Then written language appeared with symbols and a combination of symbols to represent language, which in turn described nature. In the ever more complex human development, we invented mathematics to help solve problems, which were too complex for mere words. Then as the process unfolded, certain mathematical entities and concepts became fashionable and useful to explain and describe phenomenology. Such is the concept of field to describe the interaction between real physical entities: Electric or magnetic fields if the entities are charged particles or gravitational fields if the entities are macroscopic or celestial bodies. Then, to find out about the microcosm, we shoot particles into matter to learn about the behavior of the particles which bounce back. Then, as we become comfortable with particles, we invent more particles as entities responsible for phenomena beyond the reach of our senses. And with this clever artifice, we are able to predict more phenomena in the measure that nature's reality conforms to the properties we assign to these noetic entities. In this unfolding and ever more complex process, *we have come to a point in which we begin to confuse reality with the noetic framework*, which we have built to describe and explain nature. However, the framework becomes a very successful paradigm and an orthodoxy becomes established as the continuing success of the noetic framework is able not only to describe and explain but also to build devises ever more elaborate for the comfort of human beings. Thus the birth and development of technology becomes a reinforcement of the building blocks of the established scientific paradigm. This noetic lattice may even become so strong as to stifle further progress in the discovery of the truth of nature.

It may do this in a twofold fashion: first by directing all further inquiries in a direction away from the real questions which have to be solved, and masquerading such research as unfashionable. Secondly by providing a paradigm within, which to interpret all inquiries and reject all unorthodox interpretations. The strength of this paradigm we can call the "power of the mind." As teachers of experimental physics, we have often seen it at work in students. It is an inner directing authority to reject false data, which does not conform with established theory, *but also to reject true data and obtain false results, which confirm wrong hypothesis they mistakenly believe to be true*. The "power of the mind" builds theories upon theories, all solidly founded in sound experimental results in an ever-stronger lattice until nature's reality and the noetic paradigm becomes fused into a single entity in the minds of scientists. Furthermore, science in the modern world is big business. Huge laboratories, universities, science funding institutions and their research objectives are dangerously threatened by unorthodox physics. Hence, it happened that the noetic paradigm went at some point in its development, almost a century ago, into a track that left behind another, perhaps much more adequate avenue of human seeking, as Curé emphasizes and develops in his book.

At the turn of the new millennium, we are witnessing the birth of a new paradigm. A new generation of physicists with the courage to dissent and follow their own convictions is a huge challenge. To reinterpret and rebuild the whole framework is no easy challenge. We have a whole century to catch up on. However, the price is that humankind is not bonded by only one interpretation of space and time, and the stars become within our grasp. Curé's book , *Einstein on Trial*, is the introduction to a new interpretation of the *Principles of Natural Philosophy*. This new interpretation will guide us to an incredible implosive renaissance in every field of knowledge.

Eduardo D. Greaves, Ph.D.
Caracas, Venezuela. July 23rd, 2000.

CHAPTER 1

ONTOLOGICAL PRINCIPLES

"A physical science made to the model of mathematics would be empty, could not inform us about the physical world."

H. Reichenbach

Introduction.

Newton's preface to the first edition of *Philosophiæ Naturalis Principia Mathematica (Principia)* [1], written in Latin, was signed by Newton at Cambridge, Trinity College, May 8, 1686. On that occasion he wrote:

> "I heartily beg that what I have here done may be read with forbearance; and that my labors in a subject so difficult may be examined, not so much with the view to censure, as to remedy their defects."

The only major mistake we may find in Newton's *Mathematical Principles of Natural Philosophy* is precisely in the title of his masterpiece. It is difficult to understand why Newton thought that Natural Philosophy should be based on mathematical principles instead of metaphysical (ontological) principles. The main purpose of this book is to remedy the "defects" of the Principia consisting of the omission of different fields of applications of Newton's dynamical methodology. These fields are Gravitodynamics, Electrodynamics, Atomic dynamics (quantum mechanics), Nuclear dynamics and Ergodynamics of moving vortices of condensed energy.

1.1 Why Mathematical Principles of Natural Philosophy?

One can wonder why Newton said that Natural Philosophy rests on Mathematical principles. If we understand philosophy as the love for wisdom, and if we understand *wisdom* as the search for *first principles* and *first causes*, then the original title of Newton's *Principia* is incorrect. However, Einstein would disagree with us. The title of Newton's *Principia is* correct if what Einstein [2, p. 274] once said is correct:

> "Our experience hitherto justifies us believing that nature is the realization of the simplest conceivable mathematical ideas."

We first observe that Einstein only *believes*. He does not *know* that Nature, the world of things, is the *realization* of our mathematical *thoughts*. In the second place, Einstein's use of the word "realization" was very unfortunate. If we etymologically consider the word "realization," that is derived from the Latin word *res*, meaning *thing*, then Einstein is telling us that Nature *is* the actualization of our mathematical thoughts. We are convinced that Einstein did not mean this. What Einstein meant was that our mathematical thoughts can very adequately *describe* the motion of *things* in Nature. Otherwise, we must conclude that Einstein decided that *to be and to think is the same thing.* The same idea was advanced by Parmenides 26 centuries ago. This is the reason Einstein ended the paragraph, from which we extracted the previous quotation, by saying: "*In a certain sense, therefore, I hold it true that pure thought can grasp reality, as the ancients dreamed.*" Now Einstein is talking about *grasping reality;* he is talking about apprehending reality or the external world of things. We cannot *identify* Nature with our thoughts. Ontologically, this is an impossible identification. Nature cannot *be* thoughts. Nature *is that it is,* absolutely *independent* from any human *thought.* Thus, if Nature *is* human mathematics, then Newton's title is correct. Nevertheless, we contend that Nature *is not* human mathematics, and therefore, Newton's title is incorrect.

From another point of view, Einstein would agree that the title of Newton's *Principia* is *incorrect* if what Einstein [2, p. 233] said in another instance is correct:

> "How can it be that mathematics, being after all a product of human thought which is independent of experience, is so admirably appropriate to the objects of reality? Is human reason, then, without experience, merely by taking thought, able to fathom the properties of real things?
>
> "In my opinion the answer to this question is, briefly, this: as far as the propositions of mathematics refer to reality, they are not certain; and as far as they are certain, they do not refer to reality."

As mathematical propositions are analytical judgments, and every analytical judgment is true, according to the logical principle of identity, therefore, every mathematical proposition is true or certain. Einstein now is telling us that the use of, or the application of mathematical propositions to reality, is not guaranteed to be true or to be certain. Thus, *"nature is - not - the realization of the simplest conceivable mathematical ideas."* Therefore, this is the Einsteinian dilemma (violation of the ontological principle of non-contradiction): (1) Nature is mathematics. (2) Nature is not mathematics. Only one of the previous contradictory statements can be true. Being that mathematics is the science of forms and quantities, it is foolish to identify Nature with mathematics. Hence, Newton's title is *incorrect,* because the foundations on which Natural Philosophy must rest cannot be mathematical principles. Then, why did Newton choose *mathematical principles* as the foundation for his natural philosophy? Newton must have faced the same Einsteinian dilemma we pointed out before. Now we propose the following question as a historical problem.

Did Newton choose the title "Mathematical Principles of Natural Philosophy," instead of "Metaphysical Principles of Natural Philosophy," because he did not want to face the *Inquisition?* In the introduction to Book III of his *Principia,* Newton emphasized his choice very clearly:

> "In The Preceding Books I have laid down the principles of philosophy; principles not philosophical but mathematical: such, namely, as we may build our reasoning upon in philosophical inquiries. These principles are the laws and conditions of certain motions, and powers of forces, which chiefly have respect to philosophy; but, lest they should have appeared of themselves dry and barren."

Newton had better places in his *Principia* to emphasize the foundations of his natural philosophy if he was stressing mathematical principles instead of metaphysical ones. Indeed, in the preface he wrote; *"... and therefore I offer this work as the mathematical principles of philosophy."* It is strange that Newton wrote the preface almost one year before the *Principia* was published. Westfall, in his magnificent biography of Isaac Newton, *Never at Rest* [3, p. 459], immediately below the previous quotation, wrote: *"He insisted on the word 'mathematical'."* Westfall, however, does not explain why Newton insisted on having *mathematical principles* in the title of his *Principia.* Newton's Natural Philosophy is based fundamentally on the *metaphysical* concepts of *absolute* space and *absolute* time. Science is one of the many manifestations of the culture of a given social group. We believe the political, social and religious circumstances surrounding Newton in the month of March 1687, may explain the choice Newton made for the title of his masterwork. At that time, England was under the reign of James II, who wanted to restore Catholicism to England. A terrible crisis fell upon Cambridge University when Newton had practically finished his Book III, of the *Principia.* In Chapter 11: *Revolution,* Westfall [3, p. 469], narrates the shameful events happening at Cambridge. Westfall writes:

> "The crisis had built up gradually, James understood that control of the

universities was essential to his goal of reconverting England to Catholicism . . . If we judge by the date on which Halley received Book III, Newton left Cambridge a week before the manuscript was completed. Moreover, the crisis, in which he had assumed a leading role, still hung over the university; he could not have imagined that the matter was settled. One gains the impression that Newton said enough on 11 March that he found it advisable to disappear. He returned to Cambridge only in time to prepare for the trip to London."

Newton must have realized the consequences of his leading role in politics if the Restoration succeeded. He also was well acquainted with the works of Galileo and Descartes and their fates as natural philosophers. Newton must have known the terms imposed by the *Inquisition* on Galileo. The Italian natural philosopher was supposed to retract the heliocentric conception on philosophical grounds. The *Inquisition* allowed Galileo to maintain the heliocentric conception *only as a mathematical hypothesis*. Here is the clue as to why Newton decided to name the principles of his natural philosophy *mathematical principles*. Newton also must have known why Descartes exiled himself to Sweden after Galileo's trial. Newton was an extraordinary scholar, and it is very doubtful that he did not know it is *impossible* to *identify* accidents of an entity with the *substance* of the same entity. Had the Restoration succeeded and had Newton called his masterpiece *Metaphysical Principles of Natural Philosophy,* we would still be teaching Aristotle's Physics and the Ptolemaic System of the Heavens.

Recently Galileo was vindicated by the Catholic Church. This was a gigantic step to wisdom. The present Pope addressed the Pontifical Academy of Science, on October 31, 1992, on biblical hermeneutics. In general, *hermeneutics* is the study of the methodological principles of interpretation and explanation. This Papal lecture appeared in *L'Osservatore Romano*, No. 44, 4 November, 1992. We will cite three remarkable thoughts expressed by Pope John II in his Papal lecture. The Pope, referring to Galileo's time, said:

1. "The majority of theologians did not recognize *the formal distinction between Sacred Scripture and its interpretation*, and this led them unduly to transpose into the realm of the doctrine of the faith a question which in fact pertained to scientific investigation."

In his lecture of 1992, the Pope also referred to Saint Augustine, saying:

2. "If it happens that the authority of Sacred Scripture is set in opposition to clear and certain reasoning, this must mean that the person who *interprets Scriptures* does not understand it correctly."

The Pope also alluded to Pope Leo XIII, 1894, who echoed the Augustinian precept:

3. "Truth cannot contradict truth, and we may be sure that some mistake has been made either in the interpretation of the sacred words or in the polemical discussion itself."

In any intellectual activity, either theological or scientific, any proposition is the result of an interpretation. In the three quotations we selected from the Pope's lecture in 1992, we see clearly the Papal message to theologians: *"If there is discrepancy between Science and Sacred Scripture, most likely it is because of a misinterpretation of the Sacred Scripture."* But this is one face of this epistemological coin. The other face of the coin, not expressed by the Pope, should read as follows: *"If there is discrepancy between Sacred Scripture and Science, most likely it is because of a misinterpretation of the Natural Phenomenon."* The rest of this book precisely is dedicated to unveil the many misinterpretations we have made in science, and particularly in physics. Thus, this extraordinary lesson of Pope John II is not only for theologians but also for scientists. The great majority of scientists should recite the Indian poet Tagore's paraphrased lines: "We wrongly read the world, and then we say the world has deceived us." At the end of his lecture, the Pope quoted Einstein saying: "The most incomprehensible thing about the universe is that it is comprehensible." We believe the universe is only a little bit comprehensible by the human mind.

1.2 Natural Philosophy.

It is our opinion that natural philosophy is the application of ontological and logical principles to empirical knowledge in order to establish rational relationships among the irrational sensorial data. The word onto-logy is derived from the Greek words *on, ontos,* meaning "being," a form of the verb *einei*, meaning *to be*. The other root word is "logy." Today's meaning of "logy" is the "study of" or the "science of." Thus, *ontology* is the study of the *being* of *entities*. Naturally, the problem of the *being* of entities, material or immaterial, is still an open problem in philosophy. Some philosophers, like Gabriel Marcel, do not consider "the *being* of entities" a problem. Marcel considers it a mystery, i.e., an object the human mind is unable to comprehend. The ultimate goal of the human mind is to apprehend the *Being of all entities*. This presumptuous objective has been in the hands of theologians for many centuries.

Any theory is a set of principles, definitions, logical rules to establish inferences and a vast collection of conclusions. The axiomatic foundations of classical ontology rest on four principles: the principle of *identity*, the principle of *noncontradiction,* the principle of the *middle excluded,* and the principle of *causality.* These principles are applied to the *world* of *things* or *reality* (Latin: *res* = thing). They are applied to *nature* in the hope of unveiling the first principles and the first causes of the ever-changing reality. They are applied to the entire universe to uncover the common *substance*, which provides existence to the plurality of entities. These principles are applied to all different entities to discover their ultimate common *essence.* In the 20th century, thinkers were hypnotized by the forms of things, and they geometrized reality. Also in this same century, we have completely ignored the ontological fact that things are made from *something* that must be common to all. In physics, we have gone as far as to create *rules* of covariance and invariance. In doing so, we preserve incorrect physical laws. These rules constitute *bad metaphysics.* We must go back again to study *natural philosophy.* We must further study ontology in order to discover the very nature of things and their interactions with the *rest of the universe.*

1.3 About Theoretical Knowledge.

A theory is a hypothetico-deductive structure. It is constituted, as we said above, of a set of principles, definitions, logical rules and a vast collection of conclusions. What is the purpose of any theory, particularly in physics? Allow us to repeat the answer using different words.

1. The purpose of a theory is to explain the seemingly unexplainable reality.
2. The purpose of a theory is to rationalize the seemingly irrational reality.
3. The purpose of a theory is to comprehend the seemingly incomprehensible reality.

A theoretical structure can be compared with a building. There is first a solid bulky foundation. On top of this foundation, the remainder of the structure is built. The very foundation of any theory is made out of *principles.* On the first floor, we have the definitions and the logical rules. The upper floors are built with the many conclusions that are "manufactured" by combining *rationally* the principles, definitions and previous conclusions. When the theoretical building is finished, the ugly, sturdy foundation is covered to hide the logical shame contained in it. The rest of the theoretical building glows majestically, and is illuminated with the light of *human reason.* We now have the feeling of being truly superior beings. This feeling of superiority among most scientists is based on pure epistemological ignorance. Let us see how this is possible. What is a principle? The word "principle" is derived from the Latin word *principium* (plural: *principia),* which means *beginning, fountainhead, original or initial state.* In philosophical dictionaries, we find that "the truth of principles cannot be proven." In ancient times, philosophers said that the truth of a principle is *evident* in itself, like the truth of an *axiom.* On the other hand, they claimed that the truth of a *postulate,* though necessary, is not *evident* in itself. Whether *evident* in itself or not, the truth of a principle, axiom, postulate or *dogma* cannot be proven. It cannot be logically *deduced* nor can it be *rationalized.* Thus, principles, axioms, postulates and dogmas are *irrational statements.* Now we can expand on the purpose of any theory, scientific or not.

1. The purpose of a theory is to explain the seemingly unexplainable reality, starting with unexplainable principles.
2. The purpose of a theory is to rationalize the seemingly irrational reality, starting with irrational principles.
3. The purpose of a theory is to comprehend the seemingly incomprehensible reality, starting with incomprehensible principles.

This is the sad truth about the beginning of any scientific theory. A scientific theory is a dogmatic, inflexible, biased, opinionated, hypothetical deductive structure. The consequence is that its defenders are fanatics, maniacs or just plain extremists. A knowledgeable scientist is a person of *faith,* one that knows he has to *believe* in the truth of the principles. Thus, if ignorance is not an excuse, then, from an ethical point of view, a religious person deserves more respect than a scientist, because a religious person *knows* that he *believes,* while a scientist *believes* that he *knows.* These considerations over the years forced the mind of this author to invent two definitions: one of philosophy and the other one of science. Philosophy is an anarchic set of opinions (dogma is derived from the Greek word *dokein,* meaning opinion), while science is a set of average (mediocre) thinkers who follow the opinions of one intellectual master for some time. This state of affairs brings a dilemma for dissidents. As we do not appreciate followers, too much, it is surprising that we still would like to have our own group of followers share what we believe is a better physics.

The sad truth about theoretical science is that it is a logical fraud. The fraud is in the very foundations of any scientific theory. The glowing rationality of the theoretical structure vanishes in the blackness of irrationality of its principles. There is no rational escape from this irrational black hole in our minds. However, only those who have created theories know the existence of another *tao (method, path).* They alone apprehend the essence of the principles of a new theory. This path is not logical, not rational nor is it irrational. It is *trans-rational.* It is a path which leads humans to experience an altered state of their minds. Over the years, we realize that university professors, especially physicists, with their usual "rational" arrogance, were at ease with the word "trans-rational" but not with the words *mystical experience.* A mystical experience is beyond, it surpasses rational comprehension. In this trans-rational state of the mind, the creator knows that he knows without any syllogistic exercise. The creator grasps the essence of things: material or immaterial, visible or invisible. The creator is a true ontologist. Einstein [2, p. 289] knew this trans-rational *method* well when he wrote about the genesis of his GRT:

> "In the light of knowledge attained, the happy achievement seems almost a matter of course, and any intelligent student can grasp it without too much trouble. But the years of anxious searching in the dark, with their intense longing, their alternations of confidence and exhaustion and the final emergence into the light- only those who have experienced it can understand that."

The universal statements (principles), grasped by the few creators in physics, correspond to the Kantian *a priori* synthetic judgments. "These fundamental concepts and postulates"- Einstein [2, p. 272] says, "which cannot be further reduced logically, form the essential part of a theory, which reason cannot touch."

Now we will write about the metaphysical or ontological principles and, of course, about Natural Philosophy. Again we will express our opinion that Natural Philosophy is the application of ontological principles to natural phenomena. Once the ontological speculation is done, we must translate this philosophical cogitation into the universal language of mathematics. This translation is necessary to avoid metaphysical obscurities, which confuse our discourse and add *nothing* new to the mathematical structure of natural laws. Finally, the true natural philosopher must report experimental verifications of his endeavors in Natural Philosophy. If they cannot present experimental evidence of their enterprise in Natural Philosophy, at least they should propose experiments to empirically corroborate the theoretical conclusions. If we do not follow this path, we will exhibit incurable philosophical verboseness. We will present a farrago of mathematical equations, about symbols, without any essential meaning. This will cause an accumulation of superfluous experimental results without any sign of progress in science. The accumulation of experimental facts does not constitute a rational science, but a collection of unrelated facts. Ortega-y-Gasset alluded to irrelevant experimental activity in the 20th century. This was expressed in his essay *La Rebelion de las Masas*. Erwin Schrödinger, *the Schrödinger of Wave Mechanics,* was so impressed by this essay that he translated Ortega-y-Gasset's conceptions of the new *barbarian scientist* who "endangers the survival of true civilization." In 1952, Schrödinger [4], in his powerful little book *Science and Humanism - Physics in our Times,* page 6, wrote Ortega-y-Gasset's concepts about the *average* scientist and his experimental work:

> "He is a person who, of all the things that a truly educated person ought to know of, is familiar only with one particular science, nay even of this science only that small portion is known to him, in which he himself is engaged in research. He reaches the point where he proclaims it a virtue not to take any notice of all that remains outside the narrow domain he himself cultivates, and denounces as *dilettantist* the curiosity that aims at the synthesis of all knowledge.
>
> "It comes to pass that he, secluded in the narrowness of his field of vision, actually succeeds in discovering new facts and in promoting his science (which he hardly knows) and promoting along with it the integrated human thought - which he with full determination ignores. How has anything like this been possible, and how does it continue to be possible? For we must strongly underline the inordinateness of this undeniable fact: experimental science has been advanced to a considerable extent by the work of fabulously mediocre and even less than mediocre persons."

Most of the university professors practice this irrelevant activity today. They do these superfluous exercises because they are afraid to lose a prestige they have never had. They publish many inconsequential "scientific" papers per year as they do not want to perish individually. In doing so, they are causing science to perish as a whole. This behavior is not new. The late Professor Richard S. Westfall, from Indiana University, in his book *The Construction of Modern Science* [5, p. 105], reminds us that in the 17th century:

> "Not only were the universities of Europe not the foci of scientific activity, not only did natural science have to develop its own centers of activity independent of the universities, but the universities were the principal centers of opposition to the new conception of nature which modern science constructed."

In spite of the present *Scientific Inquisition,* which burns the modern Giordanos in the *flames of silence,* new knowledge will gain new adepts. Unfortunately, they, too, will probably create another Scientific Inquisition in the new present century. As misery loves company, let us rejoice and love others for having the courage to be free thinkers. Let us admire them for having a philosophy, a natural philosophy, which someday will find a *true university* with which to plant this new epistemological seed.

1.4 Ontological Principles.

Ontology offers four principles to natural philosophers: (1) The principle of Identity, (2) the principle of Noncontradiction, (3) the principle of the Middle Excluded (in other languages it is called the principle of the Third Excluded), and (4) the principle of Causality (in logic it is called the principle of Sufficient Reason or the Great Principle according to Leibniz). The ontological principles refer to the *being* of things. Logical principles refer to *judgments.* Ontological and logical principles have the same names except the fourth principle as we noted above.

Principle of Identity.
Every thing is identical to itself.

Symbolically "A is A," or "A≡A." Thus, the notion of *identity* implies *unity.* Identity is *idem* plus *tas* or *tatum,* a Latin suffix that corresponds to the English suffix *-ty,* to indicate a quality or condition of *idem* that means the same. Identity, therefore, is the quality or condition of being the same. Identity is absolute or essential *sameness.* Identity is *oneness.* To identify is to reduce the incomprehensive plurality of reality to one *being*. At this point, perhaps, David Bohm would have asked: What is the unbroken wholeness of the entire universe?

Material things constantly change. In spite of the changes, the physical object still is the *same.* We still recognize that in *that* object *something* remains the same. The form of the object may change, the color and a long list of its *accidents* (qualities, properties, attributes) may change, but we can still identify *that* object with itself. The *being* of things is not corrupted by the passage of time or by the changes of its accidents. If we can identify *that* object with itself, in its chaotic storm of changes, it is because our mind is able to differentiate between what is accidental in *that* object and retain what is essential. This, however, is not an easy task. What is the *Being* of all entities? If the ability of establishing new relationships is to think, then to think requires first to identify. If we have three entities, A, B, and C, it would be insane to ask, what is the relationship between A, B, and C? However, if we identify certain common aspects (accidents) in A and B, say a and b, and in B and C, say b and c, then we can think. These two identifications are symbolically represented by: $a \equiv b$ and $b \equiv c$. Now we need logical rules to truly think rationally. From the two previous identities we *cannot deduce* that $a \equiv c$. Of course the statement $a \equiv c$ is extremely *evident* in itself, but we *cannot deduce* it. We need another principle: the *transitive principle* that is not an ontological one. Whenever a conclusion is based on the application of two principles, two definitions, or a principle and one definition, along with the use of the transitive principle, we have only one answer to give to the embarrassing question: Why does $a \equiv c$? The only honest answer is *just because.* Thus, we can think because our mind is capable of identifying. Parmenides, 26 centuries ago, taught us that *to be and to think is the same thing.* The exact essence of this Parmenidean assertion was repeated by Einstein [2, p. 274] last century. Einstein said:

> "Our experience hitherto justifies us believing that nature is the realization of the simplest conceivable mathematical ideas."

To say that *nature is the fulfillment of our mathematical thoughts* seems to be shocking at first sight. For years, this author reacted violently against this Einsteinian statement. The best argument against it was provided by the epistemologist Hans Reichenbach [6]:

> "To regard mathematics as the ideal which the physical sciences

> should try to approximate means misunderstanding the nature of mathematics. A physical science made to the model of mathematics would be empty, could not inform us about the physical world . . . The philosopher, who in the twentieth century still attempts to derive knowledge from reason, has lost his most potent support, the support by the mathematician."

When we trace the origin of mathematics, we must succumb to the fact that *geometry* came to the minds of the first Sumerian and Babylonian philosophers from the pebbles on the road. Geometry also descended from the starry heavens to the minds of the first ancient mathematicians. Philosophers began to count the stars in the heavens, while the common people counted mundane things to continue with their everyday business. Initially, the human mind had pedestrian and celestial encounters with mathematics. A stone thrown into a pond of water showed humans the perfection of diverging circumferences. Mathematics is *part* of nature and *precedes* any human mind. We have gone so far into the *abstract* realm of mathematics that we have forgotten the ordinary origin of it. The Pythagorean theorem was used centuries before Pythagoras deduced it formally. Einstein's ontological conception of nature coincides with Parmenides' assertion: *"To be and to think is the same."* Centuries later Descartes discovered a corollary of Parmenides' statement: *"I think, therefore I am."*

For years, this author did not understand the last part of an Aristotelian assertion: "Things are what they are, and not different." This author began to ask his students-- Why are things what they are, and not different?-- "Because they are what they are"— was their answer. This is equivalent to saying "Just because." No one learns anything with this kind of answer. Why are not the three atoms of a water molecule in a straight line? Because there are *natural* laws which force the atoms to adopt a *unique* geometrical configuration. The shape or the *form* of things is dictated by the natural laws of interaction between the atoms. Yes! The form of natural things can be grasped with our mathematical thoughts. Better yet, the mathematical structure of natural laws is the fulfillment of our mathematical thoughts. Thus, we have to come to the conclusion that Einstein is right and he is also wrong. He is right from a formal point of view, and he is wrong from an essential point of view. Mathematically Einstein is right. Ontologically Einstein is wrong. Reality, which is the world of *things,* is not only form, but most importantly is *substance.* We use here the word "substance" in a philosophical sense. Unfortunately, today the word "substance" has a strong chemical connotation. Consequently, we will use the word *essence* instead of the word *substance* as it was used before the birth of the science of chemistry.

We can ask again: What is the *essence* of the unbroken wholeness of the entire universe? We will dare to answer this question in chapter 5. The answer will not be given by a philosopher but by a natural philosopher. Philosophers, in general, do not know physics nor mathematics. Philosophers, in general, do not *do* physics nor mathematics. In summary, the ontological principle of *identity* allows us to *unify* the overwhelming diversity of the world of things, of reality projected in the human mind.

Principle of Non-contradiction.
Nothing can be and not be at the same time.

Another way to enunciate this principle is to say that it is *impossible to affirm* that *A is* and *A is not,* simultaneously and in the same sense. E. Mach used this principle when he said the universe is one with stars and not a universe without stars. In this way, Mach ontologically refuted Newton in relation to two globes united by a rope. Newton imagined the globes revolved around their center of mass in a universe without stars.

Principle of the Middle Excluded.
Everything has to be or not to be.

Another way to express this principle is to say that it is *impossible to deny* that *A is* and *A is not,* simultaneously and in the same sense. The following two propositions are contradictory: (1) "an electron *is* a particle" and (2) "an electron *is not* a particle." According to the principle of noncontradiction, it is impossible to *affirm,* simultaneously, that "an electron is a particle" and that "an electron is not a particle." On the other hand, according to the principle of the middle excluded, it is impossible to *deny,* simultaneously, that "an electron is a particle" and that "an electron is not a particle." D.H. Freedman in his article *Weird Science,* (Discover, p. 62, November 1990), writes:

> "At the cornerstone of quantum mechanics is the bizarre-sounding truth that bits of matter and energy sometimes behave like particles and sometimes like waves- depending on how you measure them- but try as you might, you will never observe both characteristics at the same time. The very act of your observation will cause the object of your attention to

assume a single, mundane identity."

Freedman refers to the four ontological principles in a very *weird* manner. At least he respects the ontological principles of noncontradiction and middle excluded. In the last sentence, Freedman resorts to the principle of causality to "deduce" that a mere human observation is the *cause* of the *object* adopting a mundane identity. One wonders what that mundane *thing* is that identifies the essence of the object. The quotation finishes with the identification of a quantum entity with a pedestrian thing.

But Freedman goes further when he dares to uncover the ontological ignorance of physicists. He continues writing:

> "Physicists have long been obsessed with this apparent limitation, and in the past seven decades many have designed clever experiments to catch something behaving like a particle and as a wave simultaneously."

These experiments are not clever at all. No person with a sane faculty of thinking can fall in such an ontological absurdity in designing such experiments. Is it possible to observe an entity that *is* and, simultaneously, *is not*? It could only be possible in a *weird science* because the weird scientists are totally *ignorant* of ontology. Freedman's first quotation contradicts the second one. The second quotation is definitively more offensive to our intelligence. It is even offensive to common sense. Why does this *dualism* exist in quantum physics? We think this quantum dualism is a necessary consequence of the unawareness of the primordial cosmic energy field. An electron carries an "atmosphere" of energy superimposed upon the pre-existing primordial cosmic energy field. The electron is a corpuscle which remains identical to itself, but its intrinsic rotation and motion in the primordial cosmic energy field modifies the energy content in its environment. This energy modification sets a density wave of energy which eventually generates a diffraction pattern. The diffraction phenomenon is due to energy waves and momentum waves generated by the motion of the electron, as a particle, in the cosmic energetic medium. The particle remains a particle and the momentum wave remains a momentum wave, as we will prove in chapter 2. There is no ontological electronic metamorphosis. The pilot or matter wave of de Broglie has an *identity.* It is not a ghost wave but an energy-momentum wave.

Principle of Causality.

Every effect has a cause.

This principle, over the centuries, has *created* vast philosophical literature which is impossible to summarize in a few lines. There are many ways to enunciate this principle. Everything has a cause. Nothing happens in this world without a cause. Nothing can be without a cause. Nothing can stop being without a cause. Everything that *be*comes is *be*cause of a cause. Everything that *be*gins must have a cause. Any effect is the actualization of a cause.

Aristotle distinguished four different causes: material, formal, teleological (final) and efficient. The latter was kept in science during the renaissance. An efficient cause is the agent which produces some change in a certain environment. Even today, however, the teleological cause is still used in Biology. For Galileo, *the necessary and sufficient condition of the emergence of something is the efficient cause.* An extensive discussion of the principle of causality is found in *"Causality. The place of the causal principle in modern science"* by Mario Bunge [7].

Newton's principle of action and reaction can be only used if we resort to the principle of causality. Forces act on *material* bodies. These material bodies are the original *cause of the force.* Every good student of Newton's theory of dynamics knows that his principle of action and reaction is *not valid* in a noninertial reference system. Einstein falsely accused Newton of having hypostatized absolute space *as the cause* of centrifugal forces. This is because Einstein was a very poor student of the *Principia.* Bertrand Russell, on the other hand, considers the principle of causality as another useless relic in modern science. However, he was not a theoretical physicist. To *deduce* the mathematical structure of the gravitostatic force, it is definitively necessary to use the principle of causality. Newton did not have a mystical vision of the gravitostatic force when "the apple fell on his head." If we use Kepler's observational laws along with Newton's theory, in a rational way, we will then be amazed to discover that the mathematical structure of Newton's gravitostatic law *contains* the inertial masses of the interacting celestial bodies and not the so-called gravitational masses. We will demonstrate this incredible conclusion in chapter 6. By the way, Newton, in his *Principia,* very seldom used the word *mass*. Why do misconceptions remain for such a long period of time in science? Perhaps, because scientific and philosophical books are usually *copies* of other books, repeating these misconceptions, over the years, without any critical analysis. Now, we need to add two more ontological principles. The principle of reality and the principle of inseparability.

Principle of Reality.
An external world exists, independent of any human mind.

Ideas are mental images, are intellectual icons, are phenomena (luminous mental images) in the human mind. These sensorial mental appearances are *caused* by an external world of *things* (reality, from the Latin word *res*, meaning thing, material thing). On the other hand, we have learned over the centuries that the human mind can *conceive* new noumenal (mental) entities which are not caused by the external reality but by the human mind itself, and they are called *concepts*. We will leave the statement of the principle of reality in the hands of Einstein. In this respect, Einstein [2, p. 266] was absolutely clear when he asserted that:

> "The belief in an external world independent of the perceiving subject is the basis of all natural science. Since, however, sense perception only gives information of this external world or of 'physical reality' indirectly, we can only grasp the latter by speculative means."

To speculate does not only mean to guess. The speculative process to which Einstein refers, is a synthesis of all the ontological, logical and mathematical principles plus intuitive and imaginative abilities of the human mind.

Principle of Inseparability.
Any material entity is inseparable from the rest of the universe.

We may call this principle, "the principle of the unbroken wholeness." In physics this principle establishes that it is impossible to separate two interacting material entities from the rest of the universe in order to study the interaction between them. We have been doing this epistemological dichotomy in physics since Newton's time, except when we decided to analyze the periodical and secular perturbations exerted on one planet by the other planets. The perihelic motion of planet Mercury is an excellent example of the action on Mercury by a collective planetary potential. In chapter 2, we will use this principle to identify the quantum collective potential, and in chapter 6, we will again use this principle to identify Mach's Principle with the cosmic collective potential. Finally, in chapter 7, we will use again this principle of inseparability to analyze the concept of *consciousness*.

1.5 Application of Ontological Principles.

We think the best way to apply the ontological principles is to go back into the history of science and philosophy. If we do this, we may critically retrace the generation and degeneration of concepts. Now we wonder how we can apply the principle of identity to natural events and natural things. How do we apply the statement "any thing is identical to itself." Over the years, a good natural philosopher realizes that the *unreasonable* conclusions of human scientists force them to call *one and the same entity* with different names and attributes that only correspond to aspects of the whole entity. The history of physics is an excellent place to initiate *identification* of "essentially" different things, events and attributes or qualities of things. A good example is the "essential" difference between inertial mass and gravitational mass. We will analyze this ontological confusion in chapter 6. In the other extreme, we find some people who have no problem in *identifying* two totally different entities such as a corpuscle and a wave. The history of physics is a great source to apply the ontological principles, and to discover many mathematical obscurities without any physical and ontological essences.

1.6 Rules of Reasoning in Philosophy.

In Book III of the *Principia,* Newton established methodological and philosophical rules to guide our thoughts in matters of natural philosophy. We simply reproduce them with some short comments.

Rule I. "We are to admit no more causes of natural things than such as are both true and sufficient to explain their appearances."

Implicitly, in this Rule I, it is a fundamental ontological consequence. To reach this conclusion, we need the use of psychology, biology and the ontological principle of cause and effect. This is a conclusion about the cause of mental appearances. It is a conclusion about the existence of *reality*. It is a conclusion about the existence of external *things* which are the causes of the internal mental effects. The mental effects are the *appearances or images* in the human mind. In other words, it is an ontological conclusion about the existence of a *real world* which seems to be independent of any human mind. This independence of reality, from any human mind, is not contained in the principle of cause and effect. This independence of *reality,* from the human *subject,* i.e., this so-called *objectivity,* is not contained in the axiomatic structure of ancient ontology. This *objectivity* constitutes an *act of faith*, i.e., an extra axiom or principle. For this reason, we extended the ontological principles up to six principles. The fifth ontological principle is *the principle of reality or objectivity.* This principle of reality, as any other principle in any theory, is a dogma of faith about the truth we believe it contains. Any scientific theory only provides relative knowledge. The scientific knowledge is *relative* to the truth contained in the dogmas of faith (principles, axioms, postulates) constituting the foundations of any theory, scientific or not.

The degree of abstraction, in modern physics, has superseded the obsolete degrees of Aristotelian abstraction. Modern speculations, in quantum physics for example, requires far more mental capacity to abstract from the *appearances* projected in the human mind. Even the realm of quantum physics is presently deprived of a quantum reality, though more and more physicists are becoming *true* natural philosophers in the present. Aristotle's physics was kindergarten physics, of which the great ancient philosopher assigned a modest first degree of abstraction. Aristotle was right in his epoch. When Newton introduced the concept of *force,* he *ontologized* the kinematics of Galileo. Newton fused kinematics into philosophy when he established that the *force* is the cause of the change of *motion* (motion = m**v**) of a body. Afterwards, Mach and the *positivist* physicists became frightened and hesitated to mention the word *cause* or to ask *why*. Physics is a *speculative* science; physics is natural philosophy; physics is metaphysics today if we consider the high level of abstraction required in quantum physics or field theories. By the way, philosophers have always done philosophy out of physical theories. We have reached such a high gnoseological level of abstraction in science that it is ridiculous not to bring the sciences back to philosophy. All of the sciences abandoned philosophy during the explosive renaissance. Now is the time to initiate an implosive renaissance, but we must be careful to remember that the mental activity of human beings is based on unexplained principles, on acts of faith. Let us carefully read what Einstein [2, p. 266] said: "*We must always be ready to change these notions—that is to say, the axiomatic basis of physics.*" In chapter 7, we will come back to this subject. Physics today is far beyond the old Aristotelian Metaphysics. Physics today, as a matter of fact, is Meta-Metaphysics. Any metaphysician or theologian, in the present, who knows nothing about modern physics cannot write anything about Meta-modern-physics. In the present, their writings are stagnant in the archaic Aristotelian-St.Thomas Aquinas Metaphysics. To put it in other words, let us quote the contemporaneous Jesuit priest and astronomer George Coyne:

> "I'm trying to understand God's universe. As a man of faith, I believe it is God's Universe. But that's faith. I mean, I can't prove that

to any one else that this is God's universe. That's simply my faith. But once I have that faith, then all my scientific research helps to enrich that faith . . . In fact, if I bring God in to explain what I cannot explain scientifically, I think that is one of the greatest sins ever committed against God and against myself."

This sin is caused by the infernal beast of "ignorance." Siddhartha Gautama, the Buddha, many centuries ago considered this same subject. D.T. Susuki [2*] wrote in the 20th century, saying that:

"Gautama felt as though a prison which had confined him for thousands of lifetimes had broken open. Ignorance had been the jail keeper. Because of ignorance, his mind had been obscured, just like the moon and stars hidden by the storm clouds. Clouded by endless waves of deluded thoughts, the mind had falsely divided reality into subject and object, self and others, existence and non-existence, birth and death, and from these discriminations arose wrong views—the prisons of feelings, craving, grasping, and becoming. The suffering of birth, old age, sickness, and death only made the prison walls thicker. The only thing to do was to seize the jail keeper and see his true face. The jail keeper was ignorance . . .Once the jail keeper was gone, the jail would disappear and never be rebuilt again."

Rule II. "Therefore to the same natural effects we must, as far as possible, assign the same causes."

This rule is an application of the same ontological principle of causation. This rule is not always true. However, its *converse* is always true: the same causes always produce the same effects if the exact same initial conditions are present. Newton's rule requires the application of the first principle of ontology, which is the principle of *identity.* First we must identify the different *effects* among them. The incapacity to identify seemingly different effects have created *fictitious* problems in physics and biology through the ages.

Rule III. "The qualities of bodies, which admit neither intensi-

> fication nor remission of degrees, and that are found to belong to all bodies within the reach of our experiments, are to be esteemed the universal qualities of all bodies whatsoever."

Newton illustrated this rule in the similar dynamical behavior of planets with respect to the sun as well as the behavior of satellites with respect to planets. The rule relies heavily on experiments. However, the last judge of our theoretical speculations should not be left completely to *experiments*.

> If there is contradiction between a theoretical conclusion and the corresponding experiment to verify it, then the *interpretation* of the conclusion and/or the interpretation of the experiment has to be critically reviewed.

Any experiment requires at least one instrument. We have never met a scientist who has built an instrument and does not know what to measure with it. Instruments measure only what the scientists want to measure, i.e., all instruments are *biased* by preconceptions or by the *dogmatic* hypothetical-deductive structure of any theory. We should never trust the outcome of any experimental result by 100%.

Rule IV. "In experimental philosophy, we are to look upon propositions, inferred by general induction from phenomena, as accurately or very nearly true, notwithstanding, any contrary hypotheses that may be imagined, till such time as other phenomena occur, by which they may either be made more accurate or liable to exceptions."

A typical example of the nonapplication of this rule is provided by Kepler's laws and Newton's establishment of his gravitostatic law. Leverrier *corrected Kepler's first law*. By means of astronomical observations, Leverrier discovered the anomalous motion of planet Mercury in 1859. The whole world before and after Einstein, blamed Newtonian dynamics for being *powerless* in explaining the perihelic rotation of planet Mercury. In 1917, Einstein [8, p. 102] strongly criticized Newton's Classical Mechanics. He wrote:

"[Relativity theory] has already explained a result of observation in astronomy against which Classical Mechanics is powerless."

Brahe's astronomical observations, performed in the second half of the 16th century, even before the invention of the telescope, plus Kepler's own astronomical observations, were the *inaccurate* data used by Kepler to establish his astronomical laws. Later, Newton used these laws, along with his classical dynamics, in order to obtain the Newtonian gravitostatic force. Obviously, Newton's law of gravitation *inherited* all the *imperfections* of the astronomical observations. After Leverrier's discovery of Mercury's perihelic motion, no one applied Rule IV to this new problem. Also, few, if anyone, studied the theoretical solution to this problem that was *contained* in the *Principia.* Remember this was published in 1687, and has been available to physicists all this time. Exactly 300 years after its publication, Phipps [9, p. 322 f] published his *Heretical Verities.* He rediscovered a formally identical gravitational force contained in Newton's *Principia* under Proposition XLIV, Theorem XIV in Book I. Even today, few physicists know that Newton has *two gravitational laws.* All the details of 304 years of ignoring Newton's explanation of planetary perihelic motion are described in J.C. Curé's paper of 1991 [10]. Einstein's assertion is not true. It constitutes false testimony against Newton's classical dynamics. Einstein was an extremely poor student of the *Principia.* Present physicists are much more inclined to offer *thought or virtual experiments* instead of *factual experiments.*

1.7 Logical Principles.

Any science, in order to become a *rational* intellectual discipline, must not violate the principles of Logic. Logical principles are the *dogmatic* first judgments which every science needs. They are needed in spite of the undeniable fact that *logical principles* are *irrational* assertions, as we have previously pointed out. The names of the first three logical principles are the same names as the first three ontological principles. The fourth logical principle is named principle of Sufficient Reason. From an axiological point of view, a *judgment* is a proposition that has only *two values: true* or *false.* Thus, western logic is *bivalent* logic.

Principle of Identity.
Every analytical judgment is true.

An analytical judgment is a statement in which the predicate is implicit in the subject. Thus, every thesis of a mathematical theorem is an *analytical judgment.* Hence, every mathematical proposition is true after the extraction of the predicate from the subject of the sentence which expresses the thesis of the theorem. Every proposition in mathematics *must* be *proven* logically by the use of the corresponding mathematical *axioms* and previous proven theorems. More than one epistemologist has said that mathematics or any theory is nothing but a mere gigantic tautological structure. Any judgment is a conclusion of a syllogism. In the judgment, the middle term disappears. Different criticisms have been raised against the principle of identity in logic. To say that "A is A," according to Goblot, does not constitute a judgment because "to know that A is A is to know nothing." The logical principle of identity never was intended to be a gnoseological principle. No one acquires any knowledge if someone says "I am that I am." This judgment is absolutely *true,* from a logical point of view, because the statement is a *perfect* tautology. It is absolutely *true,* because it is an analytical judgment. The frustration of critics is rooted in gnoseology, not in logic.

It is a shame that today's physics textbooks are not written following the terminology and organization of textbooks on geometry. This geometrical methodology was adopted by Newton in his *Principia.* Every conclusion in theoretical physics should be presented in the form of a *conditional proposition,* like a mathematical theorem: *If P, then Q.* Students will then truly appreciate the logical reasoning of their teachers, and the powerful logical structure of the scientific theory they are studying. In the present, very few teachers are appreciated in this respect.

Experimental science can only grow by the accumulation of *little* or particular numerical tables obtained in laboratories. These numerical tables are usually presented in the form of a graph. If the scientist knows enough mathematics, he will adjust a mathematical equation to the empirical numerical table. On the other hand, theoretical science is the science which *interconnects* little empirical laws. To develop or unfold analytical judgments, to deduce conclusions or theoretical laws, we require the existence of a theory. Any theory needs *principles,* and no principle can be an analytical judgment. Principles are not *deduced logically.* On the contrary, a set of principles allows us to deduce particular conclusions. Principles are, according to Kant, *a priori synthetic judgment.* This type of judgment is the only kind that makes science progress. In a synthetic judgment, the predicate is not implicitly contained in the subject. From a gnoseological point of view, a synthetic judgment unveils new knowledge through the predicate of the proposition. This knowledge cannot be extracted by *analysis* from the subject of the judgment. The truth or falsehood of a synthetic judgment cannot be established by logic. It is established by *experience* with the *things* of reality. The validation of a synthetic judgment is an ontological-experimental activity. Once the truth of these synthetic judgments has been *scientifically* established, they can be thrown into the syllogistic machinery to draw new conclusions or new judgments. Logic does not care, in the least, about the ontological or physical contents of synthetic judgments. Logic is only interested in the logical values: true or false, of synthetical judgments.

A priori synthetic judgments are apprehended, are grasped, and are reached by a human mind when that particular mind is in an altered state. It is a transrational act of the human mind. It is an absolute noetic activity. It is an intuitive leap of the mind in which logic and previous experiences do nothing because they cannot be involved. Newton knew this mental process, as well as Maxwell, Einstein, Schrödinger, Poincaré, Pascal, Descartes and a few others. *A priori* synthetic judgments are universal statements which help the creation of theories. Once the theory is created, the development of the theory is constituted by a large set of analytical judgments.

Principle of Non-contradiction.

Any self-contradictory judgment is false. Two contradictory judgments cannot both be true.

These are the statements of the principle of *noncontradiction* in logic. The first one refers to one self-contradictory judgment. The second one refers to two judgments, one contradicts the other. The principle of identity establishes that every analytical judgment is true, but does not establish anything about synthetic judgments. Now, the principle of contradiction says something about two contradictory synthetic judgments. It establishes that both cannot be true. It opens the possibility that both judgments may be false, or that one is true and the other synthetic judgment is false. This *uncertainty* is resolved by the introduction of the principle of the *middle excluded* as we will see in a moment. Different statements of the principle of logical noncontradiction have been advanced over the years. "The same subject does not admit contrary predicates at the same time." "Affirmation and negation cannot both be simultaneously true when they refer to the same subject."

Principle of the Middle Excluded.

Two contradictory judgments cannot both be false.

This principle eliminates the *uncertainty*, as we mentioned above, by the principle of noncontradiction, when applied to two contradictory synthetic judgments. The possibility that both contradictory judgments are false is resolved by this principle of the middle excluded. The application of the last two logical principles to the two contradictory synthetic judgments shows that one judgment *necessarily* is true and the other is logically false. Logic, however, cannot help us in deciding which synthetic judgment is true or false. This decision falls in the hands of the corresponding science.

The principle of the middle excluded has been eliminated from the axiomatic foundations of some classical logic. Brouwer and his followers deny the validity of this logical principle in mathematics. Heyting, also, created a new logic without the principle of the middle excluded. Brouwer and Heyting are advocates of mathematic and logic intuitionism which does not accept demonstrations by reduction *ad absurdum.* In the context of this logic, Bell's theorem, about *nonlocality* in quantum physics, is not valid because it proceeds by reduction to absurdity.

Principle of Sufficient Reason.
Every true judgment or false judgment is true or false because of some reason.

The four logical principles say something about the truth or falsehood of judgments. Leibniz called the principle of sufficient reason the *great principle*. The *reason* why the fourth principle is the "great principle" is because, before we apply the other principles, we have to give the reason(s) why the judgments are true or false. The only time we never give a reason why a judgment is true is when the judgment represents a principle, an axiom, a postulate or a dogma. In this case, the childish answer *just because* is the most appropriate one. Why, in Euclidean geometry, is the shortest distance between two points the straight segment between the points? Just because! A principle cannot be *deduced* from anything. The truth of any principle is accepted blindly. The acceptance of the truth of a principle is an act of *faith*. It is an *irrational* act in order to initiate the so-called *rational* mental activity.

1.8 Why Should Scientists Study Philosophy?

Physicists or any other scientist should study philosophy because philosophy is the foundation and the scaffolding of any scientific theory. Scientists of the past 20th century made a virtue of practicing the most miserable and pernicious of all philosophies which consisted of ignoring, altogether, philosophy. The great majority of scientists from the 20th century prostituted, with their positivism (which is bad philosophy) and pragmatism (which is another bad philosophy), the noble science of Natural Philosophy. Every scientist should study philosophy because philosophy is a lantern; otherwise, the poor minds of all scientists will wander, like a meta-mathematician, in an obscure labyrinth of forms and quantities, devoid of substance.

Bertrand Russell in his concise book, *The Problems of Philosophy*, has a wonderful message for *the next generation of scientists* of the 21st century. For mediocre reasons, we are forced to paraphrase two wonderful paragraphs written by this great man of the 20th century.

Russell begins by saying that human beings, who have no interest in philosophy, go through life in a portable jail which are their minds. This mental individual jail is built by prejudices generated by uncritical common sense, by personal and national dogmatic beliefs. These humans, especially scientists of the 20th century who have no conception of philosophy, see the world with their hands, and therefore, everything is familiar to them. Russell says they never see anything unfamiliar in their familiar world. Russell gives many reasons why one should study philosophy. Philosophy teaches one to question everything. It perhaps never shows the last answer of each question, but it indicates what is possible, and provides the material to build the scaffolding of new metaphysical and physical speculations about the universe. Russell sees in this human capacity of formulating questions, the most important characteristic of human beings. Philosophy, according to Russell, reduces the intolerant arrogant authority, (especially of scientists without any drop of philosophy in their minds), which predisposes their insubstantial understanding against theorization or speculation. We hope, in the near future, scientists will realize that the bread we eat today is manufactured with the techniques of yesterday; which in turn are based on the technology of previous years; which in turn is based on the science of decades ago; which in turn is based on the seemingly impractical philosophy of centuries ago. Unfortunately, in the last century the concept of "practical" (pragmatic) has become a synonym of "fact," which in turn became a synonym of "statistically significant." The collectors of scientific facts are not interested, in the lest, to look for relationships between the so-called "hard facts." They do not know Bolztmann's statement: "The most practical thing is a good theory."

Do we need new theories in physics? The answer is obvious, but as Einstein said once: "It's not enough to know we need a new theory." Perhaps we need a *Metascience*, as Paul Von Ward has suggested in his encyclopedic and visionary book *Solarian Legacy: Metascience & a New Renaissance* [11]. In the opinion of the author of *Einstein on Trial or Metaphysical Principles of Natural Philosophy*, we need a nexological or holistic science. A science that goes *beyond* present science as Von Ward has predicted. *Einstein on Trial* seems to be the foundations of the fulfillment of Von Ward's visionary *neo-renaissance*.

Conclusions

In this chapter, we saw two extraordinary methodologies: One is a Philosophical Methodology with two sets of principles: the substantial or ontological principles and the formal or logical principles. Without these sets of philosophical principles, it is impossible to identify things, to distinguish when concepts correspond to things in the real world or when to determine the genesis of things. The other set of philosophical principles, i.e., the formal principles of logic, are essential for thinking properly. The meaning is to rationalize the raw sensorial experiences and the raw transrational experiences.

Thus, the philosophical methodology teaches human beings to create theories which are theatrical plays about noumenal entities and mental events, or about natural entities and real events. The hypothetico-deductive structure which deals with noumenal experiences is called *a philosophical theory* or *metaphysics*. The other hypothetico-deductive structure which deals with natural experiences is called *natural philosophy* or *physics*, or *natural science*. In chapter 7, we will come back to this subject and propose a reclassification of philosophy.

The other extraordinary methodology is Newton's dynamic methodology. This methodology teaches us to design physical theories about the force of interaction between two material bodies and the effect on them caused by ensembles of many other material bodies. Thus, in this chapter, we have presented the two most powerful methodologies needed for our attempt to comprehend the seemingly incomprehensible world. Let us never forget that this theoretical construct is just a mental *interpretation* of reality, the world of things, as we perceive and imagine it.

In relation to Einstein, we must emphasize that he was one of the few creative physicists of the 20th century who was always aware of the philosophical and epistemological terrain on which he was stepping. On diverse occasions, Einstein behaved as an oriental poet by making "contradiction" the main character of his epistemological statements. This author believes that Einstein, like Picasso, wanted to make light of these matters. We conclude, in this chapter, that Einstein is not guilty of lacking knowledge in philosophy and philosophy of science. On the contrary, he is one of the few exceptions, among 20th century physicists, who was cultivated in philosophy.

References

1 I. Newton, *Mathematical Principles of Natural Philosophy and His System of the World* (Univ. of California Press, Berkeley, CA, 1960)

2 A. Einstein, *Ideas and Opinions* (Crown Publishers, Inc., NY, 1982) [On the Method of Theoretical Physics] p. 270

2* D.T. Susuki, *The Essentials of Zen-Buddhism: Selected from the Writings of Daisetz Teitaro Susuki* (Greenwood Press Publisher, Westport, Connecticut, 1973)

3. R.S. Westfall, *Never at Rest* (Cambridge University Press, Cambridge, New York, New Rochelle, Melbourne, Sydney, 1980)

4. E Schrödinger, *Science and Humanism: Physics in Our Times* (Cambridge at the University Press, 1952)

4 R.S. Westfall, *The Construction of Modern Science* (Cambridge University Press, Cambridge, New York, New Rochelle, Melbourne, Sidney, 1977)

6. H. Reichenbach, The Philosophical Review (Cornell University Press, Ithaca, NY, 1948) p. 338, 339

7. M. Bunge, *Causality. The place of the causal principle in science* (Cambridge, MA, Harvard University Press, 1959)

8. A. Einstein, *Relativity, the Special and General Theory*. (Mathuen, London, 1920)

9. T.E. Phipps, Jr. *Heretical Verities: Mathematical Themes in Physical Description* (Classic Non Fiction, Urbana, IL, U.S.A.)

10 J.C. Curé, Galilean Electrodynamics, **2**, no. 3, 43-47 (1991) [The Perihelic Rotation of Mercury by Newton's Original Method]

11. Paul Von Ward, *Solarian Legacy: Metascience & a New Renaissance* (Oughten House Publication, www.oughtenhouse.com)

CHAPTER 2

NEWTONIAN QUANTUM MECHANICS

TABLE OF CONTENTS

"Great spirits have always encountered violent opposition from mediocre minds"

A. Einstein

Introduction.

This chapter is an essay about the ontologico-mathematical foundations of quantum mechanics and the *quantum hidden potential.* As an essay, it is a preliminary dissertation which is to be modified later, if necessary. One of the main objectives of this essay is to solve some fundamental problems about natural philosophy such as - is there a quantum reality? Other questions we plan to answer are the following: What is the identity of the quantum hidden potential? Is it possible to deduce Schrödinger's equation from Newton's dynamics? What is the identity of de Broglie's matter waves? Can radioactivity be explained logically and ontologically? As the questions refer to fundamental problems, we need to investigate the roots of natural sciences which are nothing but their philosophical foundations. After doing this, we have what Galileo called *experimental philosophy* and what Newton called *natural philosophy*.

This author's understanding of *natural philosophy* is the ontological and epistemological analyses of natural phenomena. These analyses must be done in order to apprehend the first principles and first causes of the natural phenomena. After a long metaphysical speculation on the subject, these speculations should be *translated* into the language of mathematics. Finally, the philosophical conclusions, expressed mathematically, should be tested experimentally to verify or falsify the corresponding theory of natural philosophy.

2.1 What is the Quantum Potential?

We should note that the question requires an ontological answer, in other words, the question requires an answer about the essence of the quantum potential. According to our present knowledge, we can say nothing about the being of the quantum potential, simply because we know nothing about the ontological identity of such an entity. In the present, all we know is the mathematical definition of the quantum potential energy Q* [1]:

$$Q^* = - [\hbar^2/(2m)](\nabla^2 R)/R \qquad (2.1)$$

where R is a real function. As the quantum potential was first discovered mathematically, the problem of its physical interpretation is still unknown, like the wave function in Schrödinger's equation. The first concept about the quantum potential was that it was a "hidden variable." In 1962, David Bohm [2], when writing about hidden variables, wrote that there are "*a further set of variables, describing the state of new kinds of entities existing in a deeper subquantum mechanical level and obeying qualitatively new types of individual law.*" What about the physical origin of the quantum hidden potential? In 1987, Hiley and Peat [3, p.12] wrote this:

> "To Bohm the quantum potential arises formally from the mathematics and, in order to demonstrate the logical consistency of the whole approach, it is unnecessary to seek a deep explanation of the potential's physical origin."

Unfortunately, Bohm was not interested in the ontological identification of the quantum potential. Bohm concentrated on providing an ontological-causal meaning to quantum mechanics in general. Nevertheless, Bohm knew about the *unbroken wholeness or collectiveness of the quantum potential*, as we will soon see, but first let us see the mathematical origin of the quantum potential.

2.2 Mathematical Origin of the Quantum Potential.

The mathematical or formal origin of the quantum potential is as old as Schrödinger's equation; both were established in 1926, but the quantum potential was not recognized as such at the time. Schrödinger's equation is given by:

$$-[\hbar^2/(2m)]\,\nabla^2\psi + U\psi = \hbar\, i\, \partial\psi/\partial t \qquad (2.2)$$

In 1926, Madelung [4] introduced the following substitution: $\psi = \alpha\exp\{i\beta\}$ in the previous equation. In 1952, Bohm [1], not being acquainted with the work of Madelung, introduced the same substitution in Schrödinger's equation, with α=R(x,y,z,t) and β=S(x,y,z,t)/$\hbar$:

$$\psi = R\exp\{iS/\hbar\} \qquad (2.3)$$

We call the functional substitution, given by eq. (2.3), the Madelung-Bohm substitution. R and S are real functions. In mathematical circles, S is Hamilton's characteristic function. In Analytical Mechanics, S is known as *action.* After eq. (2.3) is introduced in eq. (2.2), the real part provides a generalized Hamilton-Jacobi's equation:

$$\partial S/\partial t + (\nabla S)^2/(2m) + U + Q^* = 0 \qquad (2.4)$$

where Q^* is the *quantum collective potential energy* given by eq.(2.1). From now on, we will call eq. (2.4) Hamilton-Jacobi-Bohm's (HJB) equation. The imaginary part, after Madelung-Bohm's substitution is introduced in Schrödinger's equation, provides the continuity equation for the quantum probability density $R^2 = \psi\psi^*$, given by:

$$\partial R^2/\partial t + \nabla\cdot[R^2\, \nabla S/m] = 0 \qquad (2.5)$$

Eq. (2.5) can also be written as:

$$2m\, \partial R/\partial t + R\, \nabla^2 S + 2\, \nabla R\cdot\nabla S = 0 \qquad (2.6)$$

Madelung interpreted HJB's equation in terms of Euler's hydrodynamic equation. Max Jammer [5a, sec. 2.3], in his book *The Philosophy of Quantum Mechanics*, offers a good summary of Madelung's interpretation. In the same reference, Jammer quotes the work of A. Isaakson, in 1927, from the Polytechnical Institute of St. Petersburg, Russia. Isaakson studied the modification which we have to introduce into Hamilton-Jacobi's equation in order to deduce Schrödinger's equation. Jammer also mentions the work published in 1927 by A. Korn, who claimed that Schrödinger's equation should be interpreted as a hydrodynamic equation of a viscous compressible fluid. Let us recall from Analytical Dynamics the total energy E and linear momentum **p** are given by:

$$E = -\, \partial S/\partial t \qquad (2.7)$$

$$\mathbf{p} = m\mathbf{v} = \nabla S \qquad (2.8)$$

Korn also looked for a hydrodynamical interpretation of quantum mechanics, and then proposed to modify eq. (2.4) by introducing the term $\varepsilon\nabla^2 S$ in place of Q^*:

$$-E + (\nabla S)^2/(2m) + U + \varepsilon\nabla^2 S = 0$$

The last term, according to Korn, is a correctional term to the classical Hamilton-Jacobi equation. The last term in Korn's equation corresponds to a particular case of the quantum potential energy Q^*, given by eq. (2.1). The case $\varepsilon = 0$ corresponds to classical dynamics, while ε K0, although very small according to Korn, corresponds to known results of quantum mechanics. In 1976, this author [6], not knowing the work of Korn, proposed the same formal substitution for Q*, but specifying the factor ε equal to [-(ℏ i/(2m)]. With this last substitution in the last equation, we immediately *linearize* it, deducing Schrödinger's equation as we will show later.

The attempts to provide a physical reality to quantum mechanics did not stop with these works of the late 1920s. The hydrodynamical models of quantum mechanics, which followed Madelung's initial concepts, were created in order to have a quantum theory describing events of real entities at an atomic level. The official quantum theory describes the probability of occurrence of the results determined at a laboratory macroscopic level. Important contributions in this ontological or realistic approach were done by N. Rosen (1945), O. Buneman (1950-55-56-1970), T. Takabayasi (1950-52-53-56), M. Schönberg (1954), D. Bohm and J.P. Vigier (1954), H.W. Franke (1954) and many others. The interested reader will find an excellent bibliography in chapter Two of *The Philosophy of Quantum Mechanics* by Max Jammer, already quoted [5a]. In the work of Schönberg, the quantum potential, given by eq. (2.1), is caused by an internal stress in the fluid which depends on the derivatives of fluid density. The nature of this quantum fluid was always undetermined. However, a simple dimensional analysis of the units, in which the terms of the HJB equation are measured, immediately show the energetic nature of them.

The HJB equation is not a hydrodynamic equation but an *ergodynamic* equation. The nature of the medium, in an interpretation of quantum ergodynamics, obviously, is *energy*. When the quantum theory was emerging in the second half of the1920's, the concept of a continuous cosmic energetic medium, pervading the entire universe, was not in the minds of any physicists at that time. Even Einstein did not bring forth his resuscitated formal *ether* of 1920 [6, and chapter 5 in this book]. At Leyden University, Einstein had revived the concept of the *luminiferous ether* in a somewhat unknown lecture he delivered in May 1920. As we will see in chapter 5, Einstein identified the cosmic ether with the components of the metric tensor of his general relativity theory. This Einsteinian identification was completely *unessential*. Without a medium filling up the vacuum, like the cosmic continuum of pure energy in the theoretical foundations of quantum theory, there are many quantum absurdities found today in the innumerable interpretations of the quantum formalism. Dirac [7], in 1951, and Bohm and Vigier [8], in 1954, tried to reintroduce the concept of ether in quantum mechanics. The work of Bohm and Vigier is particularly interesting because their interpretation of the *vacuum fluctuations* is ontologized by the fluctuations of the ether.

Finally, we should mention the work of Louis de Broglie [9], who, in 1927, attempted to provide to quantum mechanics, or better yet, to *wave mechanics*, the "physical reality" it never had. The "theory of the double solution" of de Broglie in 1926-1927 was indeed a program to develop a nonlinear theory to explain the *generation* of particles by fields. As de Broglie admitted, this was a theoretical concept which belonged to Einstein. In 1927 in Brussels, de Broglie presented his very incipient theory of the "double solution," referring mainly to the concept of the guidance wave, also called matter wave, pilot wave, ghost wave or empty wave. With so many names for one and the same entity, it is evident that no one knew anything about the very nature or essence of this cryptic wave. The whole concept of de Broglie's was rejected in the congress, and de Broglie himself abandoned his own theoretical program to create a nonlinear wave mechanics [9]. Louis de Broglie was a highly intuitive spirit, and his incredible ineffable leaps made it difficult to follow his thinking. He always wrote that the mass of an elementary particle was a mathematical singularity in the amplitude of a solitary nonlinear wave. Today, this solitary nonlinear wave is known as a *soliton*. One important center, where the sophisticated nonlinear mathematics of the soliton's theory is being studied, is in St. Petersburg, Russia. The anti-reductionist will not be too happy when the *onto-mathematicians,* of this new century, fulfill the dream of de Broglie's theory of the double solution. Eventually, the mass of an electron will be shown as a lump of concentrated energy in a fast rotating vortex, being guided in its translational motion by an energy-momentum potential. The quantum ergodynamic paper, published by Bohm and Vigier [8] in 1954, is the prerequisite for a young natural philosopher who would like to venture his mind into the metaphysics of nature, but expressed mathematically.

2.3 Some Perplexing Comments about Quantum Mechanics.

What is quantum mechanics? Murray Gell-Man [10], a well-known quantum physicist, once said:

> "Quantum mechanics, that mysterious, confusing discipline which none of us really understands but which we know how to use."

For Gell-Man, quantum mechanics is a mystery. Now, a mystery is a "very special" problem which is beyond human comprehension, and therefore has no rational solution. A mystery is a permanently unsolved problem because there are no tools to solve a mystery. A mystery is a transrational problem like a mystical experience that is ineffable, indescribable. When Capra wrote *The Tao of Physics* [11], he did no favor to quantum mechanics when he found so many similarities to Eastern Mysticism. The favor actually was the other way around. In *The Tao of Physics*, Capra, in his opening of chapter 2, wrote that mystical "insights cannot be communicated verbally." In chapter 3, Capra quotes D.T. Susuki and W. Heisenberg saying that in Eastern mysticism, as well as in quantum mechanics, the basic difficulty is the inability to communicate experiences because of language problems. In quantum mechanics, when one finishes calculating the expectation value of the energy of a quantum state, for example, one cannot describe, *and must not attempt to describe*, the essential elements and processes of the atomic system. Thus, Quantum Mechanics is not Quantum Statistics Mathematics but *Quantum Statistics Mysticism*. This is the reason why quantum mechanics is incomprehensible. From another point of view, Capra's book, *The Tao of Physics*, is a magnificent book for western scientists to at least get acquainted with eastern mysticism, religion and philosophy.

No wonder Feynman [12], a Nobel laureate for his work in quantum electrodynamics, plainly confessed:

> "I think it is safe to say that no one understands quantum mechanics."

But Feynman went still further when he blamed nature for being absurd, instead of blaming the human mind for having forgotten that common sense is the least common of the senses. He wrote:

> "The theory of quantum electrodynamics describes nature as absurd, from the point of view of common sense. And it agrees fully with experiments. So I hope you can accept nature as she is - absurd."

In the 20th century, the temple of physics was invaded by *meta-mathematicians*. In a letter Einstein wrote to Erhenfest [29, p.9], he told him:

> "You are one of the few theoreticians who have not been deprived of their native intelligence by the mathematical epidemic."

Quantum mechanics, an impeccable quantum metamathematics, lacks *atomic reality*; it lacks a substantial story to tell about the atom. Quantum mechanics is an *incomplete theory* because it lacks atomic ontology. Obviously, there are no *ontological beings* in this formal epistemological tragedy called quantum mechanics.

From the beginning of quantum mechanics, Einstein's intuition made him one of the most frightful opponents of this new mechanics. Einstein, over the years, insisted that quantum mechanics was not a *complete theory*, because it could not describe the quantum behavior of an isolated system. *Einstein's opposition to the quantum theory* is the title of a recent paper by Deltete and Guy [13]. In the introduction of their paper, in 1989, these authors asked, what was the point of Einstein's criticism to quantum mechanics. They gave an interesting answer by successive approximation. The answer began with an ontological criticism and ended epistemologically. In other words, the answer began by referring to the domain of *reality*, and ended by referring to the *virtuality* of the world of mathematics. They began the answer by saying that the statistical theory of quantum mechanics does not tell the *real* description of the atom. These authors went on to say that the probability statement, established by the quantum formalism, by no means described the *real physical state* of individual systems. Finally, Deltete and Guy wrote the clearest statement about Einstein's criticism to quantum mechanics:

> "More precisely still, it was that the mathematics of the quantum theory, specifically the Schrödinger wave function, must be regarded as describing an *ensemble* of systems and cannot plausibly be regarded as describing the state of an *individual* system. This is the essence of Einstein's criticism."

Few authors can summarize so magnificently Einstein's criticism to quantum mechanics. In section 8, we will come back to Einstein's criticism of the quantum mechanical description. In 1987, Hiley and Peat [3, p.15], two members of the London school of thought, founded by David Bohm, reach Einstein's conclusion but from the quantum potential point of view. They said:

> "What is even more striking is that the quantum potential cannot be expressed as a universally determined function of all the coordinates of the particles. Rather it depends on the 'quantum state' $\psi(r_1 \ldots r_n)$ of the *system as a whole*."

In the long run, everyone will admit that Einstein was always right in his incisive criticism of quantum mechanics. The only point Einstein missed in quantum mechanics was in relation to a divine game of dice. We will see that "The Old Man", indeed, plays dice with the atomic world.

2.4 Ontological Principles and some Mathematical Theorems.

In this section we will briefly review the basic concepts of chapter 1. In the Western world, the word *onto-logy* was created in the 17th century to refer to the study or science of the *being* of entities, visible material entities and invisible immaterial entities. Sometimes the word *ontology* is taken as a synonym for the word *meta-physics*. In the 20th century, the use of the words philosophy and metaphysics have become a "heresy" among the positivist theoretical physicists who constitute the majority of the Physics Establishment. This is so because, the majority of theoretical physicists are not natural philosophers but meta-mathematicians who practice the most pernicious of all philosophies, which is, completely *ignoring* philosophy. In 1882, Ernst Mach [14] closed the 19th century with a destructive statement. He published his positivist book *Science of Mechanics (The Historical-Critical Development of Mechanics.)* In the preface of this book we read:

> "The present volume is not a treatise upon the application of the principles of mechanics. Its aim is to clear up ideas, expose the real significance of the matter, and get rid of metaphysical obscurities."

In the 20th century, Werner Heisenberg [15] in his book, *Physics and Beyond*, hammered Mach's positivist lesson into the minds of young physicists by quoting one of his classmates saying: *Philosophy is the systematic misuse of nomenclature specially invented for the purpose*. The author of this book strongly believes that Philosophy is the "love for wisdom," and that wisdom, as Aristotle taught, is the permanent search for first principles and first causes of entities and events. Philosophy is the scaffolding of any good scientific theory. Quantum mechanics was erected in thin mathematical air without any substantial scaffolding. For this reason, quantum mechanics is a *mystery*.

Let us now see how to solve this mystery by converting it into a problem. Unsolved problems can be solved because we have rational tools with which to solve them. A particular mystery is a permanently unsolved problem because there are no rational tools to solve it. For this reason, we must create or resuscitate the proper tools to convert a mystery into a problem and then solve it. The set of tools we will use to solve the quantum mystery is called *ontology*. Ontology, as any other rational discipline, is founded on a set of *principles* or *axioms* which are discovered by very special human beings. In case the reader did not study chapter 1, we will summarize the ontological principles by giving some very brief comments.

1. Principle of identity.
2. Principle of non-contradiction.
3. Principle of excluded middle.
4. Principle of cause and effect.
5. Principle of reality.
6. Principle of inseparability.

On the principle of identity. To say that anything is identical to itself, is to establish a good tautology which teaches no new knowledge at all. Perhaps the greatest value of this principle is to prepare the mind to look into the material plurality of the world of things and reduce the whole to one unit. The essence of the principle of identity is unity. In practice, the principle of identity helps us reduce things to one entity. Different entities, which have been assigned with different essences, have been given several different names by varied groups of humans. A typical example in physics is the unessential difference between inertial mass and gravitational mass. Any respected physicist knows that no one should test experimentally the ontological principle of identity as Eötvös did. We will elaborate on this behavior of some physicists in chapter 6.

On the principle of non-contradiction. This principle establishes that no-thing can *be* and *not be* at the same time. The principle of complementarity was invented by Niels Bohr to reinterpret a misinterpretation of experimental results. The misinterpretation was to believe that electrons, which are corpuscles, are also diffracted as if they were waves. Thus, an electron can *be a corpuscle* and also *cannot be a corpuscle*, but a wave. Some meta-mathematicians believe that the *being* of an electron is determined by the consciousness of the physicist at the time he is designing the experiment, but which he plans to run months later. This is outrageous, repugnant, or just plain detestable.

On the principle of the excluded middle. This principle establishes that everything has *to be* or *not to be*. Another way to express this principle is to say that it is *impossible* to deny that an *electron is what it is* and an *electron is not what it is* simultaneously and in the same sense. These two last principles are very useful to analyze the quantum wave-particle duality.

On the principle of cause and effect. Being that physics is the study of nature, physicists observe natural phenomena which are particular manifestations or effects of general laws. Thus, we may say that experimental physicists record natural effects in order to discover the causes which produce such effects. Hence, any effect must have an efficient cause which generates that effect. Or, any effect must have a cause which brings that effect into existence. To claim that we have an objective science, we must be sure that cause and effect are independent of any human mind.

Presently, the majority of physicists agree that electronic quantum transitions in atoms, in other words, when an electron changes its quantum state, the change is not caused by anything. Quantum transitions in the atomic level happens *just because*. We want to emphasize that the majority of quantum physicists believe the effect called "quantum transition" happens *without any cause*. This was the interpretation of Niels Bohr, founder of the Copenhagen school of thought. Schrödinger's [16] reaction to Bohr's acausal conception, in September 1926, was this: *If we are going to stick to this damned quantum-jumping, then I regret that I ever had anything to do with quantum theory*. Thus, in quantum mechanics the principle of cause and effect is not applicable. It is useless. This is so because, in quantum mechanics there is no reality. The atomic realm is empty, there is nothing (no *thing*) to speculate about, nothing to theorize about its behavior. This is intolerable, despicable, offensive to human intelligence. This attitude or epistemological effect in the minds of quantum mechanicists, of course, has a *cause*. The cause is found in ontological ignorance.

On the principle of reality. Ideas are virtualities, are icons, are phenomena (luminous images) in the human mind. These sensorial mental appearances are *caused* by an external world of *things* (reality, from the Latin word **res**, meaning thing, material thing), which are totally independent from any human mind. On the other hand, we have learned over the centuries that the human mind can become pregnant by *conceiving* new noumenal entities which are not caused by external reality but by the human mind itself; these are called *concepts*. In this respect, Einstein [17, p. 266] was absolutely clear when he asserted what we call now *the principle of reality*. Let us read it again:

> "The belief in an external world independent of the perceiving subject is the basis of all natural science. Since, however, sense perception only gives information of this external world or of 'physical reality' indirectly, we can only grasp the latter by speculative means."

To speculate does not only mean to guess. The speculative mental activity to which Einstein refers to are: all the ontological, logical and mathematical principles, plus intuitive and imaginative abilities of the human mind.

On the principle of inseparability. We may call this principle "the principle of the unbroken wholeness." This principle establishes that it is impossible to separate two material entities from the rest of the universe in order to study their interaction between themselves. We have been doing this epistemological dichotomy in physics since Newton's time, except when we decided to analyze the periodical and secular perturbations exerted on one planet by the other planets. The perihelic motion of planet Mercury is an excellent example of the action on Mercury from a collective planetary energy potential.

Some mathematical theorems. We will mention only two mathematical theorems. The first one allows us to go from a total time variation of a singular entity into a partial description of this total time variation. One part is an intrinsic time variation while the other is due to the motion of the entity in a certain medium. Thus, if **p** is the linear momentum of a particle, the theorem establishes the following relationship:

$$d\mathbf{p}/dt = \partial\mathbf{p}/\partial t + [(\mathbf{v} - \mathbf{v}')] \cdot\nabla)\mathbf{p} \qquad (2.9)$$

where ($\mathbf{v} - \mathbf{v}'$) is the relative velocity of the test particle with respect to the source particle in an inertial reference system. This is a Galilean relative velocity. The other mathematical theorem refers to the gradient of a scalar product of two vectors **a** and **b**:

$$\nabla(\mathbf{a} \cdot\mathbf{b}) = (\mathbf{a} \cdot\nabla)\mathbf{b} + (\mathbf{b}\cdot\nabla)\mathbf{a} + \mathbf{a}\mathrm{x}(\nabla\mathrm{x}\mathbf{b}) + \mathbf{b}\mathrm{x}(\nabla\mathrm{x}\mathbf{a}) \qquad (2.10)$$

In case that **a** = **b** = **v**, the previous equation is reduced to:

$$(\mathbf{v} \cdot\nabla)\mathbf{v} = \tfrac{1}{2}\nabla(\mathbf{v} \cdot\mathbf{v}) - \mathbf{v}\mathrm{x}(\nabla\mathrm{x}\mathbf{v}) \qquad (2.11)$$

In the next section, we will deduce a generalized Hamilton-Jacobi equation directly from Newton's second axiom of motion.

2.5 Ontological Origin of the Quantum Potential.

As natural philosophers, we have no right to doubt the impeccable logic of *The Mathematical Foundations of Quantum Mechanics* in the hands of John von Neumann [18], but we have every right to doubt some of the initial hypotheses or some strong *affirmations*. As a matter of fact, we must strongly disagree with an extraordinary scholar such as Max Jammer [19, p. 192] when he declares that:

> **"A logical derivation of Schrödinger's equation from classical mechanics, does not, and cannot exist."**

To say that such a derivation *cannot exist* is to express an opinion of excessive logical arrogance. Jammer forgets that physics is Natural Philosophy and not the formal science of logic nor the formal science of mathematics. Physics or Natural Philosophy *uses* logic and mathematics as rational tools, along with ontological principles. Let us write down Newton's second axiom of motion for the interaction of an electron (particle # 1) with a proton (particle # 2), belonging to a system of N particles, including the proton. For more than three centuries, we have written Newton's axiom in the following form:

$$d\mathbf{p}/dt = \mathbf{F}_{12} \qquad (2.12)$$

However, the right hand side of the previous equation is incorrect for many reasons. From a practical point of view, we may discard the interaction of particle 1 with the rest of the (N-2) particles of the system. We can also neglect the interaction of particle 2 with the rest of the (N-2) particles. From an ontological point of view, we cannot deny the existence of all these forces acting on particles 1 and 2. The action of this collection of forces may change the distance between particles 1 and 2. This collective action of the rest of the system changes the relative distance between particles 1 and 2. In consequence, the magnitude of $\mathbf{F}_{12}$, in eq. (2.12), also changes. At the beginning of the 20th century, everyone erroneously blamed Newton's second axiom of motion for being unable to explain the atomic spectra of Helium and other heavier atoms. Another inexcusable mistake committed in those years was to very naively think that the only force of interaction between particles 1 and 2 was Coulomb's force. In chapter 4, we will see there are fifteen interactive dynamical terms in the Newtonian Relativistic Electrodynamics. This is another reason, unknown until today, why in the last century everyone blamed Newton's second axiom of motion for being incorrect.

But there is another reason why the right-hand side of eq. (2.12) is incorrect. From a dynamical point of view, it is very difficult to understand how the interaction of particles j and k, belonging to the ensemble of (N-2) particles of the atomic system, can perturb the interaction of particles 1 and 2. Nevertheless, from an energetic point of view it is easy to see that particles 1 and 2 are *immersed* in the potential energy field of *all* the pairs in the remaining (N-2) particles of the system. Thus, the interaction of all pairs, which remain in the system, affect the interaction of particles 1 and 2, through the electric potential energy field of all the pairs. Of the four terms of the following equation, the last term takes care of this last consideration:

$$d\mathbf{p}/dt = \mathbf{F}_{12} + \sum_{k=3}^{N}\mathbf{F}_{1k} + \sum_{k=3}^{N}\mathbf{F}_{2k} + \mathbf{f}^* \qquad (2.13)$$

The force **f*** is the gradient of the total potential energy of all the pair interactions of the (N-2) particles of the system, at the position of particle 1 and at the position of particle 2. Now, we have to bring into our analysis the existence of the cosmic energetic medium. The electron is not moving in a void; the electron is moving in a cosmic ocean of energy where the energy density is permanently fluctuating. This is caused by the constant change in the geometrical configuration of the rest of the particles from 3 to N. This ontological description of the interaction between two atomic *electric material particles* with themselves, and with the rest of the atomic whole, tells us that the velocity **v** of the electron, and in consequence, the linear momentum **p** of the electron not only depends on time, but also on the location of the electron and proton. Mathematically, we have to conclude that the velocity of the electron depends on x, y, z, and t. Therefore, the time derivative, in the previous equation, is a total derivative, and we can use the mathematical theorem given by eq. (2.9) under the assumption that **v'** = 0. This assumption has further implications which we will explore in chapter 6. Eq. (2.13) becomes:

$$\partial \mathbf{p}/\partial t + (\mathbf{v}\bullet\nabla)\mathbf{p} = \mathbf{F}_{12} + \sum_{j=3}^{N}\mathbf{F}_{1j} + \sum_{j=3}^{N}\mathbf{F}_{2j} + \mathbf{f}^*$$

or
$$\partial \mathbf{p}/\partial t + (\mathbf{v}\bullet\nabla)\mathbf{p} = -\nabla U + \mathbf{F}^* \tag{2.14}$$

where U is the potential energy between the electron and the proton, and **F*** is given by:

$$\mathbf{F}^* = \sum_{j=3}^{N}\mathbf{F}_{1j} + \sum_{j=3}^{N}\mathbf{F}_{2j} + \mathbf{f}^* \tag{2.15}$$

The potential energy of this global force **F*** is never considered in the determination of the Hamiltonian of the atomic system. Introducing eq. (2.11), in the previous equation, we get:

$$\partial \mathbf{p}/\partial t + [1/(2m)]\nabla(\mathbf{p}\cdot\mathbf{p}) - \mathbf{v}\mathrm{x}(\nabla\mathrm{x}\mathbf{p}) = -\nabla U + \mathbf{F}^* \tag{2.16}$$

If we use eq. (2.8), which is **p** = ∇S in the previous equation, the term ∇x**p** is identically zero because the curl of any gradient is identically zero. Moving the gradient of U to the left hand side of eq.(2.16), we get:

$$\nabla\{\, \partial S/\partial t + [1/(2m)]\,(\nabla S)^2 + U \,\} = \mathbf{F}^* \tag{2.17}$$

Let us define the scalar function of the gradient as follows:

$$\{\partial S/\partial t + [1/(2m)]\,(\nabla S)^2 + U\} = -\,Q^* \tag{2.18}$$

Introducing this last equation in eq. (2.17) we get:

$$\mathbf{F}^* = -\,\nabla Q^* \tag{2.19}$$

The collective force **F*** of the atomic system is conservative. Now, let us move Q* to the left hand side of eq. (2.18):

$$\partial S/\partial t + [1/(2m)]\,(\nabla S)^2 + U + Q^* = 0 \tag{2.20}$$

The last equation is identical to Hamilton-Jacobi's equation if Q*=0. On the other hand, eq. (2.20) is identical to eq. (2.4) which is deduced purely by formal mathematical means. Eq. (2.20) proves that it is logically and ontologically possible to deduce a generalized Hamilton-Jacobi equation directly from Newton's second axiom of motion, avoiding the use of Calculus of Variations. We will call Q*/q = Q the *quantum collective potential* of the atomic system of elementary particles. The charge of one of the interactive particles is q. We will call eq. (2.20), as we did before, the *Hamilton-Jacobi-Bohm* (HJB) equation. So that nothing is overlooked, pay special attention to the fact that all the terms of HJB's equation (2.20) represent *energy*, like kinetic energy $[1/(2m)]\,(\nabla S)^2$, or the classical potential energy U. Hence, Q* represents the quantum collective potential *energy*. The great majority of quantum ontologic-physicists erroneously refer to Q* as *quantum potential*, instead of *quantum potential energy*.

Eq. (2.20) summarizes the Newtonian ontologico-mathematical analysis of the origin, and identity of the quantum collective potential energy Q*. Beside the classical potential U/q, we now have a quantum collective potential Q whose origin cannot be localized at all. Q is *nonlocal* because the source of the quantum collective potential *is everywhere and nowhere, in particular*. The identity of Q is mainly the collective electrostatic Coulomb potential. Still, we have to consider as components of the quantum collective potential Q the electrokinetic potentials. Finally, we can say that Q is the Newtonian quantum collective potential of an atomic ensemble composed of many interactive elementary particles. Later, we will explore other properties of the collective quantum potential Q in relation to Schrödinger's equation. Now, we have to identify the nature of de Broglie's *matter waves*.

2.6 The Missing Link in Classical Mechanics: Classical Wave Mechanics.

Now, our thesis is to linearize HJB's equation. In other words, our thesis is to deduce D'Alambert's equation from HJB's equation, in terms of function S. To ease the understanding of the mathematical process, let us assume that S=S(x,t), U=U(x), and Q=Q(x,t). Now, eq. (2.20) becomes:

$$\partial S/\partial t + [1/(2m)]\,(\partial S/\partial x)^2 + f = 0 \qquad (2.21)$$

where f = f(x,t) = U + Q. Taking now the second partial derivative with respect to time, and the second partial derivative with respect to x, we get:

$$\partial^2 S/\partial t^2 + v\,[\partial^2 S/\partial x\partial t] + \partial f/\partial t = 0,$$

where $v = (1/m)\,\partial S/\partial x$

and $[\partial^2 S/\partial x\partial t] + v\,\partial^2 S/\partial x^2 + \partial f/\partial x = 0.$

Now, multiplying this last equation by -v, and adding it to the previous equation, we get:

$$\partial^2 S/\partial x^2 - (1/v^2)\,\partial^2 S/\partial t^2 = \partial f/\partial t - v\,\partial f/\partial x = G(x,t) \qquad (2.22)$$

The velocity v is a function of x and t, and it also has a random nature. We will assume that v is the average velocity of the particle in the previous equation. Eq. (2.22) finishes the proof of our thesis. Eq. (2.22) is the nonhomogeneous D'Alambert equation for the propagation of *waves* S. If Q=0 and U=constant, then we have a free particle moving in a straight line. Under these circumstances, eq. (2.22) becomes a homogeneous D'Alambert equation in which the solution can be written as a monochromatic wave:

$$S = S_o \exp\{ik(x - vt)\}$$

or

$$S = S_o \exp\{2\pi i(x/\lambda - \nu t)\} \qquad (2.23)$$

where $k = 2\pi/\lambda$ (2.24)

and $\nu = v/\lambda$ (2.25)

Using equations (2.7) and (2.8) for the case S=S(x, t), we get from eq. (2.23):

$$E = -\partial S/\partial t = E_o \exp\{2\pi i(x/\lambda - \nu t) + 1/4\} \quad (2.24)$$

$$p = \partial S/\partial x = p_o \exp\{2\pi i(x/\lambda - \nu t) + 1/4\} \quad (2.25)$$

where the amplitudes E_o and p_o are given by:

$$E_o = 2\pi S_o \nu$$

$$p_o = 2\pi S_o/\lambda$$

To evaluate the constant $2\pi S_o$, we can use the experimental data of electron dispersion or momentum potential S diffraction, and calculate the wave length λ in the manner used in X-Ray diffraction. Using the experimental data published by Davisson and Germer [21], in 1927, this author determined an average value of $6.68\text{x}10^{-34}$ (Js) for the constant $2\pi S_o$. Obviously, this numeric value corresponds to Planck's constant $h = 6.63\text{x}10^{-34}$ (Js). Now the last two equations become:

$$E_o = h\nu \quad (2.26)$$

$$p_o = h/\lambda \quad (2.27)$$

The last two equations are Planck's hypothesis and de Broglie hypothesis, respectively. They correspond to the amplitude of an energy wave E and a linear momentum wave p. At this point, we must reconsider the concept of a linear momentum wave, because eq. (2.22) is a linear-homogeneous D'Alambert equation. It is here where we have to bring de Broglie's conception of a singularity in the amplitude of a *nonlinear wave*. Undoubtedly, the mathematical process of obtaining D'Alambert's equation is equivalent to *linearizing* HJB's equation. What we should do now is directly solve the nonlinear HJB equation, given by eq. (2.20), which requires the analytical form of the quantum collective potential.

To close this section, we should say that the nature or essence of the old mysterious "matter or pilot wave" of de Broglie is simply the *energy-momentum potential* S wave. It is from this potential S where "matter in motion" is actualized: $m\mathbf{v}=\nabla S$, and energy comes into existence: $E = -\partial S/\partial t$. The epistemological duality of a wave-corpuscle vanishes in a permanent ontological energy storm, or fluctuating ocean of energy where the corpuscle moves and "almost" never loses its corpuscular identity.

2.7 Schrödinger's Equation is a Particular Case of Hamilton-Jacobi-Bohm's Equation.

Up to here we have deduced a generalized Hamilton-Jacobi-Bohm's equation from Newton's second axiom of dynamics. Now is the time to deduce Schrödinger's equation from Newton-Hamilton-Jacobi-Bohm's equation. However, let us review very briefly how Schrödinger deduced his famous equation. He started with the Hamilton-Jacobi equation in his first two papers of 1926 which were published in *Annalen der Physik* [20]. Schrödinger was well acquainted with the analogy of the Hamilton-Jacobi equation and the *eikonal* equation in optics. The D'Alambert equation, in physical optics, allows us to explain all the undulatory phenomena of light. On the other hand, the eikonal equation, which is an approximation to wave phenomena, is the foundation of geometrical optics which describes the path of rays of light. Thus, it was natural to ask if it was possible to complete, in mechanics of particles, a total analogy with optics. This completion consisted of finding a D'Alambert equation in classical mechanics. In the 19th century, Hamilton was the first to produce a geometrical-intuitive analogy. Schrödinger, in his first paper of 1926, wrote:

> "Then an undulatory mechanics has to be established, and the most obvious approach to it is the elaboration of the Hamiltonian analogy into a wave theory."

This is exactly what we did when we deduced *analytically*, not analogically, D'Alambert's equation for the function S, from Newton-Hamilton-Jacobi-Bohm's equation. Max Jammer [5b, sec. 5.3], in his book *The Conceptual Development of Quantum Mechanics*, 1966, offers a magnificent historico-conceptual development of the initial steps of Quantum Mechanics. Schrödinger, in his second paper, *arbitrarily* writes down D'Alambert's equation for a wave function ξ. He then proposes the functional substitution $\xi = \psi(x,y,z)\exp\{2\pi i\nu t\}$, and this other substitution $v/\nu = h/[2m(E-U)]^{1/2}$. After these substitutions are introduced in D'Alambert's equation, Schrödinger obtains his memorable stationary equation:

$$\left[\hbar^2/(2m)\right]\nabla^2\Psi + (E - U)\Psi = 0 \qquad (2.28)$$

In the first paper, Schrödinger begins with Hamilton-Jacobi's equation and proposes the functional substitution:

$$S = k \log\psi \qquad (2.29)$$

[In reference to this last equation one is tempted to see Boltzmann's fundamental concept of the entropy "S" related to the thermodynamic statistical probability ψ.] Let us go back to the meaning of S in eq.(2.29). Once the substitution is performed, the result is:

$$L = (\nabla\psi)^2 - [2m/k^2](E - U)\psi^2 = 0$$

Finally, the variational integral of L allows one to, again, obtain the stationary Schrödinger's equation if we choose $k= h/(2\pi) = \hbar$, Planck constant h divided by 2π. In these two derivations, Schrödinger refers to Hamilton-Jacobi's equation without the quantum collective potential energy, of course. In the fourth communication of 1926, Schrödinger presented his time-dependent wave equation given by eq. (2.2) in this chapter. In all these derivations, Schrödinger begins with Hamilton-Jacobi's equation. No wonder Madelung [4], in 1926, and Bohm [2], in 1952, introducing ψ=Rexp{iS/$\hbar$) in Schrödinger's equation, recovered a *modified* Hamilton-Jacobi equation.

In the previous section, we proved that HJB's equation can be *linearized* into a nonhomogeneous D'Alambert wave equation, given by eq. h(2.22). This step formalizes the optical-mechanical analogy used by Schrödinger which was advanced by Hamilton almost a century before. An excellent treatment of Hamilton's work is presented by C. Lanczos [22, p. 264]. Now we ask - is there another way to *linearize* HJB's equation? The answer is in the affirmative. To accomplish this other linearization, this author [23], in 1976, tried the following substitution:

$$Q^* = [- \hbar i/(2m)]\ \nabla^2 S \qquad (2.30)$$

Introducing the previous equation in HJB's equation (2.20), we get:

$$[- \hbar i/(2m)]\ \nabla^2 S + [1/(2m)]\ \nabla S\cdot\nabla S + U = - \partial S/\partial t \qquad (2.31)$$

Multiplying the last equation by $[-2m/\hbar^2]e^{iS/\hbar}$, and after some mathematical work, the last equation becomes:

$$- \nabla\cdot[\nabla(e^{iS/\hbar})] + [2m/\hbar^2]\, U\, e^{iS/\hbar} = - [2m/(\hbar i)]\partial(e^{iS/\hbar})/\partial t \qquad (2.32)$$

Now let us define ψ as:

$$\psi = e^{iS/\hbar} \qquad (2.33)$$

Introducing the last equation in eq. (2.32), we get the famous Schrödinger's equation but showing, this time, its true mathematical *formal being;* Schrödinger's equation *is a linearized particular case* of Newton-Hamilton-Jacobi- Bohm's equation given by eq. (2.20). The quantum collective potential energy Q* must be a real function. This is so because we have already determined the origin and identity of Q*. However, eq.(2.30) seems to show that this quantum collective potential energy is not represented by a real function. Nevertheless, let us assume that S is a complex function:

$$S = S_R + iS_I$$

Introducing this last equation in eq. (2.30), we get:

$$Q^* = [\hbar/(2m)]\, \nabla^2 S_I - i\, [\hbar/(2m)]\nabla^2 S_R$$

Thus, the real part of Q* is very well represented by the Laplacian of the imaginary part of S. To generalize this derivation of Schrödinger's equation let us use Bohm's mathematical representation of Q* given by eq. (2.1):

$$Q^* = - [\, \hbar^2/(2m)]\, (\nabla^2 R)/R$$

Introducing the last equation in HJB's equation (2.20), we get:

$$(\nabla^2 R)/R - (1/\hbar^2)\nabla S\cdot\nabla S - (2m/\hbar^2)U = (2m/\hbar^2)\partial S/\partial t \qquad (2.34)$$

Now, in order to linearize the previous equation, we will add the following identity to zero obtained from eq. (2.6):

$$(i/\hbar)\nabla^2 S + (2i/\hbar)(\nabla R/R)\cdot\nabla S - (2m/\hbar i)(1/R)\partial R/\partial t = 0 \tag{2.35}$$

Adding the last two equations we get:

$$(i/\hbar)\nabla^2 S+(2i/\hbar)(\nabla R/R)\cdot\nabla S+\nabla^2 R/R -(1/\hbar^2)\nabla S\cdot\nabla S - (2m/\hbar^2)U = A \tag{2.36}$$

where A is given by:

$$A = (2m/\hbar^2)\partial S/\partial t + (2m/\hbar i)(1/R)\partial R/\partial t \tag{2.37}$$

Let us multiply eq. (2.36) by $R\, e^{iS/\hbar}$ The result can be written in the following form:

$$B - (2m/\hbar^2)U\, Re^{iS/\hbar} = (2m/\hbar i)[(i/\hbar)Re^{iS/\hbar}\partial S/\partial t + e^{iS/\hbar}\partial R/\partial t] \tag{2.38}$$

where B is the following function:

$$B=(i/\hbar)Re^{iS/\hbar}\nabla^2 S+(2i/\hbar)e^{iS/\hbar}\nabla R\cdot\nabla S+e^{iS/\hbar}\nabla^2 R - (1/\hbar^2)Re^{iS/\hbar}\nabla S\cdot\nabla S \tag{2.39}$$

After some mathematical work, it can be shown that B is the Laplacian of $Re^{iS/\hbar}$. In other words:

$$B = \nabla^2(Re^{iS/\hbar}) \tag{2.40}$$

Now, introducing the last equation in eq. (2.38) we get:

$$-\nabla^2(Re^{iS/\hbar}) + (2m/\hbar^2)\, URe^{iS/\hbar} = - (2m/\hbar i)\partial(Re^{iS/\hbar})/\partial t \tag{2.41}$$

If we use Madelung-Bohm's substitution $\psi = Re^{iS/\hbar}$ in the last equation, we then get, after rearranging the literal coefficients:

$$- (\hbar^2/2m)\, \nabla^2\psi + U\psi = \hbar i\, \partial\psi/\partial t \tag{2.42}$$

which is the time-dependent Schrödinger's equation. Recapitulating, we see that Schrödinger's equation, given by eq. (2.42), is a linearized particular case of the nonlinear Hamilton-Jacobi-Bohm's equation, given by eq. (2.20). On the other hand, we should not forget that HJB's equation is a holistic, or synergistic representation of *Newton's second axiom of motion* of two interacting particles which belong to an ensemble of N interactive particles. For this reason, from an ontological point of view (principle of inseparability), we must take into account the *existence* of all the forces of the total surrounding dynamic dominion. This collective dynamic action of the *whole* ensemble is represented by the quantum collective potential energy Q* of the entire system. Q* is function of all the spatial coordinates of the ensemble of N particles.

In section 2 of this chapter, we saw that in the original work of Bohm, he used the Madelung-Bohm functional substitution in Schrödinger's equation. After the substitution, the real part provides the HJB equation; while the imaginary part provides the continuity equation of the probability density R^2. In our treatment, we first deduced HJB's equation from Newton's second axiom of motion, and then, we deduced Schrödinger's equation. Bohm's treatment is essentially a *mathematico-physical* approach. Our treatment is essentially an *ontologico-physical* approach.

2.8 The Need for a New Electrodynamics in Nuclear Physics.

All the previous sections of this chapter have dealt mainly with the ontological identification and mathematical transformation of d**p**/dt in Newton's second axiom of motion, and, also, the synergistic action of the rest of the system with a holistic or quantum force **F*** acting on two seemingly isolated interacting particles. When people write down $\mathbf{F}_{12} = -\nabla U$, this means that U refers mainly to Coulomb's potential energy. Nevertheless, nature is much richer in electrodynamic potentials than only revealing a single electrostatic potential. If we truly want to rationally explain atomic and nuclear physics, instead of accumulating so much incomprehensible empirical data, we need to *create* a new electrodynamics with more than seven terms as the present Relativistic Electrodynamics has. This new Electrodynamics will allow us to write down the correct *Hamiltonian* in Schrödinger's equation in order to describe the atomic nucleus. This new Electrodynamics allows us to discover the true internal structure of the neutron. In chapter 3, we advance this new Newtonian Relativistic Electrodynamics, and in chapter 4, we obtain the same electrodynamics by logical and formal means. Also in chapter 4, we finally test this new electrodynamics in Eddington's model of the neutron. Let us comment here about the variation with distance of the orbital angular momentum of the electron. This variation is negligible in the atomic domain. However, this orbital angular momentum variation is very significant in the nuclear domain, reducing the orbital angular momentum of the electron to $\frac{1}{2}\hbar$ in Eddington's model. Without this latter knowledge, provided by the new Newtonian Relativistic Electrodynamics, we will never be able to causally explain the so-called *spontaneous* radioactivity of the unstable nuclei. But before we proceed in this line of thought, let us reiterate Einstein's criticism of Quantum Mechanics.

This criticism establishes that *Quantum Mechanics is an incomplete theory because it cannot describe a single system.* A single system could be, for example, one atom of hydrogen. Before the mathematical discovery of the quantum collective potential by Bohm, no one knew that the *energy momentum potential* S was a function of Q*. Whether HJB's equation is solved or not, the eventual solution of HJB's equation will show that S is a function of Q*. Hence, S is a function of the potential energy of an ensemble of N interactive particles. If we only consider the Coulomb potential energy among all the pairs of particles, the HJB's equation can be written as:

$$\partial S/\partial t + [1/(2m)]\,(\nabla S)^2 + U_{12} + \nabla\Big[\sum_{j=3}^{N}\phi_{1j} + \sum_{j=3}^{N}\phi_{2j} + \sum_{j=3}^{N+1}\sum_{k=j+1}^{N}\phi_{jk}\Big] = 0 \qquad (2.43)$$

where ϕ_{jk} is the potential energy field created by particles j and k, acting on particles 1 and 2. All the pairs of potential energies must be referred to the position of particle 1 and to the position of particle 2. Eq. (2.43), the HJB equation, shows clearly that S is a function of the coordinates of all the interactive particles of the ensemble. Therefore, S is a holistic function in *configuration space*. S carries the energetic information of the *whole* ensemble. Madelung-Bohm's substitution, given by $\psi = R\exp\{iS/\hbar)$, shows very clearly that Schrödinger's wave function ψ is a function of S, and consequently, is a function of the quantum collective potential energy Q*. Thus, the quantum wave function ψ depends on the holistic action of the entire ensemble on the electron of the hydrogen atom. As S, ψ is a function pertaining to *configuration space*. This property of ψ depends on Q*, and not the other way around. The reason why ψ is a configuration space variable is the collective quantum potential energy Q*, as is clearly shown in eq. (2.43).

In the Copenhagen interpretation of Quantum Mechanics, no one knew that ψ depended on Q*, simply because no one knew about the existence of the quantum collective potential energy Q*. In the London interpretation of the same Quantum Mechanics, though they knew the *mathematical* existence of the quantum collective potential energy, they say that Q* is a function of ψ. However, in an ontological analysis of the genesis of ψ, we saw that it is the quantum wave function which depends on the collective quantum potential energy. The important point to make here is that when we solve Schrödinger's equation, in the case of the simple system of the hydrogen atom, we are dealing with an ensemble of N interacting particles. We are not dealing with a system of only *two particles*: the electron and the proton. The *mathematical* approach of Schrödinger, in 1926, though impressive in itself, was completely incapable of unveiling *the unbroken wholeness of the entire ensemble*. The entire ensemble is one hydrogen atom surrounded by a huge *collection* of other hydrogen atoms. Here we have the meaning of Einstein's criticism to Quantum Mechanics. Running the risk of being too repetitive, let us read again the comment of Deltete and Guy [13] about Einstein's criticism:

> "More precisely still, it was that the mathematics of the quantum theory, specifically the Schrödinger wave function, must be regarded as describing an *ensemble* of systems and cannot plausibly be regarded as describing the state of an *individual* system. This is the essence of Einstein's criticism."

In spite of all the academic merits of Einstein in so many different fields of physics, he was completely rejected in respect to his criticism of quantum mechanics. In 1949, Max Born [24, p.163] refers to this epoch of rejection of Einstein:

> ". . . he kept himself aloof and sceptical. Many of us regard this as a tragedy - for him, as he gropes his way in loneliness, and for us who miss our leader and standard-bearer. I shall not try to suggest a resolution of this discord."

In this chapter, we have solved that old discord. We have redeemed Einstein's criticism of quantum mechanics. Now, we can repeat Einstein's statement loudly and clearly: *quantum mechanics is an incomplete theory*! When we use Schrödinger's equation to solve the problem of the hydrogen atom, there is a short interval of time in our minds in which we visualize a "piece of quantum reality": an electron moving in space and time around the proton. To calculate the Coulomb potential energy of the electron with respect to the proton, we imagine these particles to be separated by a distance r in space. However, we immediately annihilate that "piece of quantum reality" by referring to the electron as a *point-like particle*. We mentally destroy the essence, the *onto,* of the linear momentum of the electron by transfiguring the ontological essence of the electron linear momentum into an imaginary mathematical operator. After we obtain the wave function, we dare to calculate different probability densities for the electron, like the radial probability density, for example. Where is the *ensemble* of particles which justifies the use of probabilities when we are dealing only with a very simple system of one electron and one proton? Some physicists proposed the existence of a sub-quantum level with imaginary entities which continuously collide with the electron, imitating the Brownian motion at a molecular level. Other physicists, like J.P. Vigier *et. al.* [25, p. 169], were not satisfied with the mathematical origin of the quantum potential presented by Bohm. They realized that one should look for a causal and physical explanation of the origin of the quantum potential. Hiley and Peat [3, p.12] believe that:

> "... here there are a wide variety of possibilities and Vigier *et. al.* have adopted a particular position in which they argue that the quantum potential has its origin in 'non-locally correlated stochastic fluctuations of an underlying covariant ether'."

Now, if — *the quantum potential has its origin in 'non-locally correlated stochastic fluctuations of an underlying covariant ether'* - then according to this statement, the quantum collective potential still is an ontological unsolved mystery. The above quotation describes very well the *effect* of the quantum collective potential but not its origin.

At the location of the electron, beside the classical potential, we have the action of the quantum collective potential which is not localized anywhere, and nevertheless, it acts everywhere in the ensemble of systems: hydrogen atoms. The medium in which the electron moves in space and time is not empty, is not void. It is full of energy whose density is changing constantly. In chapter 2, we demonstrated that the essence of the cosmic ether is pure energy. Our ether has ontological roots, while Vigier's ether only has formal roots because his team adopted Dirac's ether [7]. The cause of this energy, in the vicinity of the electron under analysis, is caused by the electrons and protons of the ensemble of hydrogen atoms. Now, these other atoms are in constant motion, causing, in this way, permanent fluctuations in the quantum collective potential at the site of the electron in question. Schrödinger's equation is a very abstract and sophisticated way to write Newton's second axiom of motion of the electron, belonging to one atom, which in turn belongs to a huge ensemble of other atoms. The electron in question has an erratic or stochastic motion around its proton. But on the average, it moves in a quasi-orbit around the proton. We can rewrite eq. (2.13) in the following way:

$$d\mathbf{p}/dt = -\nabla U - \nabla Q^* \quad (2.44)$$

The quantum force: $-\nabla Q^*$ is a force which is constantly changing in magnitude and direction. This quantum force may even have a greater magnitude than the classical force: $-\nabla U$, and for a brief interval of time may point in the opposite direction of the classical force. The effect in atoms would be the emission of light by an electron transition from one quantum state to another (old quantum jump of the electron from one orbit to another). In Schrödinger's time, and after Bohr created his insubstantial school of thought in Copenhagen, people became used to repeating the absurd statement that these quantum jumps were *spontaneous,* originated or generated *without any cause*. No one in those years had any clue about the ontological existence of the quantum collective potential energy Q^*. For this reason, Schrödinger [16] said what we know already:

> "If we are going to stick to this damned quantum-jumping, then I regret that I ever had anything to do with quantum theory."

The same quantum cause-effect explains the tunneling effect. Most important yet, combining this new quantum cause-effect of the quantum collective *nuclear* potential along with the inner structure of the neutron described in chapter 4, we find an impressive conclusion. We rationally and ontologically strongly reject the idiotic Copenhagean statement that *nothing causes the phenomenon of radioactivity in an unstable nucleus*. The α and β radiations are *caused* by the *quantum collective nuclear potential energy.* We believe that now we are ready to develop an intelligent nuclear theory. The next and future generations have most of the philosophical, mathematical and experimental tools to create a comprehensive ontological quantum theory of the nucleus. Bohm has shown the new *Camino de la Fisica*. Also, the quantum collective potential energy has shown "The Old Man," indeed, plays dice with the micro-world.

2.9 Some Other Causal Explanations in Quantum Mechanics.

If it were possible to determine with accuracy the position of an electron, the linear momentum of the electron would be completely undetermined according to Heisenberg's uncertainty principle. Now, if Heisenberg's principle were true, it would be unthinkable, and completely irrational, to conceive and talk about the *trajectory* of any electron. Any trajectory, at an atomic level, would imply that we know the electron's position and velocity simultaneously and at any time. But to know the electron's velocity, it would mean a violation of Heisenberg's principle, and hence, a total contradiction with our previous conclusion that the electron linear momentum is totally undetermined. Here we see very clearly that for the Copenhagen interpretation to remain logically and mathematically *sane, coherent, rational*, it must introduce ontological *insanities*, like denying almost all the ontological principles. But Heisenberg's principle is part of the Copenhagen interpretation as well as Schrödinger's equation. Heisenberg's principle belongs to *a linear theory*. However, in this chapter we have demonstrated that Schrödinger's equation is a *particular* case of a more general and *nonlinear equation*: Hamilton-Jacobi-Bohm's equation. Thus, we must question the validity of Heisenberg's principle in the context of the London interpretation, i.e., in the context of Bohm's ontological quantum theory. To decide about the validity of Heisenberg's uncertainty principle in the context of Bohm's theory, we see at least two methods to determine the trajectory of one electron. In both cases, we will assume that the position of the electron can be determined with acceptable accuracy.

1. Newton-Hamilton-Jacobi-Bohm's Method.

Step 1. To determine the energy-momentum potential function S, we must solve the nonlinear HJB's equation.
Step 2. To determine the velocity of the electron, we use the definition $\mathbf{p} = m\mathbf{v} = \nabla S$.
Step 3. Finally, we integrate $\mathbf{v}$ to determine the trajectory of the electron.

This method would be particularly powerful if the mathematical structure of the quantum collective potential energy Q^* is known. In the near future, Q^* will be expressed in terms of an electromagnetic tensor. In the present, this method only theoretically proves the invalidity of Heisenberg's uncertainty principle in the context of Bohm's theory. This new method will keep all the formalism of the present Copenhagen interpretation of quantum mechanics, but with some reservations. Through Madelung-Bohm's functional substitution, given by eq. (2.3), we can determine the Schrödinger wave function ψ. Equation (2.3) is given by:

$$\psi = R\exp\{iS/\hbar\} \qquad (2.3)$$

The wave function ψ, determined by the last equation, must be different from the wave function ψ' determined by Schrödinger's equation. The reason is that S is the solution of a nonlinear equation (HJB's equation), while ψ' is the solution of a linear equation. Perhaps here we will find an essential difference between the predictions of the Formal Linear Copenhagen school of thought and the Ontological Nonlinear London school of thought. The function R is determined from the quantum collective potential energy Q*, given by eq. (2.1):

$$Q^* = - [\hbar^2/(2m)](\nabla^2 R)/R$$

Once ψ is known, we can use, with some reservations as we said before, all the mathematical formalism of the Copenhagen quantum mechanics. However, this method # 1 shows we can determine the trajectory of an electron, and consequently, it proves that Heisenberg's uncertainty principle is not valid in the context of Newtonian-Bohm's ontological quantum theory developed in this chapter.

2. Bohm's method.

Bohm does not propose to solve HJB's equation. From equation (2.3), we can determine the probability density R^2 given by:

$$R^2 = \psi\psi^* \qquad (2.45)$$

Then, from eq. (2.3) again, we solve for S, and use R to get:

$$S = (\hbar/i) \ln \pi(\psi\ \psi^*) \qquad (2.46)$$

Introducing the last equation in $m\mathbf{v} = \nabla S$, we get:

$$\mathbf{v} = (\hbar/2im)[\psi^*\nabla\psi - \psi\nabla\psi^*]/(\psi\psi^*) \qquad (2.47)$$

Integrating the last equation, we can determine the trajectory of an electron and disprove Heisenberg's principle in the context of Bohm's ontological quantum theory. This method is vital in explaining causally the so-called diffraction of corpuscles.

As our purpose in this chapter was to investigate the ontological foundations of quantum mechanics, and the identity of the quantum collective potential, other applications of the new concepts are left in the hands of ontological physicists. We refer the reader to the very source of the causal-ontological interpretation of quantum mechanics created by David Bohm and developed as the London School of quantum mechanics. We will mention some important books which contain a vast bibliography on the subject of causal interpretation of quantum mechanics. These books include explanations of *non-locality,* derived from Bell's theorem, by using the quantum collective potential. The reader will also find realistic explanations of the diffraction of the *energy-momentum potential* S. Among these books let us start with *Quantum Implications - Essays in Honor of David Bohm,* edited in 1987 by B.J. Hiley and F. David Peat [3]. Another book, the last book which Bohm wrote with his colleague B.J. Hiley, and published in 1993 is *The Undivided Universe - An ontological interpretation of quantum theory* [26]. In the preface of this book, Hiley wrote, *"Just as the final touches were being put to the manuscript, David died suddenly.*" But he will be remembered for many centuries to come. David Bohm killed the giant of *ontological ignorance* in natural philosophy. It is now the mission of the next generations to take all sciences back to the womb of mother philosophy and initiate the neo-renaissance of Natural Philosophy. Another book is *The Philosophy of Quantum Mechanics* by Max Jammer [27]. Particularly interesting is chapter Seven titled, "Hidden Variables." Another extraordinary book for dilettante and many physicists, is *Quantum Reality: Beyond the New Physics* by Nick Herbert [28]. Herbert makes a magnificent presentation of many different *quantum realities* according to different epistemologico-mathematical interpretations. Herbert, of course, includes in his book Bohm's synergistic interpretation of quantum mechanics. Further on, in chapter 7 of this book, we analyze another *quantum reality* presented by Herbert with the subtitle *Consciousness Creates Reality*. Another exceptional book we highly recommend is *Quantum Paradoxes and Physical Realities* by Franco Selleri [29]. In Selleri's mind, we have Galileo's *Experimental Philosophy*, i.e., Selleri's book is rich in philosophical speculations and historical accounts. His speculations are expressed later in the language of mathematics. Selleri's book also contains a vast bibliography on experimental physics, showing himself as an authentic modern Mediterranean Natural Philosopher. Obviously, we left aside many other books from the previous list. The reason is that this author did not have the time to study all of them. Nevertheless, the title of the following book written by B. d'Espagnat [30], seems extremely appropriate for the themes of this chapter. The book is *In Search of Reality*. A paper also written by d'Espagnat [31] deserves our attention: "Nonseparability and the tentative description of reality." Finally, we should mention the essay written by Bernard d'Espagnat *Meaning and being in contemporary physics* [32].

Conclusions

Our intentions, when establishing the ontological foundations of the quantum "hidden" potential, went far beyond what we expected to accomplish. The extra objectives of this essay are many. However, the main objective, announced in the title of this chapter, was totally achieved; *Newtonian Quantum Mechanics.* We definitively demonstrated that the quantum potential has been hidden in Schrödinger's wave function ψ. To accomplish this task, we went back to the very source of *dynamics*. We started with Newton's second axiom of motion and deduced many conclusions. One conclusion was the deduction of Schrödinger's equation from "classical mechanics," a task which had been considered an absolute impossibility. In disproving this impossibility, we demonstrated that wave mechanics is a logical consequence of Newton's equation of motion. In this deductive process, we found many corollaries or secondary conclusions which have been considered previously and to be of primary importance in quantum mechanics.

When mathematically expressing Newton's equation of motion, the "unbroken wholeness of the entire universe" of Thoth and David Bohm, we discovered the origin and identity of the quantum collective potential. At that moment, the ontological principle of inseparability was mathematically used in Newton's equation of motion. The by-product of these ontological and mathematical speculations was the *ergodynamic* equation of Hamilton-Jacobi-Bohm which contained the extra term of the quantum collective potential energy. For more than a century this extra term was always absent in the expert minds of meta-mathematicians.

The nonlinear partial differential equation of Hamilton-Jacobi-Bohm was linearized in two ways. One way shows that the energy-momentum potential S represents the essence of the pilot or matter wave of de Broglie or the ghost wave of Einstein. The name "matter wave" is absolutely unacceptable as well as the name "ghost wave." The result of this linearization is the D'Alambert wave equation for the energy-momentum potential S. This D'Alambert equation for S was the missing *classical wave mechanics* which we deduced analytically in this essay. We deduced the mathematical expressions of the amplitude of energy waves and linear momentum waves. They correspond to Planck-Einstein's hypothesis $E=h\nu$, and de Broglie's hypothesis $p=h/\lambda$, respectively. We have left the nonlinear problem of the propagating singularity of de Broglie-Einstein's solitary waves (solitons) unsolved, but now we have the equation to solve this problem. The equation is the Hamilton-Jacobi-Bohm equation. Before anyone attempts to solve this problem, he must express the quantum collective potential in terms of an electromagnetic tensor. Once this is done, we will see the formal and ontological similarities between Einstein's field equations and Hamilton-Jacobi-Bohm's field equations. The other way to linearize Hamilton-Jacobi-Bohm's equation is to use Madelung-Bohm's functional substitution. This procedure provides Schrödinger's equation and the continuity equation of the probability density.

We criticized one of the methods used by Schrödinger because it imposed an arbitrary D'Alambert's wave equation in mechanics. This was obtained by analogy with optics. In this essay, we proved that this *analogy* is not necessary anymore, because we deduced, with logico-mathematical rigor, the formal existence of a D'Alambert's *wave equation in classical mechanics* from the Hamilton-Jacobi-Bohm equation of *classical mechanics of particles*. The undeniable existence of the *primordial cosmic energy field* proves the absurdity of the so-called "particle-wave duality." An electron is always a corpuscle which moves in an energetic medium. The motion of the electron modifies the energy density of its immediate vicinity which propagates in this primordial cosmic energy field. We have also shown, in this essay, that Nature has never been absurd. Only the human mind can draw absurd conclusions when it ignores or violates the ontological principles along with the logical principles.

What about Einstein's position in respect to his criticism of Quantum Mechanics? Max Born [24] is very eloquent in answering this question:

> "Einstein himself stands aloof, critical, sceptical, and hoping that this episode may pass by and physics return to classical principles."

In this chapter, we have proven that Einstein definitively was not guilty when he accused *Quantum Mechanics of being incomplete*. This chapter proves that the hopes of Einstein, for the return of physics to classical principles, have been amply fulfilled. Among these classical principles, and perhaps the most important one, was Newton's second principle of motion. The other one was the principle of reality. The unbroken faith of Einstein on the principle of reality was the solid rock which kept him immovable in respect to his position against Quantum Mechanics. In this trial of chapter 2, Einstein was absolutely not guilty. He never suspected that, some day, Newton's conception of the universe was going to vindicate him. Einstein was unjustly rejected and ignored in Quantum Mechanics because of the ontological and epistemological ignorance of the majority of 20th century physicists. We are tempted to say that the present Quantum Mechanics is an *incomplete* theory because it is *too complete* to describe an ensemble of micro-systems. It does not deal with *one sys-*

tem, but with an ensemble of systems. *The great spirit of Einstein finally has triumphed over the violent opposition he encountered from the average minds of 20th century physicists!*

References

1. D. Bohm, *A Suggested Interpretation of the Quantum Theory in Terms of "Hidden" Variables* Part I, Phys. Rev., **85**, No. 2, 166-179 (1952). Part II, Phys. Rev.., **85**, No. 2, 180-193 (1952)

1. 2. D. Bohm, *Hidden Variables in Quantum Theory* found in the book *Quantum Theory* (Ed. D.R. Bates, Academic Press, N.Y., London, 1962), vol. **3**, p. 348

3. B.H. Hiley and F.D. Peat, *Quantum Implications, Essays in Honour of David Bohm* (Ed. by B.H. Hiley and F.D. Peat, Routledge & Kegan Paul, London, N.Y., 1987)

1. 4. E. Madelung, *Quantentheorie in Hydrodynamicher Form,* Zeitschrift für Physik, vol. **40**, 322-326 (1926)

5a. M. Jammer, *The Philosophy of Quantum Mechanics* (John Wiley & Sons, N.Y., 1974)

5b M Jammer, *The Conceptual Development of Quantum Mechanics* (Mc Graw-Hill Book Company, N.Y., 1966)

6. J.C. Curé, *On the Mathematical Principles of Natural Philosophy*, Bull. Am. Phys. Soc., **21**, 42 (1976)

7. P.A.M. Dirac, *Is there any aether?*, Nature, **168**, 906 (1951)

8. D. Bohm and J.P. Vigier, Phys. Rev., **96**, 208 (1954)

9. L. de Broglie, *Non-Linear Wave Mechanics: A Causal Interpretation* (Elsevier, Amsterdam, N.Y., 1960); *The Current Interpretation of Wave Mechanics* (Elsevier, Amsterdam, N.Y., 1964)

10. M. Gell-Man, *The Nature of Matter, Wolfson College Lecture, 1980* (Clarendon Press, Oxford, 1981)

11. F. Capra, *The Tao of Physics* (Shambala, Boston, 1991)

12. R.P. Feynman, *QED, The Strange Theory of Light and Matter* (Princeton University Press, Princeton, NJ, 1988)

13. R. Deltete and R. Guy, *Einstein's Opposition to the Quantum Theory*. Am. J. Phys., **58**, (7), July, 673-683 (1990)

14. E. Mach, *The Science of Mechanics*. (The Open Court Publishing Co., La Salle, Illinois, 1989)
15. W. Heisenberg, *Physics and Beyond: Encounters and Conversations* (George Allen and Unwin Publishers, London, 1957)
16. Schrödinger's statement in *Niels Bohr and the Development of Physics* (Editor W. Pauli, Pergamon Press, N.Y., London, 1955)
17. A. Einstein, *Ideas and Opinions* (Crown Publishers, Inc., NY, 1982)
18. J. von Neumann, *Mathematical Foundations of Quantum Mechanics* (Princeton University Press, Princeton, N.J., 1955)
19. M. Jammer, *Concepts of Mass in Classical and Modern Physics* (Harvard University Press, Cambridge, MA, 1961)
20. E. Schrödinger in Ref. 5a
21. C. Davisson and L.H. Germer, Phys. Rev., **30**, 705 (1927) [Diffraction of electrons by crystals of nickel.]
22. C. Lanczos, *The Variational Principles of Mechanics* (Dover Publication, Inc., N.Y., 1986)
23. Ref. 6
24. M. Born in *Albert Einstein: Philosopher-Scientist* (Edited by P.A. Schilpp, Open Court Publishing Company, La Salle, IL, 1988)
25. J.P. Vigier *et. al.*, in Ref. 3, p. 169
26. D. Bohm and B.J. Hiley, *The Undivided Universe: An Ontological Interpretation of Quantum Theory* (Routledge, 11 New Fetter Lane, London EC4P 4EE, and 29 West 35th Street, New York, NY, 1993)
27. Ref. 5b
28. Nick Herbert, *Quantum Reality: Beyond the new Physics* (Anchor Books, Doubleday, N.Y., 1985)
29. Franco Selleri, *Quantum Paradoxes and Physical Realities* (Kluwer Academic Publishers, Boston, London, 1990)
30. B. d'Espagnat, *In Search of Reality* (Springer, N.Y., 1983)
31. B. d'Espagnat, "Nonseparability and the tentative description of reality," Physics Report, **110**, 201 (1984)
32 Ref. 3, pp. 151-168

CHAPTER 3

COMPENDIUM OF ELECTROKINETICS AND ELECTRODYNAMICS

TABLE OF CONTENTS

"If *our choice is 'only' between Grassmann's and Ampère's electrokinetic forces, better we should choose Maxwell's electrokinetic force."*

J.C. Curé

Introduction.

In chapter 5, we deduce a wave density equation from Maxwell's field equations. The existence of this equation definitively proves the reality of a cosmic medium pervading the entire universe. This proof removes, once and for all, a serious objection to the concept of a cosmic ether. We also *identified* the essence of this universal medium as pure energy, concluding that the *being* of all entities is *energy.* We can affirm, therefore, that the universe is a cosmic energy field. Einstein confirms this assertion by saying in his GRT that: *"There can be no space nor any part of space without gravitational potentials; for these confer upon space its metrical qualities, without which it cannot be imagined at all."* Finally, we quoted Maxwell, identifying all forms of energy as *one and the same.* In this chapter, we will propose a *Parametrized Newtonian Relativistic Electro-Dynamics (PNRED).* From this PNRED, we can *obtain* many old electrokinetics and electrodynamics, including Maxwell-Lorentz-Einsteins' electrodynamics. All this work is done in a Newtonian-Euclidean-Causal [1] frame of reference. We will also derive Maxwell's electrokinetics, which is a generalization of Ampère's electrokinetics.

In the conceptual development of electrical science, we should not become fanatic advocates of any *electrokinetics* or *electrodynamics.* If we do so, there is no difference between relativists and dissidents. We must run the risk of being called a dissident among dissidents. Nevertheless, among present dissidents, we already have different "tribes" fighting for intellectual power. They cluster in different *isms*: Ampèrism, Gaussism, Grassmannism, Neumannism, Weberism, Riemannism, Clausiusism, Maxwellism-Lorentzism-Einsteinism, Ritzism, Whittakerism, Brownism, Marinovism, Wesleyism, and some minor *isms.* History teaches that humans have a strong unconscious inclination to bondage their minds to one *ism.* If this historical lesson is unavoidable, why not see the *good* elements in every *ism,* and become advocates of *eclecticism.* Today, we still have a controversy between Ampère and Grassmann's electrokinetics. We continue to deal with interactions between differential *current elements.* O'Rahilly [2, vol. 2, p. 514], talking about Ampère's force, wrote: "*... what is historically out-of-date and logically offensive is the almost universal failure to derive it from the professedly held electron theory.*" This means, that in the interaction of two electric conductors, we need to analyze four different electrodynamic interactions: electron-electron*, electron-proton*, proton-electron*, and proton-proton*. The use of the asterisk (*) is to label the electron and proton of one conductor. Wesley [3] and Assis [4] escape from this criticism.

3.1 A Brief Disquisition on the concept of "Field."

We will use the name "unilateral" to refer to the study of the field created by only one moving electric particle. Any unilateral field theory is not only a *one-sided* theory, but it is definitively *metaphysical*. J.C. Maxwell [5, p. 47], in his little book *Matter and Motion,* wrote about this one-sided viewpoint:

> "The mutual action between two portions of matter receives different names according to the aspect under which it is studied, and this aspect depends on the extent of the material system which forms the subject of our attention."
>
> "If we take into account the whole phenomenon of the action between the two portions of matter, we call it Stress. This stress, according to the mode in which it acts, may be described as Attraction, Repulsion, Tension, Pressure, Shearing stress, Torsion, etc.
>
> "But if, as in Article 2, we confine our attention to one of the portions of matter, we see, as it were, only one side of the transaction- namely, that which affects the portion of matter under our consideration- and we call this aspect of the phenomenon, with reference to its effect, an External Force acting on that portion of matter, and with reference to its cause we call it the Action of the other portion of matter. The opposite aspect of the stress is called the Reaction on the other portion of matter."

If we want to understand the *interaction* of two moving electric particles, i.e., the forces of action and reaction on the two electric particles, we must *"consider the whole phenomenon of the action between the two portions of matter."* A unilateral electromagnetic field theory explains the propagation of a modification of the energy content in the surroundings of a source electric particle. The unilateral field equations refer to a *metaphysical* interaction of an electric particle and a *geometrical* point in the environment of the said particle. At that mathematical point, in the unilateral field, there is *no* physical metric attribute of any *real* test electric particle. As the velocity of propagation of the electromagnetic field varies inversely proportionately to the square root of the energy density of the region, the presence of a real electric particle, at the mathematical point, modifies this velocity of propagation. Thus, the electrodynamic force, on a real test electric particle in a *bilateral* field theory, must be different in respect to a unilateral field theory. The number of force-terms in a unilateral electrodynamics must be, at least, half the number of force-terms in a bilateral electrodynamics. The number of potentials in a unilateral theory must also be, at least, half the number of potentials in a bilateral field theory.

An electromagnetic field theory describes the *propagation* of energy and momentum. To complete an electromagnetic field theory with an electrodynamic force theory, we have two methods. One is to *invent* a Lagrangian to deduce the force-terms of the total electrodynamic force. This force should act on a *real* test electric particle at a point in the unilateral field of the source electric particle. Maxwell used this method, himself, almost twenty years before Lorentz used the same Lagrangian method applied to Clausius' electrokinetic potential, as described by Whittaker [6, p. 393]. This so-called Lorentz force, deduced in this manner, looks completely independent from Maxwell's field equations. Assis [4] wrote recently: *"Maxwell's equations are independent of Lorentz's force. This means that Maxwell's equations could remain valid even if we have a force different from that of Lorentz."* This is true in the Lagrangian context. However, we have another method to extract the corresponding electrodynamics from a field theory. Let us be more explicit. From Maxwell's field equations, after the introduction of two potentials, we get two D'Alambert's equations for these potentials. Using the Liénard-Wiechert's retarded potentials, we can deduce the Liénard-Schwarzschild's electrodynamics. To honor historical precedence, we should point out that those retarded propagating potentials, scalar and magnetic-vector potentials, were introduced by Lorenz (spelled without a "t") in 1867. In notes written by Riemann in 1858, and published posthumously in 1867, he introduced a D'Alambert equation for the scalar potential. Axiomatically starting from these retarded potentials, we can deduce Maxwell's field equations and the corresponding classical electrodynamics. O'Rahilly [2, vol. 1, Chap. VI] has treated this subject with extensive references. The retarded potentials in this unilateral approach are the essential field variables that provide the force-terms acting on a real test electric particle. In this treatment, the electrodynamic force depends on Maxwell's field equations. The converse statement is also true. Thus, if new experiments show that Lorentz's force is incomplete, then Maxwell's field equations will not remain unaltered. Wesley, with his *bilateral* field theory, has capably shown that Maxwell's field equations are incomplete. We should seriously consider Wesley's pioneering work in this respect. From Wesley's seminal work in his book *Advanced Fundamental Physics* [3, Chap. 6], let us first mention his deep criticism of the Lagrangian method. Wesley wrote:

> "The Lagrangian requires a knowledge of the energy integral (the kinetic energy and the potential energy) to start with, so one has to essentially solve the problem before one can even state the problem by the Lagrangian method."

Theoretical physics began with Newton's *Principia*. This means, that in the beginning of Natural Philosophy, the concept of *force* preceded the concept of *vis viva* (alive force) or *energy*. Ontologically, a variation of energy in and around a material particle is the *cause* of the force acting on the particle. In 1883, Mascart and Joubert [7] repeated Maxwell's concern about the necessity of proving the existence and the identity of a universal medium. They wrote:

> "The great problem which the philosophy of science raises is to know the constitution of the single medium by which all physical phenomena may be explained."

Ten years earlier Maxwell's concern was written in his *Treatise:*

> "If we admit this medium as a hypothesis, I think it ought to occupy a prominent place in our investigations and that we ought to endeavour to construct a mental representation of all the details of its action; and this has been my constant aim in this treatise."

This is an ontological preoccupation for the ultimate *substance* or the essence of the *being* of all entities. This is a concern for the ultimate reality. In vulgar terms, it is the unveiling of the supreme "stuff." Every entity, visible or invisible, is made out of the same "stuff." This concern is not only scientific and ontological, but it is also theological. The existence of an energy density wave equation, as we will explain in chapter 5 of this book on Natural Philosophy, shows that the ultimate essence is *energy.* This *energy* is the manifestation of the ultimate *Being.* The first time a line integral of a force was done it should have brought to the mind of the mathematician a gnoseological message. It should have told him that he was dealing with a *continuous* dynamical function that *existed* at every point in the interval of integration. The integrand, which represents an infinitesimal amount of energy, is present in all the infinite points in the space of integration. From a pedagogical point of view, it is more comprehensible to deal with the line integral of an electrodynamic force though mathematically it may be very cumbersome. However, if we determine the line integral of an electrodynamic force, we end with the energy associated to every force-term of the corresponding bilateral electrodynamics. This is precisely the method followed by Wesley [3, p. 217] in the section entitled, *"Wesley's Generalization to Fields and Radiation"* of Weber's Electrodynamics. Is Wesley's electromagnetic field theory a unilateral theory? No! It is a bilateral theory because Wesley used Weber's Electrodynamics, which describes the *physical* interaction between two *real* electric particles. To continue, Weber's Electrodynamics shows the electrodynamic interaction force as a function of the *relative* distance, the *relative* velocity and the *relative* acceleration of two *real* particles. Einsteinian *relativistic* electrodynamics presents a force that does not depend on *relative* velocity nor on *relative* acceleration. Relativistic electrodynamics is not a relativistic theory. In fact, Einstein's electrodynamics, coming from a unilateral theory, is not even a physical theory. It is a beautiful logical, mathematical and metaphysical theory. Obviously, the number of potentials in Wesley's bilateral field theory has twice the number of potentials contained in Maxwell's unilateral field theory. We anticipated this result before. Is Weber's Electrodynamics the most general electrodynamics we have today? Does Weber's Electrodynamics contain, as a particular case, the Liénard-Schwarzschild's electrodynamics? We will answer these questions in the following sections.

3.2. The English Electromagnetics versus the German Electrodynamics.

Assis [4], in his *unique* and excellent book *Weber's Electrodynamics,* extensively quoted Maxwell's preface in his *Treatise.* Maxwell said the English method (Faraday and Maxwell himself) and the German method (Gauss, Weber, and others), from *"a philosophical point of view . . . are radically different."* Assis says that *"Weber's theory is compatible with what we call Maxwell's equations (namely, laws of Gauss, Ampère and Faraday) although it is completely different from Maxwell's conceptions in philosophical matters."* Wesley, on the other hand, sees a formal identification of these two seemingly antagonistic European methods. Wesley [3] says:

> "An action at a distance theory [German method] can be represented directly in terms of the force between two particles . . . or it can be represented in terms of intermediate fields [English method]. In the field representation, a particle or distribution of particles, is viewed as first giving rise to an intermediate field. It is then the field that acts on another particle thereby giving rise to the observed force. Although these two representations may evoke different images of the physical mechanisms involved; they are, in fact, mathematically isomorphic (when no time retardation is involved)."

Wesley resorts to a psychological description of what the human mind records through its sensorial interaction with *something* which does not belong to the human mind. These images, or platonic shadows in the cavern of our minds, are illusions or projections onto the screen of the dark chamber of our mind's photographic camera. These images are the result of the sensorial input into our organic hardware which saves this sensorial data in the memory of our *selves.* In our minds, we only have the illusions, the subtle energetic *effects* of *something* that ontologically must be the *cause*. *Reality,* the external world of things, is the *cause* of our sensorial input. Later we organize, systematize, formalize and rationalize the initially chaotic sets of sensorial illusions. Our own "selves" are just insignificant parts of an unbroken *real whole.* The overwhelming flood of sensorial experiences should convince us that a real universe *causes them.* These experiences are caused by a true world of *things.* These *things* have an *immanent, inherent being* common to every entity. We westerners have dogmatically accepted the Greek dichotomy between the "*ego" or "I"* and the *rest*. This dichotomy is the true illusion. The *"I" and the rest are one*. In the following quotation, Maxwell's concept is initially Oriental and Occidental at the end. Weber's concept is initially Occidental and Oriental at the end. Neither of them is wrong, because they are referring to an entity that is *the same!* In the ontological root, there is no essential difference in these two European approaches. In the preface of his *Treatise,* Maxwell expressed this *unity*:

> "I found that in general the results of the two methods coincided, so that the same phenomena were accounted for, and the same laws of action deduced by both methods, but that Faraday's methods [English electromagnetics] resembled those in which we begin with the whole and arrive at the parts by analysis, while the ordinary mathematical methods [German electrodynamics] were founded on the

principle of beginning with the parts and building up the whole by synthesis."

In the 17th century, we found the same antagonism between the English Astrodynamics of Newton and the French Ethereal Philosophy of Descartes's vortices in a plenum. This time the English natural philosophers were defending the "Deutsch Naturwissenschaft" of the 19th century. Simultaneously, the French natural philosophers were defending the English natural philosophy of the 19th century. Today, dissidents are defending the German approach while relativists defend the English approach but in higher dimensions. If we change the name *Descartes* for *Einstein* in the following quotation, we will translate Roger Cotes' complaint, in the preface of Newton's *Principia* (1687), to present dissidents' grievance:

> "Some there are who dislike this celestial mechanics because it contradicts the opinions of *Descartes,* and seems hardly to be reconciled with them. Let these enjoy their own opinion, but let them act fairly, and not deny the same liberty to us which they demand for themselves."

This pendular philosophical movement, between the *continuum of the whole and the discreteness of the parts* in a seemingly empty whole, can be traced back into the night of times. We will end this sterile dispute when we understand that elementary matter is just concentrated energy. Both ancient natural philosophy and modern natural philosophy are wrong because they are partially right. Let us select the good pieces of antagonistic natural philosophies, and create an eclectic natural philosophy. Soon we can say that - German Electrodynamics is *action-at-a-distance* through the English *energetic medium* of Electromagnetics-. Westfall [8], already quoted, has a meaningful suggestion to initiate this modern eclecticism in physics. He wrote:

> "Newton himself considered forces between particles, not as a denial of the mechanical philosophy [Descartes], but as the conception needed to perfect it. By adding a third category, force, to matter and motion, he sought to reconcile mathematical mechanics to the mechanical philosophy."

Liénard and Wiechert did this reconciliation but in a *unilateral* way. Recently, Wesley produced a better reconciliation in a *bilateral* way. Perhaps we should develop a *trilateral* treatment of the interaction of two particles. The third entity *alters* the *bi-interaction*. In chapter 2, we considered a *polylateral or nonlocal* action on the ontological unreal interaction of only two particles.

3.3 Compendium of Electrokinetics and Electrodynamics.

Before we look for a Parametrized Newtonian Relativistic Electro-Dynamics (PNRED), we will display most of the existing electrokinetics and electrodynamics. To simplify the nomenclature, we will put aside the Coulomb *electrostatic* force. After all, we will be talking about electrodynamic forces and not about electrostatic forces. In this way, we will avoid the c^2 term in the denominators. The forces are supposed to be between elementary particles. We now introduce the following terminology.

Coulomb's Electrostatic Force.

$$\mathbf{F}_C = Kqq'r^{-3}\,\mathbf{r} \tag{3.1}$$

$$K = 1/(4\pi\varepsilon_0) \tag{3.2}$$

q' is the electric charge of the source particle; q is the electric charge of the test particle. The following variables are referred to an *Inertial Reference System (IRS)*
R' is the vector position of the source particle.
R is the vector position of the test particle.

$$\mathbf{r} = \mathbf{R} - \mathbf{R}' \tag{3.3}$$

is the vector position of the test particle with respect to the source particle.
v' is the velocity of the source particle with respect to the IRS.
v is the velocity of the test particle with respect to the IRS.
a' is the acceleration of the source particle with respect to the IRS.
a is the acceleration of the test particle with respect to the IRS.

$$v^* = d\mathbf{r}/dt = d(\mathbf{R}-\mathbf{R}')/dt = \mathbf{v}-\mathbf{v}' \tag{3.4}$$

v* is the relative velocity of the test particle with respect to the source particle.

$$\mathbf{a}^* = d\mathbf{v}^*/dt = \mathbf{a} - \mathbf{a}' \quad (3.5)$$

a* is the relative acceleration of the test particle with respect to the source particle.

$$k = K/c^2 = 1/(4\pi\varepsilon_o c^2) = \mu_o/4\pi \quad (3.6)$$
$$g = g(r) = kqq'/r^3 \quad (3.7)$$

In relation to Newton's third axiom of action and reaction, we will call the force **F** the action, acting on the test particle. We will call the force **F'** the reaction, acting on the source particle. To determine the reaction force, we replace **F** for **F'**, **r** for **-r**, **v** for **v'**, **v'** for **v**, **a** for **a'**, and **a'** for **a**, in the **F** formula. The electrokinetic or electrodynamic force will be called:

Newtonian force (N), if the following two conditions are satisfied.
1. **F' = -F**
2. **F'**and **F** are in the same straight line. They are collinear

Quasi-Newtonian force (QN), if the following two conditions are satisfied.
1. **F' = -F**
2. **F** and **F'** are not in the same straight lines. They are not collinear.

Non-Newtonian force (NN) when **F' ≠ -F**

Any electrodynamic force (ED) contains an electrokinetic force (EK). In the list of forces that follow, we will use these abbreviations.

3.3.1 ELECTROKINETIC FORCES

Ampère (N, EK)

$$\mathbf{F} = g\mathbf{r}\{-2(\mathbf{v}\cdot\mathbf{v}') + 3\,r^{-2}(\mathbf{r}\cdot\mathbf{v})(\mathbf{r}\cdot\mathbf{v}')\} \quad (3.8)$$

Gauss (N, EK)

$$\mathbf{F} = g\,\mathbf{r}\{\mathbf{v}^{*2} - 3/2\,r^{-2}\,(\mathbf{r}\cdot\mathbf{v}^*)^2\} \quad (3.9)$$

Using eq. (3.4) in the previous equation, and after rearranging the terms, we get:

$$\mathbf{F} = g\mathbf{r}\{-2(\mathbf{v}\cdot\mathbf{v}')+3\, r^{-2}(\mathbf{r}\cdot\mathbf{v})(\mathbf{r}\cdot\mathbf{v}')+v^2 - 3/2\, r^{-2}\, (\mathbf{r}\cdot\mathbf{v})^2 - 3/2\, r^{-2}\, (\mathbf{r}\cdot\mathbf{v}')^2 + v'^2\} \qquad (3.10)$$

The first two terms in Gauss' force represent Ampère's force. Of all the electrokinetics, Gauss' is the richest electrokinetics. It contains square-velocity terms, and is classically *relativist* and invariant like Ampère's electrokinetics.

Grassmann (NN, EK)

$$\mathbf{F} = g\mathbf{v}\mathrm{x}(\mathbf{v}'\mathrm{x}\mathbf{r}) = g[\mathbf{v}'(\mathbf{r}\cdot\mathbf{v}) - \mathbf{r}(\mathbf{v}\cdot\mathbf{v}')] \qquad (3.11)$$

Neumann (N, EK). The name "Neumann" was suggested by Marinov [9, p. 299]:

$$\mathbf{F} = -\, g\mathbf{r}(\mathbf{v}\cdot\mathbf{v}') \qquad (3.12)$$

Whittaker (QN, EK)

$$\mathbf{F} = g[-\mathbf{r}(\mathbf{v}\cdot\mathbf{v}')+\mathbf{v}(\mathbf{r}\cdot\mathbf{v}')+\mathbf{v}'(\mathbf{r}\cdot\mathbf{v})] \qquad (3.13)$$

Aspden (NN, EK)

$$\mathbf{F} = g[-\mathbf{r}(\mathbf{v}\cdot\mathbf{v}')+\mathbf{v}(\mathbf{r}\cdot\mathbf{v}') -\mathbf{v}'(\mathbf{r}\cdot\mathbf{v})] \qquad (3.14)$$

Marinov (QN, EK)

$$\mathbf{F} = g[-\mathbf{r}(\mathbf{v}\cdot\mathbf{v}')+\tfrac{1}{2}\mathbf{v}(\mathbf{r}\cdot\mathbf{v}')+\tfrac{1}{2}\mathbf{v}'(\mathbf{r}\cdot\mathbf{v})] \qquad (3.15)$$

A summary of some previous electrokinetics is given by the following equation.

Maxwell (QN, EK)

$$\mathbf{F} = g[\mathbf{r}\{A(\mathbf{v}\cdot\mathbf{v}')+Br^{-2}(\mathbf{r}\cdot\mathbf{v})(\mathbf{r}\cdot\mathbf{v}')\}+C\mathbf{v}(\mathbf{r}\cdot\mathbf{v}')+D\mathbf{v}'(\mathbf{r}\cdot\mathbf{v})] \qquad (3.16)$$

To have $\mathbf{F} = -\,\mathbf{F}'$, we need the condition C = D. O'Rahilly [2, vol. 1, p. 107], translated the old mathematical notation of Maxwell's generalization of Ampère's force into a more modern notation given by eq. (3.16). O'Rahilly's coefficients A, B, C, D, are expressed in terms of an undetermined constant m:

$$A = -(3-m)/2 \tag{3.17a}$$
$$B = 3(1-m)/2 \tag{3.17b}$$
$$C = (1+m)/2 \tag{3.17c}$$
$$D = (1+m)/2 \tag{3.17d}$$

Giving different numeric values to m, we can reproduce some previous electrokinetic forces. With m = -1, we get Ampère's force. With m = 1, we get Whittaker's force. Obviously, we cannot expect to reproduce all electrokinetic forces assigning different values to m. There is, also, no physical reason O'Rahilly's coefficients A, B, C, and D should be related as follows:

$$m = 3 - 2A = 1 - 2B/3 = 2C - 1 = 2D - 1$$

If Maxwell's electrokinetic force is the most general one, then *experiments* should provide the numeric values of O'Rahilly's coefficients. If we disregard the condition C=D, and ignore the relationships given by eqs, (3.17), we can get Grassmann's electrokinetic force by assigning the following numeric values to the coefficients A through D. A= 1, B=0, C=0, and D= - 1.

3.3.2 ELECTRODYNAMIC FORCES

Let us recall that we put aside the Coulomb's electrostatic force in the following list of electrodynamic forces.

Weber (N, ED)

$$F = g\mathbf{r}\{\mathbf{v}^{*2} - 3/2\, r^{-2}\, (\mathbf{r}\cdot\mathbf{v}^*)^2 + (\mathbf{r}\cdot\mathbf{a}^*)\} \tag{3.18}$$

We observe that Weber's force differs from Gauss' force only in the term containing the relative acceleration. Using eq. (3.4) and eq. (3.5) in eq. (3.18), we get Weber's force as:

$$\mathbf{F} = g\mathbf{r}\{-2(\mathbf{v}\cdot\mathbf{v}') + 3\, r^{-2}(\mathbf{r}\cdot\mathbf{v})(\mathbf{r}\cdot\mathbf{v}') + v^2 - 3/2\, r^{-2}\, (\mathbf{r}\cdot\mathbf{v})^2 - 3/2\, r^{-2}\, (\mathbf{r}\cdot\mathbf{v}')^2 + v'^2 + (\mathbf{r}\cdot\mathbf{a}) - (\mathbf{r}\cdot\mathbf{a}')\} \tag{3.19}$$

The reader interested in a serious study of Weber's Electrodynamics, should consider Assis' [4] and Wesley's [3] books.

Riemann (QN, ED)

$$\mathbf{F} = g[\mathbf{r}\{\tfrac{1}{2}v^{*2}\} - \mathbf{v}^*(\mathbf{r}\cdot\mathbf{v}^*)+r^2\mathbf{a}^*] \qquad (3.20)$$

Riemann was the first one to introduce the nonradial accelerative term $gr^2\mathbf{a}^*$. This electrodynamics is classically relativist. Introducing eq.'s (3.4) and (3.5) in eq. (3.20), we get:

$$\mathbf{F} = g[\mathbf{r}\{\tfrac{1}{2}v^2 -(\mathbf{v}\cdot\mathbf{v}') -\tfrac{1}{2}v'^2\} -\mathbf{v}(\mathbf{r}\cdot\mathbf{v})+\mathbf{v}'(\mathbf{r}\cdot\mathbf{v}) + \mathbf{v}(\mathbf{r}\cdot\mathbf{v}') -\mathbf{v}'(\mathbf{r}\cdot\mathbf{v}')+r^2\mathbf{a} -r^2\mathbf{a}'] \qquad (3.21)$$

The combination of the second and fifth terms, in the last equation, provides Grassmann's force.

Clausius (NN, ED)

$$\mathbf{F} = g[\mathbf{r}\{-(\mathbf{v}\cdot\mathbf{v}')\}+\mathbf{v}'(\mathbf{r}\cdot\mathbf{v}) -\mathbf{v}'(\mathbf{r}\cdot\mathbf{v}') - r^2\mathbf{a}'] \qquad (3.22)$$

Clausius' electrodynamics is very poor when compared to Weber's or Riemann's electrodynamics. It contains, however, Grassmann's force in the first two terms of eq. (3.22). It is impossible to express Clausius' force in terms of relative velocity and relative acceleration.

Liénard-Schwarzschild (NN, ED)

$$\mathbf{F} = g[\mathbf{r}\{\tfrac{1}{2}v'^2 -3/2\, r^{-2} (\mathbf{r}\cdot\mathbf{v}')^2 -\tfrac{1}{2}(\mathbf{r}\cdot\mathbf{a}') -(\mathbf{v}\cdot\mathbf{v}')\} + \mathbf{v}'(\mathbf{r}\cdot\mathbf{v}) -\tfrac{1}{2}\, r^2\mathbf{a}'] \qquad (3.23)$$

The emergence and evolution of the last equation has many progenitors in an interval of time that extends from 1835 up to 1908. The forerunners were Gauss, Riemann, Lorenz, Liénard, Wiechert, Heaviside, Schwarzschild and Ritz. The conceptual core of eq. (3.23) is represented by the retarded scalar and magnetic vector potentials. The best references, to follow this development, are the books of Whittaker [6], O'Rahilly [2], and Assis [4]. Assis, in chapter 6 of his book, compares Weber's electrodynamics with Lorentz and Liénard-Schwarzschild's electrodynamics. Eq. (3.23) clearly displays Grassmann's force when the fourth and fifth terms are combined. The second term of this equation insinuates a vestige of Ampère's force. The first term of eq. (3.23) is found in Riemann and Weber's electrodynamics. Edwards *et. al.* [10] tested experimentally the term proportional to v'^2. Again, in Assis' book [4, Sec. 6.6], the interested reader will find an *encyclopedic* treatment and bibliography on the subject of an electrokinetic force, dependent on v'^2, acting on a static charge. In this respect, Wesley [3, p. 257] discusses Curé's experiment (to be described in chapter 4) that uses a Millikan Apparatus to observe the interaction of a permanent magnet with static charges.

Ritz's (NN, ED)

$$\mathbf{F} = g[\mathbf{r}\{A'\mathbf{v}^{*2}+B'r^{-2}(\mathbf{r}\cdot\mathbf{v}^*)^2 -\tfrac{1}{2}(\mathbf{r}\cdot\mathbf{a}')\}+ C'\mathbf{v}^*(\mathbf{r}\cdot\mathbf{v}^*) -\tfrac{1}{2}r^2\mathbf{a}'] \tag{3.24}$$

where $A' = - A/2 = (3-m)/4$ (3.25a)

$B' = - B/2 = - 3(1-m)/4$ (3.25b)

$C' = - C = - (1+m)/2$ (3.25c)

The constants A, B, and C are given by eqs. (3.17). O'Rahilly did the most exhaustive study on Ritz's electrodynamics [2. vol. 2, Chap. 11]. It is a shame that Ritz does not have *relative acceleration* in his "ballistic" theory of electrodynamics given by eq. (3.24). Now, let us expand eq. (3.24) by using eq. (3.4) for the relative velocity $\mathbf{v}^*$:

$$\mathbf{F} = g[\mathbf{r}\{A'(v^2+v'^2) -2A'(\mathbf{v}\cdot\mathbf{v}')+B'r^{-2}((\mathbf{r}\cdot\mathbf{v})^2+(\mathbf{r}\cdot\mathbf{v}')^2) -2B'r^{-2}(\mathbf{r}\cdot\mathbf{v})(\mathbf{r}\cdot\mathbf{v}') - \tfrac{1}{2}(\mathbf{r}\cdot\mathbf{a}')\} + C'\{\mathbf{v}(\mathbf{r}\cdot\mathbf{v}) -\mathbf{v}'(\mathbf{r}\cdot\mathbf{v}) -\mathbf{v}(\mathbf{r}\cdot\mathbf{v}')+\mathbf{v}'(\mathbf{r}\cdot\mathbf{v}')\} -\tfrac{1}{2} r^2\mathbf{a}'] \tag{3.26}$$

Ritz's electrodynamics has 12 force-terms. Rearranging these terms, we can write:

$$\mathbf{F} = \mathbf{F}_1+\mathbf{F}_2$$

$$\mathbf{F}_1 = g[\mathbf{r}\{A'v'^2 -2A'(\mathbf{v}\cdot\mathbf{v}')+B'r^{-2}(\mathbf{r}\cdot\mathbf{v}')^2 -\tfrac{1}{2}(\mathbf{r}\cdot\mathbf{a}')\} - C'\mathbf{v}'(\mathbf{r}\cdot\mathbf{v}) -\tfrac{1}{2} r^2\mathbf{a}'] \tag{3.27}$$

$$\mathbf{F}_2 = g[\mathbf{r}\{A'v^2 -2B'r^{-2}(\mathbf{r}\cdot\mathbf{v})(\mathbf{r}\cdot\mathbf{v}')+B'r^{-2}(\mathbf{r}\cdot\mathbf{v})^2\}+C'\{\mathbf{v}(\mathbf{r}\cdot\mathbf{v}) - \mathbf{v}(\mathbf{r}\cdot\mathbf{v}')+\mathbf{v}'(\mathbf{r}\cdot\mathbf{v}')\}] \tag{3.28}$$

If we arbitrarily assign to A', B' and C' the following numeric values:

$$A' = \tfrac{1}{2} \qquad (3.29a)$$
$$B' = -3/2 \qquad (3.29b)$$
$$C' = 1 \qquad (3.29c)$$

The force $\mathbf{F}_1$, given by eq. (3.27), becomes *identical* to Liénard-Schwarzschild's electrodynamics. This theoretical conclusion is always demanded from an alternate theory which pretends to replace a so-called *well-established theory.* The new theory must contain the old theory as a particular case. Weber's electrodynamic theory, unfortunately, does not contain the mathematical structure of Liénard-Schwarzschild's electrodynamics. Nevertheless, Weber's electrodynamics is fundamentally *relativist*. Perhaps a combination of these two electrodynamics might bring a better electrodynamic theory, as we will see later. Only new *experiments* can decide the reality of some or all of the force-terms of eq. (3.28). After we derive from this plurality of electrodynamics a Parametrized Newtonian Relativistic Electrodynamics, we will come back to this subject. Meanwhile, we can rearrange the force-terms of Ritz's electrodynamics, eq. (3.26), in this other way:

$$\mathbf{F} = \mathbf{f}_1 + \mathbf{f}_2$$

$$\mathbf{f}_1 = g[\mathbf{r}\{-2A'(\mathbf{v}\cdot\mathbf{v}') - 2B'r^{-2}(\mathbf{r}\cdot\mathbf{v})(\mathbf{r}\cdot\mathbf{v}')\} - C'\mathbf{v}(\mathbf{r}\cdot\mathbf{v}') - C'\mathbf{v}'(\mathbf{r}\cdot\mathbf{v})] \qquad (3.30)$$

$$\mathbf{f}_2 = g[\mathbf{r}\{A'(v^2+v'^2) + B'r^{-2}((\mathbf{r}\cdot\mathbf{v})^2+(\mathbf{r}\cdot\mathbf{v}')^2) - \tfrac{1}{2}(\mathbf{r}\cdot\mathbf{a}')\} + C'\{\mathbf{v}(\mathbf{r}\cdot\mathbf{v}) + \mathbf{v}'(\mathbf{r}\cdot\mathbf{v}')\} - \tfrac{1}{2}\, r^2\mathbf{a}'] \qquad (3.31)$$

Using eqs. (3.25) in eq. (3.30), we derive exactly Maxwell's electrokinetic force given by eq. (3.16). Ritz's force also can accommodate Clausius' electrodynamics given by eq. (3.22).

G.B. Brown (NN, ED)

Brown's [11, p. 46] electrodynamic force is formally derived from Ritz's electrodynamics for the particular value m = 5 in eqs. (3.25). Brown's force is given by:

$$\mathbf{F} = g[\mathbf{r}\{-\tfrac{1}{2}\,\mathbf{v}^{*2} + 3\,r^{-2}(\mathbf{r}\cdot\mathbf{v}^*)^2 - \tfrac{1}{2}\,(\mathbf{r}\cdot\mathbf{a}')\} - 3\mathbf{v}^*(\mathbf{r}\cdot\mathbf{v}^*) - \tfrac{1}{2}\, r^2\mathbf{a}'] \qquad (3.32)$$

Brown writes: "*This approximate formula turns out to be the same as that arrived by Ritz, based on ballistic ideas . . . The derivation of our formula (6) from results of experiment and not involving any theory, as has been done in this treatment, seems therefore to be preferable."* Brown, like Bacon and Newton, believes that universal or general statements can be *induced* from singular or particular experiments. Brown initiates his so-called *inductive* research with Maxwell's generalization of Ampère's force, thinking that even Ampère's force is absolutely an empirical law.

However, Ampère's force is impregnated with abundant *theoretical* components. Maxwell [12, vol. 2, Art. 522, p. 171] tells us the truth about this "empirical" law of Ampère:

> "The only experimental fact which we have made use of in this investigation is the fact established by Ampère that the action of a closed circuit on any portion of another circuit is perpendicular to the direction of the latter. Every other part of the investigation depends on purely mathematical considerations depending on the properties of lines in space. The reasoning therefore may be presented in a much more condensed and appropriate form by the use of the ideas and language of the mathematical method specially adapted to the expression of such geometrical relations - the Quaternions of Hamilton."

The fascinating aspect of Brown's book [11], titled *Retarded Action-at-a-Distance*, is the method he uses to figure the numerical coefficients of eq. (3.32). This equation refers to the interaction of elementary electric particles in the *microcosmos*. Brown uses astronomical data of the perihelic rotation of planet Mercury, in the *macrocosmos,* to find the numeric coefficients of his eq. (3.32). Brown suggests a formal *cosmic analogy,* a cosmic isomorphism: *The mathematical structure of the gravitodynamic forces that rule the motion of celestial bodies, is the same as the electrodynamic forces that rule the motion of electric elementary particles.* This formal *identity* of the mathematical structure of gravitodynamic and electrodynamic forces brings a formal *unification.* If a field theory is developed out of Ritz's electrodynamics, as Wesley did with Weber's electrodynamics, then we will have a formally *unified field* theory. This author is convinced that Einstein created this formally unified field theory represented by the field equations of General Relativity Theory. As we will mention in chapter 5, we should keep Einstein's field equations, but reinterpret the ontological, geometrical and physical background.

Spencer-Gauss (N, ED)

$$\mathbf{F} = g[\mathbf{r}\{½\ \mathbf{v}^*\cdot\mathbf{v}^*\} - \mathbf{v}^*(\mathbf{r}\cdot\mathbf{v}^*) + ½\ r^2\mathbf{a}^* + O(1/c^3)] \qquad (3.33)$$

Recently, in a private communication with Dr. Domina E. Spencer [13], this author translated the mathematical notation of the original equation into our present notation. Eq. (3.33) applies to low velocities of the interacting electric particles. The original equation, from which eq. (3.33) is obtained, contains more force-terms. Using eqs. (3.4) and (3.5) in eq. (3.33), we get:

$$\mathbf{F} = g[\mathbf{r}\{½\ (v^2+v'^2) - (\mathbf{v}\cdot\mathbf{v}')\} - \{\mathbf{v}(\mathbf{r}\cdot\mathbf{v}) - \mathbf{v}'(\mathbf{r}\cdot\mathbf{v}) - \mathbf{v}(\mathbf{r}\cdot\mathbf{v}') + \mathbf{v}'(\mathbf{r}\cdot\mathbf{v}')\} - ½\ r^2\mathbf{a}' + ½\ r^2\mathbf{a}] \qquad (3.34)$$

It is interesting to compare Spencer-Gauss' electrodynamics, eq. (3.33), with Ritz's force given by eq. (3.26). The main purpose of this chapter is to compare each electrokinetics and electrodynamics with the rest of them. The conclusion is that each of them has parts of the truth contained in the others.

Feynman (ES & ED)

In this case, we must include in the electrodynamics the electrostatic force. Feynman's force [14, vol. 1, p. 28-2 and vol. 2, p. 21-1] is given by:

$$\mathbf{F} = Kqq'[\mathbf{s}s^{-3} + sc^{-1}d(\mathbf{s}s^{-3})/dt + c^{-2}d^2(\mathbf{s}s^{-1})/dt^2] \qquad (3.35)$$

The retarded position vector is $\boldsymbol{s}$ at time (t-s/c). We can express eq. (3.35) in terms of *present* parameters instead of retarded ones. Feynman, Leigthon and Sands [14, vol. 1, p. 28] comment on this electrodynamics:

> "For those purists who know more (the professors who happen to be reading this), we should add that when we say that (28.3) is a complete expression of the knowledge of electrodynamics, we are not being entirely accurate . . . so we shall avoid the puzzle for as long as we can."

To find the solution of this *puzzle,* we must convince ourselves of the true existence of a *cosmic energy field*. Once our minds are aware of this cosmic energy field, our *intuition* will clearly show that the presence of a test particle in the energy field of a source electric particle, superimposed to the background cosmic energy field, *necessarily* altered the energy content in the neighborhood of the test particle. Another anti-ontological concept we must eliminate is the absurd idea that an electron is a point-like particle. An electron, and any elementary particles, are probably rotating *toroids* of energy. This is what many physicists are suggesting today. Particularly important is the toroidal or ring model of the electron advanced by Bergman and Wesley [15]. The concept of vortices of energy, which evolve to stable toroids, will expel from physics the conception of *negative and positive* electric charges. Some day in the future, we will have an *Ergodynamics of Moving Toroids of Energy.* Meanwhile, let us attempt to establish a *bilateral* electrodynamics.

3.4 Parametrized Newtonian Relativistic Electrodynamics.

The previous collection of electrodynamics suggests that we may write down a *general electrodynamics* containing all the force-terms. These force-terms are partially contained in most of them. We should call this general electrodynamics *Parametrized Newtonian Relativistic Electro-Dynamics (PNRED),* which is given by the following equation:

$$\mathbf{F} = g[\mathbf{r}\{\alpha\mathbf{v}^{*2}+\beta r^{-2}(\mathbf{r}\cdot\mathbf{v}^*)^2+\gamma(\mathbf{r}\cdot\mathbf{a}^*)\}+\delta\mathbf{v}^*(\mathbf{r}\cdot\mathbf{v}^*) + \varepsilon r^2\mathbf{a}^*] \tag{3.36}$$

The numeric values of the undetermined parameters α, β, γ, δ, and ε, should be obtained experimentally. Introducing eqs. (3.4) and (3.5) in eq. (3.36) we get:

$$\mathbf{F} = g[\mathbf{r}\{\alpha v^2+\alpha v'^2 -2\alpha(\mathbf{v}\cdot\mathbf{v}')+\beta r^{-2}(\mathbf{r}\cdot\mathbf{v})^2+\beta r^{-2}(\mathbf{r}\cdot\mathbf{v}')^2-2\beta r^{-2}(\mathbf{r}\cdot\mathbf{v})(\mathbf{r}\cdot\mathbf{v}')+\gamma(\mathbf{r}\cdot\mathbf{a}) -\gamma(\mathbf{r}\cdot\mathbf{a}')\} + \delta\{\mathbf{v}(\mathbf{r}\cdot\mathbf{v}) -\mathbf{v}'(\mathbf{r}\cdot\mathbf{v}) -\mathbf{v}(\mathbf{r}\cdot\mathbf{v}')+\mathbf{v}'(\mathbf{r}\cdot\mathbf{v}')\}+\varepsilon r^2\mathbf{a} -\varepsilon r^2\mathbf{a}'] \tag{3.37}$$

Eq. (3.37) contains Liénard-Schwarzschild's electrodynamics if we make $\alpha = \gamma = \varepsilon = ½$, $\beta = -3/2$, and $\delta = 1$. It also accommodates Maxwell's electrokinetics. Now, if we make $\alpha = 1$, $\beta = -3/2$, $\gamma = 1$, and $\delta = \varepsilon = 0$, then eq. (3.36), the PNRED, becomes identical to Weber's electrodynamics.

Is it possible to *deduce* eq. (3.37) in the context of Newton's theory of classical dynamics? The answer is an affirmative one, but we need to add two extra axioms to Newton's theory of classical dynamics. We will present this deduction in the next chapter.

3.5 Gravitodynamics and Geometrodynamics.

In the 19th century, when the only "known" force-term in astrodynamics was the gravitostatic term of Newton, physicists and astronomers created *gravitodynamics,* by analogy, with existent electrodynamics. Recently, André K. T. Assis [4] published an unusual book on the subject of *electrodynamics and gravitodynamics.* For the interested reader, it is essential to read section 7.5 titled *Weber's Law Applied to Gravitation.* There is only one historical note we would like to add to this section of Assis' book. We believe the first gravitodynamics was created by Newton under the title Proposition XLIV, Theorem XIV, Book I, of his *Principia* [16]. In this Proposition, Newton developed the mathematical procedure to solve the problem of the perihelic rotation of planet Mercury. This was almost two hundred years before Leverrier discovered it astronomically. To get a Parametrized Newtonian Relativistic Gravitodynamics, let us introduce the following terminology:

$$K^* = G = 1/(4\pi\varepsilon^*) \tag{3.38}$$

where G is the universal gravitational constant; ε^* is the gravitostatic permmitivity of the cosmic energy field.

$$k^* = K^*/c^2 = 1/(4\pi\varepsilon^* c^2) = \mu^*/4\pi \tag{3.39}$$

where μ^* is the gravitodynamic permeability of the cosmic energy field.

Newton's Gravitostatics

$$F = -K^* m m' \mathbf{r}/r^3 \tag{3.40}$$

Newton's Gravitodynamics (Parametrized Newtonian Relativistic Gravitodynamics)

$$\mathbf{F} = g^*[\mathbf{r}\{\alpha v^2 + \alpha v'^2 - 2\alpha(\mathbf{v}\cdot\mathbf{v}') + \beta r^{-2}(\mathbf{r}\cdot\mathbf{v})^2 + \beta r^{-2}(\mathbf{r}\cdot\mathbf{v}')^2 - 2\beta r^{-2}(\mathbf{r}\cdot\mathbf{v})(\mathbf{r}\cdot\mathbf{v}') + \gamma(\mathbf{r}\cdot\mathbf{a}) - \gamma(\mathbf{r}\cdot\mathbf{a}')\} + \delta\{\mathbf{v}(\mathbf{r}\cdot\mathbf{v}) - \mathbf{v}'(\mathbf{r}\cdot\mathbf{v}) - \mathbf{v}(\mathbf{r}\cdot\mathbf{v}') + \mathbf{v}'(\mathbf{r}\cdot\mathbf{v}')\} + \varepsilon r^2\mathbf{a} - \varepsilon r^2\mathbf{a}'] \tag{3.41}$$

$$g^* = -k^* m m'/r^3 \tag{3.42}$$

Einstein's Geometrodynamics

Einstein [17, p. 79], in his book *The Meaning of Relativity* writes: "*We shall accordingly have to assume . . . that the motion of a material particle, under the action only of inertia and gravitation, is described by the equation*":

$$d^2x_\mu/ds^2 + \Gamma^\mu_{\alpha\beta}(dx_\alpha/ds)(dx_\beta/ds) = 0 \qquad (3.43)$$

Einstein, after the introduction of different types of approximations arrives at the following equation:

$$d[(1+\sigma)\mathbf{v}]/dt = \nabla\,\sigma + \partial\mathbf{A}/\partial t + (\nabla\,x\mathbf{A})x\mathbf{v} \qquad (3.44)$$

The last equation is a consequence of *linearizing* the nonlinear Einstein's field equations which provide the components of the metric tensor (gravitostatic and gravitodynamic potentials) to be used in the Christoffel symbol $\Gamma^\mu_{\alpha\beta}$. The right-hand side of eq. (3.44) clearly represents Lorentz's force with a negative sign. The point we want to make here is that Einstein's geometrodynamics, given by eq. (3.43), along with his nonlinear field equations, may lead to Newton's gravitodynamics given by eq. (3.41). We should say "may lead," because we only have indirect evidence of this possibility. If this is the case, then we have the extraordinary possibility of reinterpreting Einstein's field equations. Is Newton's gravitodynamics a *magic algorithm* that gives all the solutions to the "vacuum" field equations of GRT? At this moment we do not know. In 1975, when this author visited Prof. J.A. Wheeler, he gave this author a preprint he was planning to present in the Seventh International Congress on General Relativity and Gravitation. In this preprint Wheeler wondered about this magic algorithm:

> "Section 7. <u>From Exact Solutions to all Solutions</u>? Kinnerley's comprehensive and systematic survey of what we know about exact solutions of the equations of general relativity invites the question whether there does not exist some magic algorithm that will give <u>all</u> of the solutions of the vacuum field equations. Why do I still have hope that such an algorithm will be discovered? Did I not bet Rainer K. Sachs $5 at Les Houches in 1963 that by 1973 one would know how to get all vacuum solutions? And did I not send a last minute cable to Roger Penrose asking if he could see a way to save me?

And in the end did I not have to pay?"

Can we deduce field equations from the *bilateral* Newtonian Gravitodynamics? Perhaps, Wesley can do this gargantuan mathematical work, because he already has deduced electrodynamic potentials from Weber's electrodynamics. In this new work, we must consider *all* force-terms including those proportional to the square of the velocities **v** and **v'**. Anyway, these are problems for the next generation of natural philosophers. In what follows, we refer to indirect evidence in only three cases in our attempt to identify Newton's gravitodynamics or electrodynamics with Einstein's geometrodynamics.

3.6 Probable Experimental Evidence of Forces Proportional to v'^2/c^2.

1. On Edwards' Effect.

As we mentioned before, Assis [4, sec. 6.6] treats this subject in an excellent way. We would like to add that in 1982, Curé [18], using an analogy with GRT, attempted an explanation of Edwards' Effect. If this experimental effect is independently verified without any further doubt, then Einstein's GRT will prove that Einstein's SRT is wrong because it is incomplete. By the way, SRT is sometimes called *Restricted* Theory of Relativity. *Limited* is a synonym of "restricted," so why would one prefer to use a *limited* theory instead of a *general* theory?

If we use the PNRED given by eq. (3.37) to analyze the action between a steady current I in a circular coil of radius R, and cross-section A on a particle of charge q at rest on the axis of the coil at a distance z from the plane of the coil, then the axial force of action is given by:

$$F_z = 2\pi kqI^2(\alpha-\gamma)Rz/[|\rho_e|A(R^2+z^2)^{3/2}] \qquad (3.45)$$

where $|\rho_e|$ is the absolute value of the conduction electron density in the coil. In the case of Liénard-Schwarzchild's electrodynamics, the axial force is zero along the axis of the coil where $\alpha = \gamma = 1/2$. This axial force is also zero in Weber's electrodynamics where $\alpha = \gamma = 1$. This result confirms the conclusions of Assis [4, p. 165].

2. On Marinov's Claim of the Non-conservation of Angular Momentum.

If in Maxwell's electrokinetic force eq. (3.16), we make A = -1, B = 0, and C

= D = ½, we get Marinov's latest electrokinetics [9], given by eq. (3.15). The important point to notice in Marinov's force or any other electrokinetics or electrodynamics is the *non-radial component*. This non-radial component makes the orbital angular momentum, of a revolving particle around another, variable. Keeping C undetermined,
in Marinov's electrokinetics, the result for the specific orbital angular momentum of a revolving particle is:

$$r^2 d\theta/dt = h(1 - K/r) \tag{3.46}$$

with K constant and equal to $2Cqq'/(4\pi\varepsilon_0 c^2 m')$. A similar result is valid for PNRED. By using GRT in the study of the perihelic rotation of planet Mercury, we obtain a completely isomorphic result. We would like to mention a very curious *attitude* in relation to a renowned relativist, the late Professor C. Møller [19], in two editions of his otherwise excellent book *The Theory of Relativity*. In the edition of 1952, page 350, we read:

> "However, the left-hand side of (18) cannot, in general, be interpreted as angular momentum, since the notion of a 'radius vector,' occurring in the definition of the angular momentum, has an unambiguous meaning only in a Euclidean space."

This last quotation disappeared entirely from the new edition of the same book in 1974. Eq. (18) in Møller's book is given by eq. (3.46), but written in the following way:

$$r^2 d\theta/dt/(1 - K/r) = h$$

in which h = constant. In the gravitational case $K = 2GM/c^2$. What was the justification of erasing the above quotation from Møller's book? Eddington [20, p. 89] gave the justification of this *omission* many years before Møller. Eddington, writing about the perihelic rotation of planet Mercury, says:

> "We have to be on our guard against results of this latter kind which

would only be of interest if the radius-vector were a directly measured quantity instead of a conventional coordinate. The advance of perihelion is a phenomenon of a different category."

It is understandable there must be a change in scale for the radius vector. What is not understandable, is a change in the mathematical form of a physical law, when going from an abstract tetradimensional Riemannian spacetime to an astronomical heliocentric reference system. GRT offers a unique transcendental abstract theory, which is metaphysical, and a multitude of physical theories depending on the initial and boundary conditions, which essentially, are inaccessible. When going to the spacetime dimensions of a terrestrial laboratory, we think we must critically analyze Marinov's experiments which, according to him and Einstein, violate the conservation of orbital angular momentum.

3. On DePalma's and Tate's Effects.

DePalma's effect [21] and Tate's effect [22, 23] are related to *rotating* discs and coils, respectively. In one sense, DePalma's effect resembles Rowland's experiment. With DePalma's rotating flywheel, we have no free electric charges on the flywheel. Therefore, there is no magnetic field in the surroundings of the rotating cylinder or flywheel. DePalma, however, claims to have experimentally detected some *inertial anisotropy.* He found this effect, in the neighborhood of the rotating cylinder, to be proportional to the mass, radius, and most importantly, to the square of the angular velocity of the cylinder. DePalma measured a *shift* in a timekeeper device when it is on the axis of the rotating cylinder. It is our contention that DePalma's effect is *real,* but it is caused by an electrodynamic force proportional to v'^2 as we will see in this section.

In 1968, J.B. Tate [22, 23], at the University of Houston, conducted Edwards' experiment six years before Edwards himself. In what follows, we will refer to Tate-Edwards' effect. The difference between these two experiments is that Tate put the coil into rotation, while Edwards used a nonrotating bifilar superconducting coil. In order to save present day incomplete electrodynamics, Tate's experiment avoids the *ad hoc* explanation offered by Bonnet [24] to Edwards' results.

DePalma's effect may be caused by the square of the electric currents produced by the positive and negative charges of the rotating disc. On the other hand, Tate-Edwards' effect depends on the action of the square of a steady electric current on static charges. If in eq. (3.37) we make $\mathbf{v} = 0$, and replace q' by dq', then the electrodynamic field d**E** is given by:

$$d\mathbf{E} = kdq'r^{-3}[\mathbf{r}\{\alpha v'^2 + \beta r^{-2}(\mathbf{r.v'})^2 - \gamma(\mathbf{r.a'})\} + \delta\{\mathbf{v'}(\mathbf{r.v'})\} - \varepsilon r^2\mathbf{a'}] \qquad (3.47)$$

where **a'** is the centripetal acceleration of the source electric charges, proportional to v'^2, and in consequence proportional to the square of the angular velocity, which is DePalma's empirical determination.

Some people may see this chapter as an interesting taxonomic work. Though it is true that this chapter contains a collection and a classification of many Electrokinetics and Electrodynamics, these people might have missed the most important conclusion of this chapter. The most important inference is represented by eq. (3.36). This equation shows the *Parametrized Newtonian Relativistic Electro-Dynamics,* which summarizes the best force-elements of a compendium of most of the existing Electrodynamics up to the end of the 20th century. This new Electrodynamics, which we present in this chapter, shows the existence of many more electrokinetic forces beyond the orthodox Grassmann's force. Some of these new forces depend on the square of the velocity of the test particle. SRT contains the same type of force at the cost of having a variable mass. This new Electrodynamics does not need to have a variable mass because it contains a force term proportional to v^2/c^2. This new Electrodynamics contains another new force proportional to v'^2/c^2, which is not contained in SRT-Electrodynamics. The mathematical structure of this new electrodynamic force is contained in GRT. Thus, the next generation will use a modified version of GRT to prove that SRT-electrodynamics is incomplete.

Conclusions.

In this chapter, we presented a taxonomic work on different electrokinetics and electrodynamics. In electrokinetics, we have force-terms which only depend on relative velocities. In electrodynamics, we have force-terms which depend on relative velocities and relative accelerations. All energy-potentials are electrokinetic energy-potentials depending only on relative velocities. The Lagrangian of these electrokinetic energy-potentials provide different electrodynamics. Very few authors distinguish between electromagnetics and electrodynamics. Einstein is one who clearly distinguishes between the two. Electromagnetics is a set of field equations with which to study the propagation of electromagnetic waves. Electrodynamics is a set of force-terms between two electric interacting particles. This chapter, however, presents not only a classification of many electrodynamics, but an eclectic Parametrized Newtonian Relativistic Electrodynamics.

We corrected some historical statements about the absence of forces in Maxwell's electromagnetic theory. We clearly demonstrated that Lorentz's force was deduced by Maxwell almost twenty years before Lorentz. We also deduced Maxwell's electrokinetic force. Dissidents are still deliberating as to whether Ampère's law or Grassmann's law is the proper one to use in electrodynamics. In this author's opinion, neither of the two given choices are the accurate ones. What we need to test experimentally is Maxwell's electrokinetic force given by eq. (3.16). From a theoretical point of view of the Parametrized Newtonian Relativistic Electrodynamics, we need the four terms of Maxwell's electrokinetics. In chapter 4, we will come back to the subject of Lorentz's force paternity.

We considered force-terms proportional to v'^2/c^2, and concluded that the experimental verification was long overdue; either that or experimental rejection of this type of electrokinetic force. The magnitude of this type of force is extraordinarily small at laboratory levels. For this reason, we have to design new experiments at an atomic level where the magnitude of a force proportional to v'^2/c^2 is more significant. These new atomic effects may be unveiled through a new Hamiltonian function in quantum mechanics, based on the Newtonian Relativistic Electrodynamics.

In this chapter, Einstein might be considered guilty of having thought that Newton's second axiom of motion was wrong and Lorentz's force was right. Einstein never considered the other possibility that Newton's second axiom of motion was correct, and Lorentz's force was incomplete. After careful consideration, this author felt that Einstein was not the true guilty one, but instead, all the followers of SRT of the 20th century were the guilty ones.

References

1. M. Bunge, *Causality. The place of the causal principle in science* (Cambridge, MA, Harvard University Press, 1959)
2. A. O'Rahilly, *Electromagnetic Theory* (Dover Publications, Inc., NY, 1965)
3. J.P. Wesley, *Advanced Fundamental Physics* (Benjamin Wesley - Publisher, Weiherdammstrasse 24, 7712 Blumberg, West Germany, 1991)
4. A.K.T. Assis, *Weber's Electrodynamics* (Kluwer Academic Publishers, Dordrecht, Boston, London, 1994)
5. J.C. Maxwell, *Matter and Motion* (Dover Publications, Inc.— from 1877)

6. E. Whittaker, *A History of the Theories of Aether and Electricity* (Thomas Nelson and Sons Ltd., London, NY, 1951)
7. E. Mascart and J. Joubert, *Treatise on Electricity and Magnetism* (2 volumes, London, 1883-1888)
8. R.S Westfall, *The Construction of Modern Science* (Cambridge University Press, Cambridge, New York, New Rochelle, Melbourne, Sidney, 1977)
9. S. Marinov, International Symposium on New Energy (Ed. Maurice L Albertson, Room 203 Weber Building, Colorado State University, Fort Collins, CO 80523, 1994)
10. W.F. Edwards, C.S. Kenyon, and D.K. Lemon, Physical Review D, **14**, 922-938 (1976)
11. G.B. Brown, *Retarded Action at a Distance* (Cortney Publications, 95/115 Windmill Road, Luton Bedfordshire, Great Britain, 1982)
12. J.C. Maxwell, *A Treatise on Electricity and Magnetism* (Dover Publications Inc., NY, 1954)
13. P. Moon, D.E. Spencer, A.S. Mirchandaney, U.Y. Shama, and P.J. Mann (Private communication on "The Gaussian Electrodynamic Force," 1994)
14. R.P. Feynman, R.B. Leighton, and M. Sands, *The Feynman Lectures on Physics* (Addison-Wesley Publishing Company, Inc., Reading, MA; Palo Alto, London, 1964)
15. D.L. Bergman, and J.P. Wesley, Galilean Electrodynamics, **1**, 63 (1990)
16. I. Newton, *Mathematical Principles of Natural Philosophy and His System of the World* (Univ. of California Press, Berkeley, CA, 1960)
17. A. Einstein, *The Meaning of Relativity* (Princeton University Press, Princeton, NJ, 1922)
18. J.C. Curé, Phys. Lett. B, **116**, 158 (1982)
19. C. Møller, *The Theory of Relativity* (Oxford at Clarendon Press, 1952)
20. A. Eddington, *The Mathematical Theory of Relativity* (Cambridge at the University Press, 1960)
21. R.L. Dione, *Is God Supernatural?* (Bantam Books, Inc., New York, NY, 1976), p.148
22. J.B.Tate, *A Study of the Relativistic Interaction of Moving Charges* (Thesis presented to the Faculty of the Department of Physics, The University of Houston, 1968)
23 Tate. J. B., "The scalar potential developed by a spinning current-carrying solenoid". J. Phys D: Appl. Phys. 7 2166-2168 (1974).
24. G. Bonnet, Phys. Lett., **82**, 465-467 (1981)

CHAPTER 4

NEWTONIAN RELATIVISTIC ELECTRODYNAMICS

TABLE OF CONTENTS

"For those purists who know more (the professors who happen to be reading this), we should add that when we say that (28.3) is a complete expression of the knowledge of electrodynamics, we are not being entirely accurate . . . so we shall avoid the puzzle for as long as we can." [Underline added]

R.P. Feynman

Introduction.

When Feynman writes, "*we are not being entirely accurate,*" he is saying that present Electromagnetic-Electrodynamic theory is not entirely **correct**, factual, faithful, reliable, exact, meticulous, precise, unequivocal, genuine, truthful. In a final analysis, all physicists, including Feynman and this author, of course, have been teaching the present **incorrect** Electromagnetic-Electrodynamic theory for more than a century. Feynman is, perhaps, one of the few honest physicists of the 20th century. Therefore - why do we keep on wasting time and intelligence proving that Maxwell's incorrect (incomplete) field equations, are **covariant** with respect to some spacetime transformation? Therefore - why do we not extend the **obsolete** Lorentz's Electrodynamics we are still teaching in this new century, in this new millennium? Nevertheless, the essence of this chapter was motivated by a serious accusation made by Einstein against Newton's Dynamical Methodology of his *Principia* [1], 1687. On page 300 of *Ideas and Opinions* [2], Einstein was extremely adamant referring to Newton's theoretical method. Einstein writes: "*this theoretical method is deficient in so far as the laws of force cannot be obtained by logical and formal considerations, so that their choice is a priori to a large extent arbitrary.*" In the preface of Newton's *Principia*, his dynamical methodology is very clear. His dynamical methodology consists of the following steps.

1. Study the motion of bodies. The study of the kinematics of bodies will allow the serious student to determine the mathematical structure of the acceleration of the bodies. This first step, in Newton's methodology, establishes all the kinematical *empirical* laws of the motion of bodies. Einstein, in his shocking accusation, saying that the theoretical method of Newton is *deficient*, forgot that physics or Natural Philosophy was in the past, an experimental or empirical science. Newton proposed to open all the windows of the working place in order to observe Nature. Einstein proposed to close all the windows of the working place in order not to observe Nature. Einstein wanted to deduce the mathematical structure of all the forces of Nature by the power of his mathematical mind; he did not want to degrade his concepts about the behavior of Nature with sensorial empirical data. Empirical facts were annoying elements for Einstein's mind.

2. Use the Second Axiom of Motion. Once the mathematical structure of the acceleration of the bodies is determined from the experimental data of observing actual moving bodies, Newton's method, through his second axiom of motion, established that the mathematical structure of the forces acting on bodies is equal to the product of the mass of the bodies times the acceleration of the corresponding bodies. With this methodology and Kepler's laws, Newton *deduced* his law of gravitostatic. Einstein falsely accused Newton's theoretical method by saying it was incapable of obtaining, "deducing", the laws of force "by logical and formal considerations." This accusation is highly unfair. Einstein [3, p. 272], when he claims that a theoretical construct is a free invention of the human intellect, denies the very essence of physics which is *experimental science*. Einstein, with this accusation against Newton, challenged the whole community of classical physicists to deduce the mathematical structure of all the force-terms of a new Gravitodynamics and a new Electrodynamics by purely logical and formal considerations.

Very few physicists have mastered Newton's *Principia.* In 1980, R.S. Westfall's book *Never at Rest* [4, chap. 1], presented Isaac Newton from a modern academic perspective. Westfall presented *the discovery of a new world*, made by the Lucasian professor in the second half of the 17th century. After many millennia the world received the new dynamical methodology: the first treatise *On the Gravitodynamics of Moving Celestial Bodies.* From a historical point of view the *Principia* is described by Westfall [5, chap. 10] with unusual mastery. In chapter 6, we will expose in detail another accusation of Einstein [6, p.258] concerning the incapacity of Newton's dynamics to explain the experimental equality between inertial mass and gravitational mass. Another accusation of Einstein [7, p.102] against Newton's classical mechanics is in relation to the so-called anomalous motion of planet Mercury. In chapter 6, we will discuss this latter accusation at length, based on a paper by this author [8], published in 1991. Newton's identification between the cosmic ether and absolute space was denied by Einstein [9, p. 276]. In chapter 5 we initiate the objection of this accusation, and we will continue this disapproval.

In this chapter, we accept Einstein's challenge in the name of Newton. Our goal is to deduce eq. (3.36), the Parametrized Newtonian Relativistic Electrodynamics, in a totally hermetic way, i.e., not considering, at all, any external observation of Nature. We will enter the hermetic realm of our minds, and there we will speculate as to what the mathematical structure of gravitatodynamic and electrodynamic forces should be.

4.1 Newton's Dynamical Methodology.

Newton's dynamical methodology is clearly established in the preface of his *Principia*:

> "... for the whole burden of philosophy seems to consist in this -from the phenomena of motions to investigate the forces of nature, and then from these forces to demonstrate the other phenomena;"

We will now expand on the subject of the Introduction. According to Newton's methodology, we must first study the kinematics of a moving body with the sole purpose of determining the mathematical structure of the *acceleration* of the body. Then, by using

Newton's second axiom of motion, we multiply the *constant* mass of the body by the kinematically determined *acceleration* to establish the mathematical structure of the force acting on the moving body.

It is obvious that, by direct observation, no one has ever determined the acceleration of a moving elementary electric particle in atoms. Because of this impossibility, Newton's dynamical methodology is useless in molecular, atomic and nuclear physics. Its initial success was in astrodynamics. It is unfair to blame Newtonian dynamics for the initial failure of atomic physics in the hands of Niels Bohr. Bohr used Coulomb's force. Coulomb's force was not determined by using Newton's dynamical methodology. When V. Bush [10] used Weber's electrodynamics, in 1926, to study the physics of the hydrogen atom, he used an electrodynamics *not* determined by the use of Newtonian methodology. However, even today, physicists blame Newtonian dynamics for its failure in atomic and nuclear physics. The total blame should fall upon the wrong electrodynamics, which physicists are still using. Even Relativity Theory kept the right-hand side of Newton's second axiom of motion intact:

$$d(m\mathbf{v})/dt = \Sigma_j \mathbf{F}_j \qquad (4.1)$$

The fundamental problem is still in the right-hand side of the previous equation of motion. J.C. Maxwell [11. p. 105] describes Newtonian methodology in perfect terms:

> "The process of dynamical reasoning consists in deducing from successive configurations of the heavenly bodies, as observed by astronomers, their velocities and their accelerations, and in this way determining the direction and the relative magnitude of the force which acts on them.
>
> "Kepler had already prepared the way for Newton's investigation by deducing from careful study of the observations of Tycho Brahe the three laws of planetary motion which bear his name."

The most clear and accurate exposition of the empiric-logical *deduction* of Newton's gravitational law is offered by Max Born [12, p. 63]. The *necessity* of the application of the ontological principle, of cause and effect, through the axiom of action and reaction, is lucidly presented by Born. On the other hand, Einstein [13, p. 300] decided to point out a serious *defect* in Newton's dynamical methodology:

> "Classical mechanics is only a general scheme; it becomes a theory only by explicit indication of the force law (d) as was done so very successfully by Newton for celestial mechanics. From the point of view of the aim of the greatest logical simplicity of the foundations, this theoretical method is deficient in so far as the laws of force cannot be obtained by logical and formal considerations, so that their choice is *a priori* -to a large extent arbitrary. Also Newton's gravitation law of force is distinguished from other conceivable laws of force exclusively by its success."

When Einstein said that Newtonian dynamics "is only a general scheme," he could not have said it better. Newton's dynamics is a *methodology* to create theories. Newton's elaboration of the force law of gravitation was not *a priori* nor arbitrary. Newton, from the observed and measured phenomena of astronomical motions: -Kepler's observational laws investigated the acceleration of the planets and natural satellites. Later, he used his second axiom of motion to establish the mathematical structure of the centripetal force acting on the planets. Newton is very clear when he renamed this centripetal force calling it *gravitational* force (Scholium to Proposition V. Theorem V, Book III, in his *Principia*). But Newton wrote his work in Latin. He *identified* the centripetal force acting on a planet with the *gravitas* (weight) of the planet with respect to the sun. Newton brought the pedestrian concept of weight to heaven. There is no doubt we speak tautologically when we say, the weight of a body is the *gravitational* pull of the earth on the body. What we are saying is, the weight of the body is the weight of the body.

Einstein accused Newton's dynamical methodology for being experimental, for being observational, for being realistic in the sense of observing *things* in the starry nights. To perform all these activities is to do Natural Philosophy. Not for Einstein! He would have preferred that Newton deduce his gravitational law from the *aprioristic* olympus of transcendental symbolism. But Newton was English. Newton was a natural philosopher and a mathematician who created the mathematics he needed. In the following, we will attempt to do what Einstein wanted Newton to do. We will deduce a Newtonian Relativistic Gravitodynamics without looking up to the wandering lights of heaven. We will use modern Newtonian axiomatic to formally and logically deduce a general electrodynamics and a general gravitodynamics.

4.2 Extension of Newton's Axioms

To preserve the old numbering system of Newton's axioms we will start with axiom number zero.

Axiom 0. Ontological principle of existence and substance of absolute space.

Absolute space exists and its substance is pure energy.

Axiom 1. Axiological principle.

Newton's theory of dynamics is valid with respect to absolute space.

Later, we will extend the validity of this axiom to *inertial reference systems*, moving with constant velocity, with respect to absolute space. Now we can say that Newton's theory of dynamics is valid with respect to absolute space, and with respect to inertial reference systems. It is insulting to the reader's intelligence to say that Newton's dynamics *is not valid* with respect to *accelerated* reference systems. But we must insist that:

> Newton's theory of dynamics is *not valid,* regardless of whether the reference system is linearly accelerated or is in rotation with respect to absolute space.

As we will see later, this obvious elementary conclusion is necessary in order to defend Newton against the false accusations made by Einstein in 1916. Even Einstein knew that the third axiom of Newton's dynamical theory was invalid in accelerated reference systems, as we will see in chapter 6

Axiom 2. Physical principle of motion.

Definition: The mass m of a body is given by $m = a/a_s$, where a is the acceleration of the body in question and a_s is the acceleration of a *standard* body of unit mass, when it interacts with the test body. Both accelerations are measured with respect to an inertial reference system. If the previous ratio has different numeric values associated with different directions, then the reference system is not inertial. In chapter 6, we will discuss an operational definition of an inertial system and the concept of mass in more detail.

Definition: Linear momentum of a body is the product of its mass times its velocity. Axiom 2 establishes that:

The time variation of the linear momentum of a body is equal to the resultant force acting on the body.

$$d(m\mathbf{v})/dt = \Sigma_j \mathbf{F}_j \tag{4.1}$$

Axiom 3. Physical principle of action and reaction.

*The action **F** and the reaction **F'** in the interactions of two bodies are collinear, equal and opposite:*

$$\mathbf{F} = - \mathbf{F'} \tag{4.2}$$

(1) Strong principle of action and reaction (Assis [14, p. 22]), or *Newtonian* forces of action and reaction are *collinear* and satisfy eq. (4.2).

(2) Weak principle of action and reaction (Assis [14, p. 22]), or *Quasi-Newtonian* forces of action and reaction are *non-collinear* and satisfy eq. (4.2). Now we will introduce the other axioms.

Axiom 4. Principle of cosmic analogy, or principle of mathematical isomorfism.

The same mathematical structure of the forces which govern the motion of celestial bodies, have the forces that govern the motion of charged elementary particles.

Axiom 5. Principle of mathematical form. (cosmic mathematical isomorphism).

The mathematical structure of the forces of interaction between two particles in motion is directly proportional to the mathematical struc

ture of the inertial accelerations: Coriolis, centrifugal, and Euler; and inversely proportional to the relative separation of the particles.

Coriolis acceleration: $2\omega x(D\mathbf{r}/Dt) = 2\omega x\mathbf{v}$ (4.4a)
Centrifugal acceleration: $\omega x(\omega x\mathbf{r})$ (4.4b)
Euler's acceleration: $(D\omega/Dt)x\mathbf{r}$ (4.4c)

where the operator D /Dt, as it is well known, does not operate on the unit vectors of the noninertial reference system. The interaction force **F**, according to *axiom* 5, is given by:

$$\mathbf{F} = C[K_1 2\boldsymbol{\omega}x(D\mathbf{r}/Dt) + K_2\boldsymbol{\omega}x(\boldsymbol{\omega}x\mathbf{r}) + K_3(D\boldsymbol{\omega}/Dt)x\mathbf{r}]/r \qquad (4.5)$$

In the above equation, C is a physical parameter to be determined later, as well as the proportionality constants K_1, K_2, and K_3. $\boldsymbol{\omega}$ is an angular velocity to be determined in what follows.

4.3 Comments on the Axioms.

We should make no distinction between axiom, principle, postulate or dogma. Rather, we should emphasize their common characteristics of being general *irrational hypotheses.* Irrational because there is no possibility to *deduce* them logically or rationally from any other field of knowledge. No matter how *evident* the statement of an axiom is, it is still logically un-deductible. This is like the truth of a postulate. Only through an act of *faith* can scientists *believe* in the *truth* of axioms. Most of the time, scientists are completely unaware that they are *strong believers* just as religious people are. Now, let us make some comments on the modified and extended axioms of Newton.

On Axiom 0.

In the past, absolute space was identified with the luminiferous ether which was assumed to pervade the entire universe. It is said that Einstein was responsible for the elimination of the ether from physics. He felt it was a "superfluous" concept in his special relativity theory. To Einstein, it was a must to get rid of the ether which was identified with absolute space. It was a must because, Newton defined the latter saying that *"in its own nature, without relation to anything external, remains always similar and immovable."* This attribute of *immobility* of absolute space was an extremely embarrassing characteristic that violently contradicted Einstein's special relativity theory.

From an ontological point of view, the cosmic ether was void of any nature or essence. Every physicist in the past spoke of the ether but no one knew anything about its very nature. Besides, from a kinematical point of view, the cosmic ether was a total paradox, not to say absurdity. Bradley's astronomical aberration showed the ether was not dragged at all when the earth went through it; therefore, physicists concluded that an ether wind must exist. Michelson-Morley's experiment, which was intended to measure the ether-wind in order to determine the absolute velocity of planet earth, showed that the cosmic ether was totally dragged by the earth. To make things more embarrassing, Fizeau showed experimentally that running water with a speed considerably less than the speed of the earth around the sun, dragged the ether partially!

Einstein, indeed, had good reasons to eliminate such "nonsense" called the cosmic ether. However, the root of these absurd conclusions is due to the total ignorance of the ontology of the cosmic ether, as we will see in chapter 5. Running the risk of repeating ourselves, we must insist on this ontological point of absolute space. Einstein [15, p. 19-23], in 1920, finally pointed out a clue as to how to grasp the essence of the cosmic ether. Talking about general relativity theory, he said:

> "What is fundamentally new in the ether of the general theory of relativity as opposed to the ether of Lorentz consists in this, that the state of the former is at every place determined by connections with the matter and the state of the ether in neighboring places, which are amenable to law in the form of differential equations; whereas the state of the Lorentzian ether in the absence of electromagnetic fields is conditioned by nothing outside itself, and is everywhere the same . . . There can be no space nor any part of space without gravitational potential . . . From the present state of the theory it looks as if the electromagnetic field, as opposed to the gravitational field, rests upon an entirely new formal motif, as though nature might just as well have endowed the gravitational ether with fields of quite another type, for example, with fields of a scalar potential, instead of fields of the electromagnetic type . . . according to the general theory of relativity space is endowed with physical qualities; in this sense, therefore, there exists an ether. According to the general theory of relativity, space without ether is unthinkable."

Thus, according to Einstein, there is no place in the universe without gravitational potential, without a certain density of gravitational energy. If we do not consider the geometrodynamical potentials of general relativity theory, except the gravitostatic potential in the neighborhood of a celestial body of mass M, then at a point distant r from the center of the body, the density of gravitostatic energy ρ at that point is given by:

$$\rho = \rho^* + GM^2/(8\pi r^4) \qquad (4.6)$$

where ρ^* is the cosmic background energy density, or the so-called *Zero-Point-Energy-Density.* Now we can say that the universal matter creates a cosmic ocean of gravitostatic energy in which it is immersed. Keep in mind that energy is one and the same, independent of the nature of the source. Maxwell [16], in this respect, was very clear:

> "In speaking of the energy of the field . . . I wish to be understood literally. All energy is the same, whether it exists in the form of motion or that of elasticity, or any other form."

Therefore, we do not need to qualify the universal gravitostatic energy. The universal matter creates a cosmic field of energy in which it is immersed. The essence of the old luminiferous ether is plain energy. The state of this cosmic energy field, Einstein would have said, "*is determined by connections with the matter and the state of the ether in neighboring places.*"

The gravitostatic energy of the earth is superimposed to the average interstellar energy density, and the gravitostatic energy of the Milky Way is superimposed to the average galactic energy density. In this way, the *cosmos* is linked by the unbroken wholeness of the entire primordial field of energy. Now, we can identify *absolute space* with the cosmic primordial energy field.

On Axiom 1.

Corollary V in Newton's *Principia* extends the validity of axiom # 1 to coordinate-reference systems (inertial reference systems) which move with constant velocity in respect to absolute space. The statement of this corollary V corresponds to the so-called "classical relativity":

> "The motions of bodies included in a given space [inertial reference system] are the same among themselves, whether that space is at rest [absolute space], or moves uniformly forwards in a straight line without any circular motion."

If **F*** is the force acting on a body of mass m^* with respect to absolute space S^*, then with respect to an inertial reference system S moving with constant velocity u^* with respect to S^*, and after using the Galilean Transformations: $x = x^* - u^*t^*$; $y = y^*$; $z = z^*$; $t = t^*$, with the additional mass invariance $m = m^*$, we conclude the dynamical invariance $\mathbf{F} = \mathbf{F}^*$. Corollary V establishes that it is *impossible* to detect a rectilinear uniform motion of an inertial reference system by performing dynamical experiments inside the said inertial system.

On Axiom 2.

Axiom # 2 is well known and is called the *equation of motion*. The so-called *principle of inertia* is not a principle, nor a postulate nor is it an axiom. It is simply a *law*, i.e., a logical consequence, a theoretical conclusion of axiom # 2. The *inertia law* corresponds to a theorem in mathematics expressed in the form: *If P, then Q*. Thus, the statement of the *law of inertia* is the following: If the resultant force acting on a body is zero, then the velocity of the body is constant. The reasoning of the demonstration of this law of inertia can be shortened considerably if we proceed mathematically. The following lines only have a pedagogical purpose. Students of physics will appreciate the solid formal-logical-*geometrical* structure of physics.

Hypothesis: $\sum_j \mathbf{F}_j = \mathbf{0}$

Thesis: $\mathbf{v} = \mathbf{constant}$

Demonstration:
Introducing the previous hypothesis in axiom # 2, given by eq. (4.1), we get:

$d(m\mathbf{v})/dt = \mathbf{0}$

$m\mathbf{v} = \mathbf{constant}$

being $m \neq 0$, but constant

then $\mathbf{v} = \mathbf{constant}$

Q.E.D.

The **constant** in the above demonstration can be zero implying that the body is at rest. If the **constant** is different from zero, then the two properties of the vector **v**, magnitude and direction, are constant. Constant magnitude in the velocity means uniform motion. Constant direction of the velocity means rectilinear motion. We believe that a good physics teacher should spend at least one hour lecturing only on the law of inertia. For example, the law of inertia has definite ontological implications. It has to do with the problem of *existence* of forces acting on a body. It also has to do with the concept of *absolute time*. Thus, the existence or nonexistence of a resultant force acting on a given body in motion depends on the accuracy of the time device we use. The congruence of two geometrical rectilinear segments presents no fundamental problem, but the congruence of two intervals of time presents serious practical problems. Only an act of faith can bring peace to our minds when we try to decide the congruence of two intervals of time. We decide to believe that absolute time flows uniformly, independent of any real motion of bodies. In investigating the nonexistence of a resultant force on a given body, we have to kinematically determine that the velocity of the moving body is uniform. However, to determine the *uniformity* of the body's motion, we must believe in a *uniform* rate of the flow of time. Today, we believe atomic clocks run *uniformly.* In the past we believed planet earth rotated uniformly, but it did not. These considerations must have forced Newton to define an *ideal or absolute time.* Newton wrote in his *Principia: "Absolute, true, and mathematical time, of itself, and from its own nature, flows equally without relation to anything external, and by another name is called duration."* We must be sure relativists use clocks which do not run in an accelerated way, but uniformly. Thus, the existence or nonexistence of a resultant force, acting on any material body, depends heavily on the accuracy of the time device we use. A *real* clock should approximate closely to an *ideal* clock to tell *absolute time. If it does not,* physics is not possible.

We may not find one single physics textbook in which the author states that a terrestrial laboratory is a highly *noninertial* reference system in which Newton's dynamics is not valid. Every teacher of physics or authors of physics books establishes that due to the small angular velocity of planet earth, a terrestrial laboratory can be considered a *good inertial reference system.* On the other hand, every physicist would agree that a laboratory accelerated at 9.8 m/s^2, with respect to the distant stars, would constitute a very bad inertial reference system. As a matter of fact, they would agree that such an accelerated reference system is definitively *non-inertial.* The same physicists would also agree that, according to Einstein's Equivalence Principle, every terrestrial laboratory is equivalent to an accelerated laboratory with respect to the distant stars. Thus, physicists can live with two contradictory judgments:

(1) Every terrestrial laboratory is a good inertial reference system.
(2) Every terrestrial laboratory is *not* a good inertial reference system.

The last two statements obviously are contradictory. Only one of them must be true. For more than three centuries we have used Newton's dynamics in terrestrial laboratories. According to the Principle of Equivalence (chapter 6), any terrestrial physical laboratory has been equivalently accelerating with respect to stellar space since the day it was built. Therefore, any terrestrial laboratory constitutes a *non-inertial* reference system. However, we have determined many forces of interaction inside terrestrial laboratories. How is this possible? The explanation is contained in Corollary VI in Newton's *Principia* [17, p. 21]:

> "If bodies, moved in any manner among themselves, are urged in the direction of parallel lines by equal accelerative forces, they will all continue to move among themselves, after the same manner, as if they had not been urged by those forces."

In modern notation, the proof of this Corollary is as follows: let **F** and **F'** be the forces of action and reaction in the interaction of bodies B and B', having masses m and m', respectively. Let *R* and *R'* be the position vectors of bodies B and B', respectively, with respect to the laboratory. Finally, let **-a*** be the laboratory acceleration with respect to absolute space. The equations of motion inside the laboratory are:

$$m\mathbf{a} = \mathbf{F} + m\mathbf{a}^*$$
$$m'\mathbf{a} = \mathbf{F'} + m'\mathbf{a}^*$$

or

$$\mathbf{a} = \mathbf{F}/m + \mathbf{a}^*$$
$$\mathbf{a} = \mathbf{F'}/m' + \mathbf{a}^*$$

Subtracting the second from the first equation we get:

$$d^2(\mathbf{R} - \mathbf{R'})/dt^2 = \mathbf{F}/m - \mathbf{F'}/m'$$

Calling $\mathbf{r} = \mathbf{R} - \mathbf{R'}$, and using axiom # 3: $\mathbf{F'} = -\mathbf{F}$, we have:

$$\mu d^2\mathbf{r}/dt^2 = \mathbf{F}$$

where μ is the reduced mass equal to $mm'/(m + m')$. The previous equation finishes the demonstration of Newton's Corollary VI. For years engineers have used Newton's second axiom of motion in noninertial reference systems, linearly accelerated or in rotation, in spite of the axiological principle given by axiom # 1. The only requirement to use Newton's equation of motion inside a noninertial reference system is to *add* to the *real forces,* acting on a body, the so-called *inertial forces, or pseudo forces or fictitious forces*. Borrowing from geometrical optics the terms real and virtual images, we would like to propose the name *"virtual forces"* to refer to fictitious or inertial forces.

The ontological difference between real forces and virtual forces is that real forces are *caused* by the interaction of *material* bodies, and they obey the principle of action and reaction. Virtual forces, on the other hand, are *not caused* by the interaction of material bodies, but by *relative accelerated motion* on only *one* body. Therefore, a virtual action force has no virtual reaction force, i.e., virtual forces do not obey Newton's principle of action and reaction. Virtual reactions do not exist in the material universe. Hence, Newton's dynamics is *not valid in noninertial reference systems.* Of course, from a practical point of view, we can use Newton's equation of motion to design machines inside noninertial systems. No one has the right to ask, not even Einstein, as we will now see, the *absurd* question about the existence of the reaction to a virtual force.

On Axiom 3.

We have just seen that the principle of action and reaction *is not valid* in noninertial reference systems that are linearly accelerated or in rotation with respect to absolute space. This was known to Newton and to every good student of Newton's *Principia.* Einstein, who never missed an opportunity to criticize Newton's theory of classical dynamics, raised false testimony against Newton's dynamics. In 1916 Einstein [18a, p. 112] wrote:

> "In classical mechanics, and no less in the special theory of relativity, there is an *inherent epistemological defect* which was, perhaps for the first time, clearly pointed out by Ernst Mach." [Italics added]

This *inherent epistemological defect* is indeed an ontological defect. It has to do with the fourth ontological principle of *cause* and *effect* through the use of the principle of action and reaction. Einstein, after a clever and long argumentation which ignored the *invalidity* of Newton's principle of action and reaction in rotating references systems, accused Newton of *hypostatizing* absolute space in order to make absolute space the *cause* of the centrifugal forces acting on a rotating *"gedanken" planet of water.* Newton never would have concluded such an aberration. Finally, Einstein concluded that absolute space cannot be the cause of the centrifugal force. Einstein said:

> "It is therefore clear that Newton's mechanics does not really satisfy the requirements of causality in the case under consideration."

Einstein *knew* that Newton's principle of action and reaction was not *valid* with respect to rotating reference systems in relation to absolute space. The following quotation proves that Einstein knew that Newton's dynamics was invalid in rotating or accelerated reference systems. Einstein [18b, p. 140], in the last edition of his book *The Meaning of Relativity*, writes:

> "Had one tried to explain to Newton the equality of inertial and gravitational mass from the equivalence principle, he would necessarily have had to reply with the following objection: it is indeed true that relative to an accelerated coordinate system bodies experience the same accelerations as they do relative to a gravitating celestial body close to its surface. But where are, in the former case, the masses that produce the accelerations?"

Newton would have asked - where are the material bodies that *cause* the *inertial force* on a body inside a noninertial system? Newton would have asked this question because he knew, as well as Einstein, that his Newtonian theory of dynamics, particularly the principle of action and reaction, was not valid in noninertial reference systems. Newton's question, imagined by Einstein, reveals that Einstein *knew* very well that *real* forces are caused by *material* bodies. Inertial forces are known by the unfortunate name of *fictitious* forces, meaning that their existence is not caused by material objects. Thus, Einstein produced two contradictory judgments. Out of the two Einsteinian quotations above, the first one is false. It would have been more acceptable for Einstein to have said that his intention was to generalize Newton's dynamics because *the laws of physics must be of such nature that they apply to systems of reference in any kind of motion*. Here we have the seed of the so-called Mach's Principle. However, Einstein never succeeded in proving that the centrifugal forces, acting on the water of Newton's bucket, were *caused* by the distant matter of the extra-galactic nebulae rotating relative to the water in the bucket.

At this point, we would like to refer to a very fundamental paper published by Hans Thirring in 1918. Thirring [19, p. 33], by linearizing Einstein's field equations, determined the forces on a small body of unit mass in the vicinity of the center of a hollow spherical shell of mass M and radius a, rotating with angular velocity ω. Thirring, in his paper, mentioned in a very subtle way that the rotation of the hollow sphere is with respect to a reference system fixed at infinity. After Thirring, relativists introduced a stationary reference system fixed at infinity to refer the rotation of the spherical shell. If relativists do not want to call *absolute space a reference system fixed at infinity*, they know, for sure, who they are deceiving. Thirring concluded his paper, saying:

> "Through a concrete example it is shown that in the gravitational field (of Einstein) produced by distant masses in rotation, appear forces that are *analogous* to the centrifugal and Coriolis forces." [Italics added]

Analogous means alike, similar: it does not mean *identical*. Let us look at the geometrodynamic accelerations, found by Thirring in 1918, with ω being the angular velocity:

$$a_x = -[8/3\ GM/(c^2R)]\omega v_y + [1/3\ GM/(c^2R)]\omega^2 x \qquad (4.7a)$$

$$a_y = [8/3\ GM/(c^2R)]\omega v_x + [1/3\ GM/(c^2R)]\omega^2 y \qquad (4.7b)$$

$$a_z = \quad 0 \quad - [2/3\ GM/(c^2R)]\omega^2 z \qquad (4.7c)$$

Thirring used $\boldsymbol{\omega} = -\mathbf{k}\omega$. If we use $\boldsymbol{\omega} = \mathbf{k}\omega$, in eq.'s (4.7), then we can write these equations in the following way:

$$\mathbf{a} = -[4/3\ GM/(c^2R)](2\boldsymbol{\omega}\mathrm{x}\mathbf{v}) - [1/3\ GM/(c^2R)]\boldsymbol{\omega}\mathrm{x}(\boldsymbol{\omega}\mathrm{x}\mathbf{r}) - \mathbf{k}[2/3\ GM/(c^2R)]\omega^2 z \qquad (4.8)$$

where M is the spherical shell's mass of radius R, rotating with angular velocity $\boldsymbol{\omega}$ with respect to a reference system *fixed at infinity* (absolute space). Let us compare the last equation with the *true* inertial acceleration $\mathbf{a}_i$ obtained from eq.'s (4.4), and referred to a reference system *co-rotating* with the spherical shell:

$$\mathbf{a}_i = -2\boldsymbol{\omega}\mathrm{x}\mathbf{v} - \boldsymbol{\omega}\mathrm{x}(\boldsymbol{\omega}\mathrm{x}\mathbf{r}) - (d\boldsymbol{\omega}/dt)\mathrm{x}\mathbf{r} \qquad (4.9)$$

In Thirring's paper $\boldsymbol{\omega}$ is constant. Therefore, Euler's inertial acceleration is zero. The gravitational radius R_g of the spherical shell is given by GM/c^2: hence, R_g/R is a pure number with no physical dimensions. Thus, if we want to claim that the *quasi-Coriolis* acceleration and the *quasi-centrifugal* acceleration, in eq. (4.8), are *identical* to the *true* Newtonian-Coriolis acceleration and the *true* Newtonian-centrifugal acceleration, given by eq. (4.9), we must equate to unity the square bracket coefficients in eq. (4.8). This procedure would introduce logical inconsistencies in the *ungrounded metaphysical obsession* of Mach and Einstein. It would have been more *reasonable* and profitable to have considered eqs. (4.8) as *authentic gravitodynamic* field intensities caused by rotating material bodies. In 1918, Lense and Thirring [20, p. 156] did some important work, determining the *gravitodynamic* force-terms on a body revolving outside a solid spherical central body. Relativists were too busy contracting tensors in Riemann's tetradimensional spacetime, and paid no attention to the solid physics buried in Einstein's field equations.

Perhaps we remember the scandal caused by Dicke and Goldenberg when they published, in 1967, the measured oblateness of the sun. Relativists and astrophysicists bitterly criticized Dicke and Goldenberg because they believed if the sun was oblate, General Relativity Theory (GRT) was doomed. Such a conclusion is based on an ontological misunderstanding of the *real* world. In 1974, relativists began to breathe normally again when Hill published new measurements of the sun's limbo, indicating no solar oblateness. However, in 1982, Hill again published new solar data showing the sun is oblate. Even in newspapers there were declarations of physicists saying that GRT is wrong. What is the cause of this fear of "general" relativists?

Einstein and his fellow followers have always been falsely proud that the relativistic explanation of the perihelic rotation of Mercury and the other planets *does not require the adjustment* of any parameter. Let us point out two objections to this relativistic claim. One is Einstein's constant $\kappa = 8\pi G/c^2$. The universal constant of gravitation G emerged from the real world through Kepler's astronomical laws. It was *adjusted* empirically! The second objection is ontological and based on the conceptions of T. Aquinas and R. Descartes about the *res extensa*. In 1915, Einstein [21, pt. 2, p. 821] solved the problem of the "anomalous motion" of planet Mercury treating the sun as a geometrical point. Einstein's solution of his field equations was approximate. The next year, Schwarzschild [22] treated the sun as a nonrotating material sphere and solved Einstein's field equations exactly. Schwarzschild proved that a nonrotating material sphere can mathematically be considered as a geometrical point. However, a geometrical point *cannot* rotate. Ontologically, it is impossible for any material object in the universe to have no geometrical dimensions. Even common sense affirms that the sun is a huge ball of fire. Astronomical observations show this ball of fire is also rotating differentially. Thus, if we start with a solar model in the shape of a rotating sphere of mass M and radius R, the model is millions of times closer to the *real* sun than conceiving the sun as a *mathematical point.* Lense and Thirring used GRT and a very real sun to explain the perihelic rotation of planets and the periplanet rotation of satellites. Introducing in Lense-Thirring's method, the quadrupole gravitational potential, we get the excess of perihelic precession Ω given by:

$$\Omega = [6\pi GM/\{Tc^2a(1-e^2)\}](1 - 2/3\,(L_o/M)/h + 1/3\,(R/h)^2\Delta) \qquad (4.10)$$

where the square bracket coefficient is the well-known Einsteinian solution to the anomalous motion of planet Mercury. L_0 is the solar intrinsic angular momentum. The constant h is the specific orbital angular momentum of the planet, and Δ is the oblateness of the sun. Thus, instead of spending so many years on this unsound idea of Mach and Einstein, about the fictitious *essence* of inertial forces, relativists and astrophysicists could have determined the intrinsic angular momentum of the sun through the analysis of the perihelic motion of the planets. They also could have calculated the intrinsic angular momentum of Jupiter, which has a more significant oblateness, through an accurate astronomical determination of the more distant perijovian rotation of its satellites.

In eq. (4.8) we have a *kind* of Coriolis' acceleration and a *kind* of centrifugal acceleration. This observation is exactly what Thirring wrote in his conclusions. Einstein's field equations, in the linear approximation, provide forces that are *similar,* never *identical,* to Coriolis and centrifugal forces. The last term in eq. (4.8) shows the existence of an axial force that is definitely not contained in the expression of the inertial acceleration given by eq. (4.9). The presence of this term, in relativistic gravitation, destroys the possibility of identifying eq. (4.8) with eq. (4.9). Einstein's GRT can perfectly survive the initial difficulties created by the oblateness of the sun, bringing real physical meaning to our knowledge of the solar system. GRT provides a totally new *gravitodynamics* if we preserve Einstein's field equations and *re*interpret its ontological, geometrical and physical background. As we can see, after these comments on Axiom 3, the ontological principle of cause and effect and the principle of action and reaction are still behind any discussion on the so-called Mach's principle.

On Axiom 4.

G. B. Brown [23] made a splendid attempt, in 1955, to unify formally the electrodynamic forces with the gravitodynamic forces in the context of Newtonian dynamics. As everyone knows, after Einstein created his field theory of General Relativity, the formal analogy of Lorentz's electrodynamic force and the Einsteinian geometrodynamic force was complete. As mentioned before, this author [24], in 1982, extended this analogy to predict the probable existence of a new electrodynamic force. This same force can be directly deduced using the Parametrized Newtonian Relativistic Electrodynamics (PNRED) presented in chapter 3, and given by eq. (3-45). The following equations represent Tate-Edwards' effect, deduced from a field theory (GRT) and also deduced from PNRED, for the case $z/R<1$:

From GRT: $F_z = (-3/2)[q/(2\varepsilon_0 c^2)](z/R^2)[\, I^2/(|\rho_e|A)]$

From PNRD: $F_z = (\alpha-\gamma)[q/(2\varepsilon_o c^2)](z/R^2)[\, I^2/(|\rho_e|A)]$

A cosmic analogy has been in the mind of every thinker since the time of Thoth. One of the profound statements, made by the great teacher Thoth, 7000 years ago was: *"As above, so below, for the fulfillment of unity."* The principle of cosmic analogy follows this line of thinking of G.B. Brown. This analogy is formal. Someday someone will discover a theory based on an essential unification of all the forces in nature. On that day we will learn that gravitational forces are, in essence, electrodynamic forces. In addition, we will also find that the weak and strong nuclear forces are also, in essence, electrodynamic forces.

On Axiom 5.

In 1969, D.W. Sciama [25] tried, in an extraordinary effort, to establish the *dynamical foundations* of GRT. He looked for the mathematical structure of *gravitodynamic* force-terms starting from the *field* theory of GRT. In 1959, Sciama [26] considered the analogy between Grassmann's force and Coriolis' force. This analogy, nevertheless, was known to Thirring in 1918, when he wrote:

> "The analogy between electrodynamics and gravitational theory (weak and in vacuum) is even greater if one observes that in the approximated integration of the quantities $g_{14}, g_{24}, g_{34}, g_{44}$, from the density and velocity of matter, are calculated in the same manner as the potentials A_x, A_y, A_z, ϕ . . . the second term of eq. (19) completely corresponds to the ponderomotive force (**E**+**v**x**B**)."

Let us now elaborate on Sciama's comparison of Grassmann's force and Coriolis' force.

$\mathbf{F} = q\mathbf{v}\mathrm{x}\mathbf{B}$

Using Bio-Savart's law in the previous equation, we get:

$\mathbf{F} = q\mathbf{v}\mathrm{x}[q'kr^{-3}\,(\mathbf{v}'\mathrm{x}\mathbf{r})]$

or $\mathbf{F} = q\mathbf{v}\mathrm{x}[q'kr^{-1}\,(\mathbf{v}'\mathrm{x}\mathbf{r})/r^2]$

Let us now define an instantaneous angular velocity $\boldsymbol{\omega}$ by the following equation:

$\boldsymbol{\omega} = (\mathbf{r}\mathrm{x}\mathbf{v}')/r^2$

Grassmann's force becomes:

$$\mathbf{F} = [\tfrac{1}{2}\, kqq'/r](2\boldsymbol{\omega}\mathrm{x}\mathbf{v}')$$

This last equation shows that Grassmann's force is directly proportional to a *kind* of Coriolis acceleration. It is also inversely proportional to the separation distance between the source charge q' and the test charge q. But what is the meaning of $\boldsymbol{\omega}$? Of course this angular velocity could be interpreted as the instantaneous angular velocity of the test charge q with respect to the source charge q'. To have this interpretation, we should replace **v'** by (**v** - **v'**) in the definition of $\boldsymbol{\omega}$. However, we must remember this is an *heuristic* exposition to invite our minds to explore the implications of axiom 5. If the analytical exploration of axiom 5 leads to theoretical laws, experimentally verifiable, then our *faith* in the truth of axiom 5 will grow, though we will never be able to demonstrate it rationally.

4.4 Logical deduction of a Parametrized Newtonian Relativistic Electrodynamics.

The redundant first part of the title of this section is to emphasize Einstein's *demand* when he criticized Newton's dynamics for being deficient. As we saw above, Einstein wrote: "*From the point of view of the aim of the greatest logical simplicity of the foundations, this theoretical method is deficient in so far as the laws of force cannot be obtained by logical and formal considerations, so that their choice is a priori to a large extent arbitrary.*" However, the main objective in this section is to deduce a Parametrized Newtonian Relativistic Electrodynamics, the same one we *"induced"* in the last chapter. There, of course, we first did a taxonomical work. This helped us to see the possibility of proposing a new general electrodynamics in the context of Newton's dynamics. Here we will resort to a Neo-Newtonian dynamics and classical Newtonian propositions or theorems related to inertial accelerations. We will introduce *interpretations* in order to draw conclusions from *axiom 5,* represented by the following equation:

$$\mathbf{F} = (C/r)\,[K_1 2\,\boldsymbol{\omega}\,\mathbf{x}(D\mathbf{r}/Dt) + K_2\,\boldsymbol{\omega}\mathbf{x}(\boldsymbol{\omega}\mathbf{x}\mathbf{r}) + K_3(D\boldsymbol{\omega}/Dt)\mathbf{x}\mathbf{r}] \qquad (4.5)$$

Let **R** and **R'** be the vector positions of a test charge q and a source charge q', respectively, with respect to absolute space S*. Let **r** be the vector position of q with respect to q'. Let S finally be a reference system bound to the source charge q'. Let the source, as well as the test charge, move with arbitrary velocities and accelerations. The reference system S, bound to the source charge q' is, therefore, a noninertial reference system. The vector **r** can be expressed with respect to S* and S as follows:

$$\mathbf{r} = \mathbf{i}(x-x')+\mathbf{j}(y-y')+\mathbf{k}(z-z') = \mathbf{R} - \mathbf{R}' \qquad (4.11)$$

where the unit vectors belonging to S are not constant in time. The coordinates x-x', y-y', z-z' are the coordinates of the test charge q, referred to the origin of S, which coincides with the source charge. Now we will use Coriolis' theorem:

$$d[\ ]/dt = D[\ ]/dt + \boldsymbol{\omega}x[\ ] \qquad (4.12)$$

Let us apply this theorem to eq. (4.11):

$$d\mathbf{r}/dt = d\mathbf{R}/dt - d\mathbf{R}'/dt = \mathbf{v} - \mathbf{v}' = \mathbf{v}^* = D\mathbf{r}/Dt + \boldsymbol{\omega}\ \mathbf{x}\ \mathbf{r} \qquad (4.13)$$

where $\mathbf{v}^*$ is the relative velocity of the test particle with respect to the source particle. The time operator D[ ]/Dt does not operate on the unit vectors of system S. From the previous equation we have:

$$D\mathbf{r}/Dt = \mathbf{v}^* - \boldsymbol{\omega}\ \mathbf{x}\ \mathbf{r} \qquad (4.14)$$

Now we have to interpret the meaning of $\boldsymbol{\omega}$ in eq. (4.5). Here we have many arbitrary choices. Any choice will constitute an extra *hypothesis.* Thus, we propose the conjecture that $\boldsymbol{\omega}$ should represent the instantaneous angular velocity of the test particle with respect to the source particle, i.e.:

$$\boldsymbol{\omega} = (\mathbf{r}x\mathbf{v}^*)/r^2 \qquad (4.15)$$

A priori we cannot justify this last equation. Only the consequences of this hypothesis will decide its usefulness. Keeping in mind that $D\boldsymbol{\omega}/Dt = d\boldsymbol{\omega}/dt$, we can now proceed to introduce eq. (4.15) and eq. (4.14) in eq. (4.5). After a lengthy mathematical development we get:

$$\mathbf{F} = Cr^{-3}[\mathbf{r}\{\alpha\mathbf{v}^{*2} + \beta r^{-2}(\mathbf{r}\cdot\mathbf{v}^*)^2 + \gamma(\mathbf{r}\cdot\mathbf{a}^*)\} + \delta\mathbf{v}^*(\mathbf{r}\cdot\mathbf{v}^*)+\varepsilon r^2\mathbf{a}^*] \qquad (4.16)$$

where $\alpha = -K_2$ (4.17a)
$\beta = 2K_3 - 2K_1+K_2$ (4.17b)
$\gamma = -K_3$ (4.17c)

$$\delta = 2K_1 - 2K_3 \tag{4.17d}$$
$$\varepsilon = K_3 \tag{4.17e}$$

If we make

$$C = kqq' = qq'/(4\pi\varepsilon_o c^2) = Kqq'/c^2 \tag{4.18}$$

eq. (4.16) is formally identical to eq. (3-36) in chapter 3. Let us now introduce eq. (4.18) in eq. (4.16), and add it to Coulomb's electrostatic force:

$$\mathbf{F} = Kqq'r^{-3}\{\mathbf{r} + c^{-2}[\mathbf{r}\{\alpha \mathbf{v}^{*2} + \beta r^{-2}(\mathbf{r}\cdot\mathbf{v}^*)^2 + \gamma(\mathbf{r}\cdot\mathbf{a}^*)\} + \delta\mathbf{v}^*(\mathbf{r}\cdot\mathbf{v}^*) + \varepsilon r^2\mathbf{a}^*]\} \tag{4.19}$$

The last equation represents the *Parametrized Newtonian Relativistic Electrodynamics.* The Greek parameters should be experimentally adjusted. Eq. (4.19) satisfies the weak principle of action and reaction, i.e., the reaction **F'** = - **F**, but **F'** is not collinear with **F**. Therefore **F** is a quasi-Newtonian force. Now, using axiom # 4, we will get a *Parametrized Newtonian Relativistic Gravitodynamics.* A great economy of thought is attained if we attach an asterisk to K, q, and q', in eq. (4.19), taking care to introduce a negative sign in front of the right-hand side of the equation. The new gravitodynamics is given by:

$$\mathbf{F}^* = -K^*q^*q^{*\prime}r^{-3}\{\mathbf{r}+c^{-2}[\mathbf{r}\{\alpha\mathbf{v}^{*2}+\beta r^{-2}(\mathbf{r}\cdot\mathbf{v}^*)^2+\gamma(\mathbf{r}\cdot\mathbf{a}^*)\}+\delta\mathbf{v}^*(\mathbf{r}\cdot\mathbf{v}^*)+\varepsilon r^2\mathbf{a}^*]\} \tag{4.20}$$

where
$$q^* = m \tag{4.21a}$$
$$q^{*\prime} = m' \tag{4.21b}$$
$$K^* = 1/(4\pi\varepsilon_o^*) = G \tag{4.21c}$$

ε_o^* can be called the gravitostatic permittivity of the cosmic energy field. We can also define a gravitodynamic permeability μ_o^* of the cosmic energy field, given by:

$$\mu_o^* = 1/(\varepsilon_o^* c^2) = 4\pi G/c^2 \tag{4.22}$$

Eq. (4.20) is the Newtonian formal answer to Einstein's criticism of Newton's dynamics. Obviously, this Newtonian relativistic gravitodynamics has to be submitted to experimental verification, in the course of which we will have the opportunity to empirically adjust the numeric values of the Greek parameters. Besides the secular variations of the perihelia of planets, there are other secular variations of orbital parameters that have been accounted for by Lense and Thirring [20]. With respect to the magnitude of forces, it is clear that eq. (4.19) offers better possibilities to detect, experimentally, new electrodynamic effects. In the next section, we will present preliminary experimental results on the probable existence of a new electrodynamic force.

4.5 Action of a Permanent Magnet on Static Charges.

In any electrokinetics derived from any Newtonian relativistic electrodynamics, we find force-terms proportional to v'^2/c^2. These types of electrokinetic force-terms were explained away by Maxwell [27, vol. 2, Art. 848, p. 850], in 1873. The reason given by Maxwell was that it *"is not capable of being experimentally tested . . ."* In 1877, Clausius, quoted by O'Rahilly [28, vol. 2, p. 589], wrote *"We accept as criterion the experimental result that a closed constant current in a stationary conductor exerts no force on stationary electricity."* O'Rahilly, in the same page, quotes Klein as saying in 1932:

> "Hitherto it has been almost a principle of faith with physicists that an electric current exerts no force on stationary charges. But it must be admitted that as yet there are no measurements in this direction, and perhaps they cannot be made owing to the extraordinary smallness of the effect."

As we mentioned in the previous chapter, in 1994 Assis [29, Sec. 6.6, p. 166], did an exhaustive bibliographic and original research on this subject. Assis proposes a serious problem of *existence*. He wrote: *"The question naturally is to know if this force exists or not."* From a theoretical point of view, the existence of this force has been known since 1846-48 when W. Weber published his Newtonian Relativistic Electrodynamics. However, this force is also contained in Gauss' electrokinetics of 1835, but published many years later. From an experimental point of view, the existence of this second order electrodynamic force is still in its infancy. J. Tate [30], in 1968, seems to have been the first one reporting experimental results on the real existence of this force. In 1974 Edwards [31], and in 1976 Edwards *et al.* [32] again reported new experiments on the existence of this new probable electrodynamic field. Why are these types of experiments meaningful or important? The *transhuman* interest in establishing the experimental evidence of this probable new electrodynamic field, beyond any reasonable doubt, is to broaden human knowledge. However, this transhuman interest has many human implications. One is to overpower the present relativistic paradigm showing that Einstein's SRT is wrong. Another implication is to attract the attention of the physicists' community by proposing *weird* explanations of Tate-Edwards' effect. After the publication of Edwards' paper, these *ad hoc* hypotheses proliferated at a greater rate than repetitions of Tate's or Edwards' experiment. We need more experimental work and less weird arbitrary explanations to save a wrong theory. These types of experiments are not *meaningless.* On the contrary, they are very important in order to search for better electrodynamic and electrofield theories. *It is important even if GRT proves that SRT is inadequate. It is important even if GRT provides the field equations for a new nonlinear electrodynamics.* It is important even if a Newtonian Relativistic Electrodynamics becomes a better theory. It is important even if we experimentally falsify the existence of this probable new electrodynamic field predicted by any of the Parametrized Newtonian Relativistic Electrodynamics. What follows is an unpublished paper this author wrote in 1987. In 1991, Wesley [33, p. 257] commented on a preprint of ~~this~~ experimental work. Later we will comment on Wesley's comments and add some recent considerations on this subject. We will present the experimental results of the action of a permanent magnet on static charges, showing the existence of a force whose magnitude was determined to be of the order of 10^{-16} N. The experiment was performed using the Millikan Apparatus.

In 1982, Curé [34] predicted the probable existence of a new electrodynamic force, by analogy, and in the context of the Linearized Theory of General Relativity. He proposed an electrodynamic experiment using the Millikan Apparatus to determine the minute force of interaction between a steady current and static charges. Previous to this prediction, some experiments had been reported on the action of steady currents on static charges [35].

The preliminary computer calculations which were done to design a solenoid in order to run the modified Millikan oil drop experiment so as to detect this minute force were very disappointing. As we decided not to use superconductors nor very high currents, in normal conductors, the weights of the designed solenoids were around 3,000 Kg. For this reason we decided to use a small permanent magnet.

4.5.1 EXPERIMENTAL DATA AND PROCEDURE

Instead of using oil drops we used latex spheres. The following data pertained to the latex drops and air. Radius = 5.05×10^{-7} m; density = 1.05×10^{3} Kg/m; mass = 5.66×10^{-16} Kg; weight = 5.55×10^{-15} N; viscosity of air in normal conditions = 1.824×10^{-5} Kg/(ms). The number N of electrons on a latex drop was determined by: W = NeV/d, where e = 1.6×10^{-19} C; V was the potential difference between the plates of the capacitor shown in Fig. 4.1; d = 0.0044 m was the separation of the plates; W was the weight of the latex drop which was balanced by Coulomb's force. The diameter of the capacitor's plates

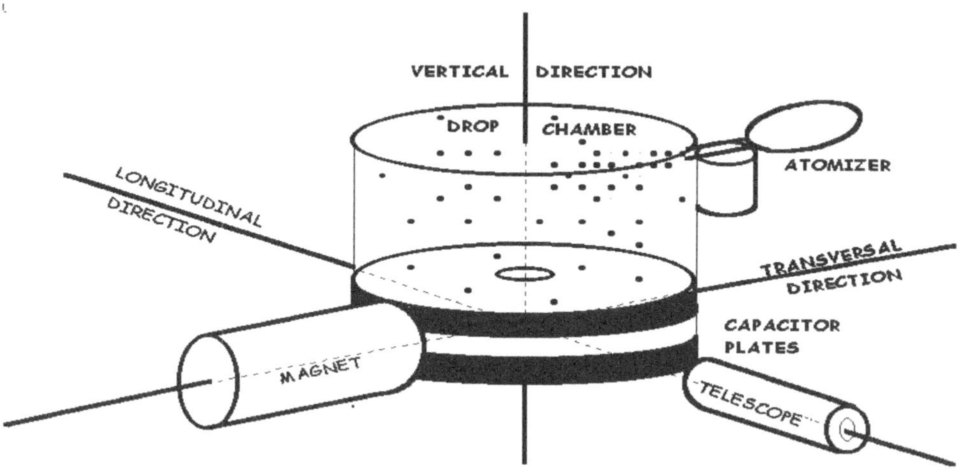

Figure 4.1 Millikan Apparatus with Magnet.

was 0.051 m and the diameter of the drop chamber was 0.057 m.

The magnet used was a cylinder formed by ten magnetic discs of ceramic. The dimensions of one magnetic disc were: diameter = 0.0254 m and height = 0.004 m. The magnetic induction B, at the center of one of the faces of the disc, was 0.38 Teslas. Once the micrometer of the eyepiece of the Millikan Apparatus was calibrated, one unit of the scale was determined to be equal to 0.326mm.

With the help of a small bubble level mounted on the upper plate of the capacitor and the three leveling screws of Millikan´s Apparatus, the capacitor was leveled with respect to two perpendicular directions shown in Fig. 4.1: a longitudinal one, coincident with the optical axis of the telescope, and a transversal one, perpendicular to the optical axis and coinciding with the axis of the magnet. The magnet's axis was made to go through the plates of the capacitor. After this operation was done, the drop chamber was installed on the Millikan Apparatus and drops were sprayed into the chamber.

Once a drop was in the view field, it was balanced by adjusting the voltage across the capacitor. However, it was necessary to further improve the leveling of the capacitor by means of the leveling screws. This operation sometimes took up to 15 minutes. Then readings of the balancing voltage, position of the drop in the view field of the microscope, and time were recorded at intervals of one to two minutes. If the position of the balanced drop did not change in the next 10 to 15 minutes, the cylindrical magnet was placed with its axis in the transversal direction, and subsequent readings were taken until the drop faded away.

4.5.2 RESULTS AND ANALYSIS OF PROBABLE SYSTEMATIC ERRORS

It is to be noted, that in every experiment, the drops could be maintained in the view field for more than one hour when no magnet was present. Nevertheless, the observation time of each drop was considerably reduced whenever the magnet was brought into the vicinity of the Millikan Apparatus. This effect corresponds to a longitudinal drift of the drop, to the extent that the drop could not be focused by the microscope anymore. This observation was corroborated by the fact that the balancing voltage had to be slightly adjusted to bring the drop along the vertical direction, back to the horizontal line of the eyepiece of the microscope.

Transversal drifts were measured for different drops along the axis of the magnet. A typical run is shown in Table 4-I. The transversal force was determined by Stoke's law. From Table 4-I, the calculated average drift speed was 1.68 microns/s, and the corresponding force was equal to 2.9×10^{-16} N. When the number of electrons on the drop was less than 30, no transversal drift was observed.

The most critical aspect of the experiment was the leveling of the parallel-plate capacitor. In the many initial runs, there was always a systematic drift of the drops of the order of 0.8 microns/s when no magnet was present. This drift persisted even when the inclination angle of the capacitor's plane, with respect to the horizontal plane, was 30 minutes of arc. Before the cause of this leveling systematic error was established, the drift was investigated assuming it was air convection currents caused by probable temperature gradients. To analyze this effect, the entire chamber of the Millikan Apparatus was covered with clay. Under these conditions, the drift persisted in the absence of the permanent magnet. Once the leveling interference was corrected, no drift was observed, even when the chamber was not covered with clay. Edge effects, of the parallel-plane capacitor, were experimentally investigated. This was done by locating the drop off the vertical axis and close to the upper or lower plates of the capacitor. Had this edge effect interfered with the main experiment, we would have observed opposing drifts along the transversal direction, but in neither case was the drop displaced in the absence of the permanent magnet.

To investigate probable electrostatic induction effects, two cylinders of aluminum and clay, of the same dimensions of the cylindrical permanent magnet, were located in the magnet position with respect to the Millikan Apparatus. In neither case did we observe any displacement of the drop.

Finally, the average approaching speed of the magnet was measured to determine the magnitude of the magnetic force on the drop. This was due to its relative motion, with respect to the magnet, when the latter was brought to its final position along the transversal direction. Its experimental value was 600 microns/s. Assuming the extreme unrealistic value of 0.38 Teslas, for the magnetic induction's components at the drop location, and the drop charged with 100 electrons, the magnetic force is approximately 100,000 times smaller than the magnitude of the measured new force.

Table 4-I. Drift of a charged latex drop in the presence of a permanent magnet.

Voltage (v)	Position (H-unit)*	Time (s)	Comments
4.80	8.5	0	
4.69	8.5	155	
4.79	8.5	280	
4.77	8.5	400	
4.77	8.5	494	
4.77	8.5	640	Magnet installed
4.59	9.0	706	
4.50	9.5	776	
4.61	9.8	900	
4.55	10.1	1020	
4.64	11.0	1125	
-----	-----	----	Drop faded away

* H-unit = Horizontal-unit = 0.326 mm

4.5.3 CONCLUSIONS

The most convincing effect of the action of a permanent magnet on static charges is the considerable reduction of time when observing the charged latex spheres in the Millikan Apparatus. Also the drops change from rest to motion when the magnet is brought to its vicinity. This is after a long waiting period while observing the drop at rest.

The observed vertical and longitudinal drifts of the drops indicate the existence of a perpendicular or radial component of the new force. This is with respect to the magnet's axis when the drop is off such axis. In addition to this component, the results reported here establish the existence of an axial component along the transversal direction shown in Fig. 4.1.

In 1994, one colleague made a critical observation of this experiment. He told this author that we should have shielded the magnet with aluminum foil grounded to earth. We repeated the experiment, according to his indications, and observed the same previous effect on negatively charged latex spheres.

As Millikan commented many years ago, the Millikan Apparatus is an extremely sensitive instrument used to detect very minute forces. The magnitude of the force reported here is of the order of 10^{-16} N. We believe that this result points to the existence of a new electrodynamic force, though, of course, more experimentation is needed to confirm this.

Preliminary calculations show that using a magnetron, surrounded by a solenoid with bifilar windings, might provide additional quantitative experimental evidence on the existence of this new electrodynamic force. Finally, if the existence of this new electrodynamic force is definitively confirmed, then it is possible that analog nonlinear electrodynamic effects are still hidden in the field equations of General Relativity Theory.

4.5.4 COMMENTS ON ELECTROKINETIC FORCES PROPORTIONAL TO v'^2/c^2

In 1991, Wesley [36, p. 257] reviewed this experiment. He estimated the equivalent current associated to the magnet equal to 16,000 A. He also estimated the radial and axial components of this new probable force using Weber's Electrodynamics:

$$F_r \approx +9\text{x}10^{-17}\ v' \text{ (dyne)} = +9\text{x}10^{-22}\ v' \text{ (N)}$$

$$F_z \approx -8x10^{-17}\ v'\ (\text{dyne}) = -\ 8x10^{-22}\ v'\ (\text{N})$$

at 1 cm off the magnet's axis and 2.85 cm from the magnet's surface; v' would be the average velocity of the electrons in a real metallic conductor. In this case, we are talking about the equivalent current which replaces a magnetic shell to produce the same magnetic field generated by a really thin magnetic disk. The observed force was equal to $2.9x10^{-16}$ N, for the empirical drift speed of the latex sphere, v = 1.68 microns/s. Wesley speculated about the average or drift speed v' of the "equivalent electrons" in the equivalent Ampère electric current associated to a permanent magnet; he wrote:

> "where the drift velocity of the negative electrons for the equivalent Ampère current for the magnet, v', remains an unknown parameter. For an observed force of 10^{-13} dynes this would mean an equivalent drift velocity of v'≈1000 cm/s, (6.87) which seems too large. But the equivalent Ampère current for Curé's magnet was large, being about 16,000 amps, and the number of equivalent electrons that should be involved remains unknown. Since v' may be regarded as merely an adjustable parameter; perhaps such a large value, Eq.(6.87) is admissible. Only repeating the experiment with an actual current loop can decide the matter."

The theoretical analysis done by Wesley is impeccable, but it has nothing to do with the experimental data and experimental results obtained by Curé. The theoretical magnitude of the radial and axial forces, of the order of 10^{-21} (N), calculated correctly by Wesley, correspond to a latex sphere located outside the plates of the capacitor in the Millikan Apparatus. The reported separation of these plates was 0.44 cm. The forces calculated by Wesley are for points *at 1 cm off the magnet's axis*. The axis of the magnet runs between the parallel plates of the capacitor which is at 0.22 cm from the surface of one of the plates of the capacitor. Thus, in science, with the best intentions, we sometimes theoretically misinterpret experimental results. Nevertheless, Wesley is right in saying that Curé's experiment should be repeated under different conditions.

Replacing a magnet by an *equivalent* current carrying loop, has been very profitable to *mathematically* calculate the magnetic induction **B** generated by elementary magnetic dipole moments embedded in a magnet. Let us propose a name for this new probable electrodynamic field proportional to v'^2/c^2 by calling it *Gauss-Weber's field.* Gauss-Weber's field, at a point on the axis of a thin cylindrical magnet or circular coil, carrying a steady current I, can be written in different ways:

$$F_z = 2\pi(\alpha\text{-}\gamma)kq\,[\,I^2/(|\rho_e|A)]\,Rz/(R^2+z^2)^{3/2} \qquad (4.23a)$$
$$F_z = (\alpha\text{-}\gamma)q\,[\,I\,/(|\rho_e|A)]\,(z/R)\,B_z(z) \qquad (4.23b)$$
$$F_z = (\alpha\text{-}\gamma)q\,v'\,(z/R)\,B_z(z) \qquad (4.23c)$$

where $B_z(z)$ is the z-component of the magnetic induction **B** given by Biot-Savart's law. The parameters $|\rho_e|$, A, and v', are all unknown quantities for a *fictitious* equivalent electric current. Rowland's equivalence seems to be totally inapplicable in the case of Gauss-Weber's field. Wesley is right when he says: *"Only repeating the experiment with an actual current loop can decide the matter."* In the case of Weber's electrodynamics, as well as for Liénard-Schwarzschild's electrodynamics, the coefficient (α-γ) is equal to zero. For Gauss's electrokinetics $\alpha = 1$, and $\gamma = 0$, hence the force field given by eqs. (4.23) is an attractive force on static electrons. In chapter 3, we saw a vast collection of electrokinetics and electrodynamics. Some of these forces provide attractive or repulsive Gauss-Weber's fields. Who is the dogmatic physicist who will tell us, from merely an *a prioristic* theoretical speculation, what the numeric Greek coefficients are in this proposed Newtonian Relativistic Electrodynamics? It is our contention that only experiments will allow us to bring some order to this *zoo* of electrodynamics.

With respect to this modified Millikan's experiment, now is the time to critically review the *interpretations* of the theoretical conclusion displayed by eqs. (4.23) and the outcome of this experiment.

1. The experiment, in the beginning, was affected by a systematic error. For this reason, the experiment should be independently executed by other experimenters in order to establish the existence or non existence of this type of error.

2. Rowland's equivalence, applicable to Biot-Savart's field, is not applicable to Gauss-Weber's field.

3. Undoubtedly, the presence of a magnet presents more interpretational problems than circuits carrying a steady current. For this reason, we propose an experiment that uses a magnetron.

Gauss-Weber's theoretical force field is an open problem we must solve sooner or later. Even GRT is implicated with its analogical prediction of Gauss-Weber's field. In 1991, Wesley concluded his comments on this modified Millikan experiment as follows:

> "In conclusion, the force observed by Curé seems to be too large and to be in the wrong direction to be a Weber velocity squared force. Nevertheless, further experiments of the Curé type using a current solenoid and accurately controlled geometry would be most desirable. It may be noted

that such experiments might also serve to further test the induction or acceleration force in the brackets on the right of Eq. (6.3) (and for a circular loop given by the last of Eqs. (6.82)."

Unfortunately, this author realized the meaning of Wesley's statement: "*at 1 cm off the magnet's axis*," in the last proofreading of the manuscript of *Einstein on Trial* on September, 2000. Thus, the magnitude of the force found by Curé is still a magnitude determined experimentally, or as Wesley would say "*it is empirically correct*." The future theoretical analyses of this empirical force, reported by Curé, is another open problem. The reader should have in mind that the purpose of Curé's experiment, with the Millikan Apparatus, was not to test Weber's force proportional to v'^2/c^2 which, by the way, is zero along the axis of the magnet.

However, a conceptual modification has occurred in the mind of Wesley. In the same reference of 1991, Wesley introduced a *neomechanics or Kaufmann mechanics.* In page 271 Wesley wrote:

"Weber-Wesley electrodynamics, where the velocity squared force is dropped, Eqs. (6.13) and (6.14), plus Kaufmann mechanics, where mass changes with velocity as $m = m_o/(1 - v^2/c^2)^{1/2}$, should be assumed true . . . This combination fits all of the presently experimental facts. Dropping the velocity squared force remains as a small theoretical flaw."

In a private communication, April 1995, Wesley [38] showed his definite renunciation of Weber's electrodynamics after years of serious intensive work in this field. It is interesting to note that Wesley anticipated the negation of Edwards' effect, published in 1992 by Lemon, Edwards and Kenyon [39]. Wesley also negated what he called Curé's effect. In his recent preprint, entitled *"Empirically Correct Electrodynamics,"* Wesley wrote:

"Unfortunately, Weber had no empirical justification for introducing both the velocity squared terms as well as the Coulomb force; because his force, Eq. (4), predicted a force on a stationary charge q due to a charge q' moving with the constant velocity v' given by

$$c^2\mathbf{F}_W = (qq'\mathbf{R}/R^3)[v'^2 - 3/2\ (\mathbf{v}'\cdot\mathbf{R}/R)^2]; \qquad (5)$$

> and this force had never been observed. Attempts to observe this very minute force for small charge velocities have failed (eg., Edwards et al. [8] and Curé [9])."

This author is convinced that the experimental attempts of Edwards, in 1974 and in 1976, were not failures, but, instead, biased theoretical misinterpretations of Edwards' experiments to save Lorentz's electro-dynamics, which does not contain a force proportional to v'^2/c^2, as Edwards' team rigorously demonstrated in 1976. With respect to Curé's experimental results, this author finds Wesley's conclusion totally erroneous.

We must recall that the velocity squared terms, in Weber's electrodynamics, is a necessary (logical) consequence of having force-terms, proportional to the square of the *relative velocity* of the interacting electric elementary particles. Wesley's criticism of Weber's electrodynamic theory is an *inductivist* criticism. In this respect, Wesley is an advocate of Newton's *hypotheses non fingo* (I do not feign hypotheses). Theoretical physics is an *aprioristic* activity of the mind, full of ontological, logical, and mathematical fundamental hypotheses called axioms. An *a posterioric* activity of the mind is a safe modest construct, which is completely incapable of predicting any future dynamical behavior of Nature. Wesley, in this respect, is not fair with Weber, nor with the outstanding development of Weber's theory in the mind of André Assis. In relation to Edwards' effect, published in 1976, Wesley wrote, in 1991:

> "Unfortunately their paper [Edwards, Kenyon, and Lemon] is so badly written that it is quite impossible to discover exactly what their experiment might have been; and a proper evaluation is not possible."

A year before, in 1990, Marinov [40, p. 114] wrote about Edwards' experiment:

> "[It] is written so badly that only a person who has [nothing] to do on this Earth would spend time to try to decipher it."

The reason why this author proposed the modified Millikan experiment, in 1982, was to avoid misinterpretations of complicated electronic circuitry, with superconducting coils, as the one used by Edwards' team. The modified Millikan experiment provides a neat *electrodynamic* action of a permanent magnet on static charges. Whoever repeats this experiment will convince himself that axial and radial forces act on the free electric charges on the latex spheres. If the next experimenter observes more than 1000 charged droplets, as this author has done, we must be sure that we will be convinced of the reality of the electrodynamic action of a permanent magnet on static charges. Another thing is to explain this action as caused by a velocity squared force-term. What we need to do is to design better experiments to verify or falsify the probable existence of a new electrodynamic force field proportional to the square of the velocity of the *source* electric particles. The probable existence of Gauss-Weber's electrodynamic force field is not a closed case. On the contrary, it is a very fundamental theoretical case which has to be decided experimentally. The Gauss-Weber force field, at laboratory scale, is almost completely negligible. This is not the case if we descend to atomic and nuclear scales. It is in the microcosmic scale where Gauss-Weber's force field manifests its incredible magnitude as we will see when we analyze Eddington's model of the neutron.

4.5.5 NESCIENCE OF EXPERTS

There is some chronological lack of knowledge, in relation to the probable existence of Gauss-Weber's electrodynamic force proportional to v'^2/c^2. There is no doubt that Edwards' team revived, in the last 25 years, the problem of the existence of Gauss-Weber's force. One important piece of knowledge which any respectful physicists *must* have is the following theoretical *truth:*

> According to Special Relativity Theory, or Einsteinian electrodynamics, Gauss-Weber's force, proportional to the square of the speed of the source particle, does not, and cannot exist when a closed circuit is involved in an experimental setup.

In 1994, an editor, and one of his referees, concluded that this author's "prediction" of Gauss-Weber's force had been *predicted* long before, in 1962, by Rosser [41] in the context of Einsteinian electrodynamics! The referee, as usual, was sarcastic and also extremely *ignorant.* Our claim of the "prediction" of Gauss-Weber's force was in the context of GRT, *not* in the context of SRT which cannot make this prediction. The illiteracy of the referee, displayed in 1994, is a consequence of his ignorance of the rigorous proof, published in 1976 by Edwards' team. This proof demonstrated the theoretical impossibility of Gauss-Weber's force existing in the context of Liénard-Schwarzschild's electrodynamics, reproduced in a more pedagogical way by Assis [42, p. 166]. Another piece of ignorance, exhibited by the referee, is contained in Rosser's paper. The referee simply did not understand the subject. In Rosser's paper we read:

> "Neglecting accelerations, Eq. (1) would have to be integrated over the velocity distribution of the electrons. It would be beyond the scope of this note to attempt to solve this problem, but it is fairly safe to conclude that the predicted resultant electric field would not be zero."

The referees of Rosser's paper did not bother to investigate the statement: "*but it is fairly safe to conclude that the predicted resultant electric field would not be zero.*" Rosser made the same mistake that O'Rahilly made in 1938. It was about the same problem, when O'Rahilly neglected the accelerations. This theoretical mistake, also, has been observed by Assis [43, p. 166]. This mistake of O'Rahilly and Rosser was taken by Bonnet [44], in 1981, to "explain" the positive Edwards' effect of Gauss-Weber's force.

Finally, we have a negative Edwards' effect, i.e., Gauss-Webers' force does not exist. As we mentioned above, Lemon, Edwards and Kenyon, in 1992, experimentally have shown that Edwards' effect is not caused by Gauss-Weber's force, but is due to strange systematic errors. Where are we standing now? Let us borrow a recent phrase. It is fairly safe to conclude, that at the entrance of the 21st century, *we do not know more than Weber and Maxwell.* The only escape from this pit of ignorance is to go back to Galileo's science: *experimental philosophy.*

4.6 On the Paternity of Lorentz's Force.

The most outstanding characteristic of *relativistic electro-dynamics (RED) or Einsteinian electro-dynamics* is not in the first part of the composite word "electro-dynamics." Einstein did not *modify* anything in the "electro"-science of Maxwell. Einstein changed *everything* in the second part of the composite word "electro-dynamics," i.e., in the concept conveyed by the word "dynamics." Einstein blamed Newton's "d(m**v**)/dt" for being wrong. This author apologizes, beforehand, to some courageous dissidents, but we are obligated to present some very critical comments about the fundamental work done by some of the few free thinkers left in this world. These criticisms have been motivated by the written work of dissident physicists. At the turn of the century, we had Lorentz's electrodynamics given by:

$$d(m\mathbf{v})/dt = \Sigma_j \mathbf{F}_j = -q\nabla\phi - q\partial\mathbf{A}/\partial t + q\mathbf{v} \times (\nabla \times \mathbf{A}) \qquad (4.24a)$$

or $$d(m\mathbf{v})/dt = \Sigma_j \mathbf{F}_j = q(\mathbf{E} + \mathbf{v} \times \mathbf{B}) \tag{4.24b}$$

For Newtonian dynamics:

$$\mathbf{p} = m\mathbf{v} = m_o\mathbf{v}, \text{ with } m_o = \text{constant} \tag{4.25}$$

For Einsteinian dynamics:

$$\mathbf{p} = m\mathbf{v} = m_o\mathbf{v}/(1\text{-}v^2/c^2)^{1/2} \tag{4.26}$$

Usually eq. (4.26) has been interpreted as the mass variation with velocity, and not as the linear momentum variation with velocity. However, from a mathematical point of view, eq. (4.26) can be rearranged in order to have:

$$m = m_o/(1\text{-}v^2/c^2)^{1/2} \tag{4.27}$$

Two recent interesting analyses about the mass variation with velocity are offered by Carl G. Adler [45, 1987], and by Lev B. Okun [46, 1989]. In both papers the same reference is made about a letter Einstein wrote to Lincoln Barnett [47] in 1948:

> "It is not good to introduce the concept of the mass $M = m/(1\text{-}v^2/c^2)^{1/2}$ of a body for which no clear definition can be given. It is better to introduce no other mass than the 'rest mass' m. Instead of introducing M, it is better to mention the expression for the momentum and energy of a body in motion."

The problem with eq. (4.27) is that it does not constitute an operational definition of mass. We cannot measure the mass of an electric particle in motion. We cannot even measure the mass of one electron *at rest.* The rest-mass m_o of an electron is an abstraction which has no real direct experimental verification. The rest-mass of an electron is indirectly *calculated* from the ratio e/m_o, expressed in terms of measured electrodynamic parameters. These parameters in turn depend on the particular electrodynamic theory we use. The most outstanding feature of Einstein's dynamics is that the linear momentum of a particle is proportional to the velocity and a function of the square of the speed of the said particle. On the other hand, in Newton's dynamics, the linear momentum is *directly* proportional to the velocity of the particle. Obviously, the *constant* of proportionality is the mass of the particle. Let us call *hybrid electrodynamics* any electrodynamics that retains Einstein's dynamics given by:

$$d[m_0(1 - v^2/c^2)^{-1/2}\mathbf{v}] / dt = \mathbf{F} \qquad (4.28)$$

where **F** obviously must be different from Lorentz's force. Before we refer to *hybrid* electrodynamics, allow us to point out a historical conceptual inaccuracy found in almost every textbook of physics in relation to Lorentz's force, given by eqs. (4.24a,b). In particular, we will quote from the book *Newton versus Einstein* by P. Graneau and N. Graneau [48, p. 139,140]:

> "There is no Maxwell force law. As discussed in the last chapter, far from unifying the force laws, Maxwell felt uncertain which of the electrodynamic laws was valid: Ampere's or Grassmann's? There is no mention in his writings of force unification. The claim that field theory unified the electric with the magnetic force is simply wrong.
>
> "Lorentz combined the Coulomb force with the Grassman force in the Lorentz force formula, long after Maxwell."

These are completely incorrect statements. *There are two Maxwell force laws!* These were published *long before* Lorentz, in Maxwell's *Treatise.* As far as this author's knowledge is concerned, the only physicist that has noticed this historical precedence to the so-called Lorentz's force is Stefan Marinov [49, p. 31]. In 1990, Marinov wrote:

> "Then Maxwell gives the Lorentz equation exactly in the form (1), so that the attribution of Lorentz's name to it is historically unwarranted."

Equation (1), in Marinov's reference, is eq. (4.24a) in this chapter. Had writers of physics books and professors of physics thoroughly studied Maxwell's *Treatise,* they never would have missed Articles 598 and 599 of Maxwell's masterpiece. Even H.A. Lorentz cannot escape this criticism. As we have mentioned in a previous chapter, Lorentz used the same method used by Maxwell. They both used Lagrange's methodology of Analytical Mechanics. Even the great historian, Whittaker, failed to mention Maxwell as the first theoretician in deducing *Maxwell*-Lorentz's electrodynamic force. Whittaker

[50, Vol. 1, Cap. XIII, p. 395] shows Lorentz's usage of Clausius' electrokinetic potential along with the Lagrangian method. Actually, Whittaker shows how Lorentz deduced Grassmann's force. Whittaker added that the so-called Lorentz force is "*in agreement with the formula obtained by Heaviside in 1889.*" Let us see where Maxwell's deduction of "Lorentz's" force is. In Volume 2, Art. 598 and 599 of Mawxell's *Treatise* [51, vol. II, p. 240, 241], we read:

> "Hence we may now disregard the circumstance that ds forms part of a circuit, and consider it simply as a portion of a moving body, acted on by the electromotive intensity E. The electromotive intensity has already been defined in Art. 68. It is also called the resultant electrical intensity, being the force which would be experienced by a unit of positive electricity placed at that point. We have now obtained the most general value of this quantity in the case of a body moving in a magnetic field due to a variable electric system ... The electromotive intensity, as defined by equations (B), may therefore be written in the quaternion form,
>
> $$\mathbf{E} = \mathbf{v} \times \mathbf{B} - \partial \mathbf{A}/\partial t - \nabla \psi \qquad (10)"$$

In the above quotation, we have changed Maxwell's notation into our present mathematical terminology. The last equation should be called *Maxwell's electrodynamic force per unit charge*. An interesting problem for a historian of science is to determine *why* Maxwell retained Grassmann's force in Maxwell's force, in spite of the fact that Maxwell, in Volume I, page 175 of his *Treatise,* referred to Ampère's force as *"the cardinal formula of electro-dynamics."*

Maxwell still has a second force! This is an electrokinetic force. It is the result of a generalization which Maxwell did to Ampère's force. In chapter 3, we referred to this other force of Maxwell. We should notice, in the previous quotation, that Maxwell abstracted from the test circuit a differential current element ds in motion. This portion **Ids** = **v**dq will later represent the granular aspect in an electron theory. Many physicists are very careless with mathematics when they refer to Rowland's equivalence by writing **Ids** = vq. An *infinitesimal* quantity cannot be equated to a *finite* quantity!

The profound change, introduced by Einstein in eq. (4.24b) after the *rejection* of Newton's concepts of absolute space and absolute time, is the remarkable consequence of the so-called variation of mass with velocity. In doing so, Einstein made acceleration depend on the square of the test particle velocity. This dependence was absolutely a new event in Maxwell's electrodynamics (Maxwell-Lorentz's force).

Before this historical moment, in the evolution of Maxwell's theory, any Newtonian Relativistic Electrodynamics provided electrodynamic force-terms proportional to the square of the velocity of the source, as well as the test particle. Since Gauss' time, this was common knowledge in Germany. Even Maxwell wrote about it in chapter XXIII, Vol. 1, of his *Treatise*. The PNRED, given by eq. (4.19), can be written as follows:

$$d(m_o\mathbf{v})/dt = \mathbf{F}_C + \mathbf{F}_{GW} + \mathbf{F}_M + \mathbf{F}_K, \qquad (4.29)$$

where

Coulomb:

$$\mathbf{F}_C = Kqq'\mathbf{r}/r^3 \qquad (4.30)$$

Gauss-Weber:

$$\mathbf{F}_{GW} = g[\mathbf{r}\{\alpha v'^2 + \beta r^{-2}(\mathbf{r}\cdot\mathbf{v'})^2 - \gamma(\mathbf{r}\cdot\mathbf{a'})\} + \delta\mathbf{v'}(\mathbf{r}\cdot\mathbf{v'}) - \varepsilon r^2\mathbf{a'}] \qquad (4.31)$$

Maxwell:

$$\mathbf{F}_M = -g[\mathbf{r}\{2\alpha(\mathbf{v}\cdot\mathbf{v'}) + 2\beta r^{-2}(\mathbf{r}\cdot\mathbf{v})(\mathbf{r}\cdot\mathbf{v'})\} + \delta\{\mathbf{v'}(\mathbf{r}\cdot\mathbf{v}) + \mathbf{v}(\mathbf{r}\cdot\mathbf{v'})\}] \qquad (4.32)$$

Kaufmann:

$$\mathbf{F}_K = g[\mathbf{r}\{\alpha v^2 + \beta r^{-2}(\mathbf{r}\cdot\mathbf{v})^2 + \gamma(\mathbf{r}\cdot\mathbf{a})\} + \delta\mathbf{v}(\mathbf{r}\cdot\mathbf{v}) + \varepsilon r^2\mathbf{a}] \qquad (4.33)$$

where
$$g = kqq'/r^3$$
$$k = K/c^2$$
$$K = 1/(4\pi\varepsilon_o)$$

We are following, in a different context, Wesley's designation of the test velocity squared force as *Kaufmann's force*. An authentic Newtonian Electrodynamics should maintain the constancy of the mass of the particles, and provide other force-terms not contained in the old Newton-Maxwell's electrodynamics. In eqs. (4.31, 32, 33), we see a formal identity between Gauss-Weber's force and Kaufmann's force. One is tempted to assert, based on symmetry, that if there are forces, dependent on the square of the test particle velocity, there must be forces dependent on the square of the source particle velocity. As we mentioned in a previous chapter, the weak point of Special Relativity Theory (SRT) is in electrodynamics and not in Einsteinian relativistic kinematics. We must never forget the origin of SRT. Whittaker [52, Vol. 2, p. 42] reminds us of that origin:

"It is clear, from the history set forth in the present chapter, that the theory of relativity had its origin in the theory of the aether and electrons. When relativity had become recognised as a doctrine covering the whole operation of physical nature, efforts were made to present it in a form free from any special association with electromagnetic theory, and deducible logically from a definite set of axioms of greater or less plausibility."

For this reason, modifications of Tate-Edwards' experiments are badly needed as well as experiments of time-of-flight of electric particles. Presently, we have enough Newtonian electrodynamics to test experimentally. One problem, which is always overlooked, is the accelerating process of electrons from a cathode by means of a positive anode. Let us reinterpret Einsteinian electrodynamics in relation to this problem. After taking the time derivative in eq. (4.24b), we get:

$$[m_o(1- v^2/c^2)^{-1/2}]\mathbf{a}+[m_o(1- v^2/c^2)^{-3/2}]va\mathbf{v}/c^2 = \mathbf{F}_C$$

In the previous equation, we see the two types of mass: transversal and longitudinal, represented by the two square brackets, respectively. The last equation can be written in another form:

$$m_o\mathbf{a} = (1 -v^2/c^2)^{1/2}\mathbf{F}_C - [m_o(1 -v^2/c^2)^{-1}av/c^2]\mathbf{v} \qquad (4.34)$$

The Newtonian force $m_o\mathbf{a}$ is not *only* proportional to Coulomb's force, but to a force proportional to v^2/c^2, plus another force that is proportional to the velocity $\mathbf{v}$, like in Maxwell's electrokinetic force. Eq. (4.34), applied to the case of an electron, being accelerated along the Z-axis by a large circular anode, can be written in the following way:

$$m_o\mathbf{a} = (1 -v^2/c^2)^{3/2}\mathbf{F}_C = \mathbf{k}(1 -v^2/c^2)^{3/2}2\pi Kq\sigma \qquad (4.35)$$

where $\mathbf{k}$ is a unit vector, and σ is the surface charge density of the anode. Eq. (4.35) shows the way Einstein's SRT introduced, in Maxwell-Lorentz's electrodynamics, force-terms proportional to v^2/c^2. In this reinterpretation, eq. (4.35) shows that in addition to the accelerating force of Coulomb, there exists another decelerating force which opposes the motion of the particle giving the impression that the mass of the particle increases with velocity. Obviously, we do not need to reinterpret Einstein's electrodynamics in Newtonian terms, because we have a vast collection of Newtonian Relativistic Electrodynamics. We also have some hybrid electro-dynamics.

4.7 Hybrid Electrodynamics(HED).

We understand by *hybrid electrodynamics,* anyone different from Maxwell-Lorentz-Einstein's electrodynamics which establishes that the mass of the particle varies with velocity.

Marinov's Hybrid Electrodynamics

Whether Marinov's concepts will eventually be proven right or wrong in matters of electrodynamics, we believe he already has a place in the future of Galileo's science. He was highly prolific and controversial, but most of all, we have to admire his love for truth. Truth, in Marinov's mind, i.e., his truth, evolves and involves at a great rate. For years he had been proposing electrokinetic theories. He had been a ferocious advocate of Grassmann's force and a retractor of Ampère's force. Recently, he had become a retractor of both of them. As far back as 1977, Marinov's HED [53, p. 46] is given in the International System of Units by the following equation:

$$d(\mathbf{p}_o + q\mathbf{A})/dt = - q\ \mathrm{grad}\phi + q\ \mathrm{grad}(\mathbf{v}\cdot\mathbf{A}) \qquad (4.36)$$

with $\mathbf{p}_o = m\mathbf{v}/(1\text{-}v^2/c^2)$ (4.37)

Besides a transposition of the sub-index zero, the last equation belongs to Einsteinian electrodynamics. At first sight, eq. (4.36) looks like a new electrodynamic equation. It is interesting to see in this equation Neumann's electrokinetic potential ($\mathbf{v}\cdot\mathbf{A}$) and, therefore, Neumann's electrokinetic force: q grad ($\mathbf{v}\cdot\mathbf{A}$). However, Neumann's force will be canceled out by a term coming from the left-hand side of eq. (4.36), once the convective operator is applied to the *total derivative* of q**A**, assuming v constant or independent of the space coordinates.

A great majority of physicists consider the magnetic vector potential **A** simply a mathematical artifice devoid of any physical reality. However, a dimensional analysis of units reveal that **A** represents a *potential linear momentum per unit charge* of the electromagnetic field. **A** is actualized, or comes to a real existence, when an electric particle of charge q moves in an electromagnetic field, changing the Newtonian or Einsteinian *inertial linear momentum* of the electric particle. This interpretation is written in Marinov's eq. (4.36), and makes intelligible the assertion that an electromagnetic field communicates linear momentum and energy to a moving electric particle. Marinov [54, p. 45] published an interesting paper on the physical reality of the specific potential linear momentum of an electromagnetic field under the suggestive title: *Is the Aharonov-Bohm Effect an Aharonov-Bohm Effect?* Eq. (4.36) can be written in this other form:

$$d\mathbf{p}_o/dt = -q\nabla\phi + q\nabla(\mathbf{v}\cdot\mathbf{A}) - qd\mathbf{A}/dt \tag{4.38}$$

which reduces to Maxwell-Lorentz's force given by eq. (4.24a). It was Clausius [55, Vol. 1, p. 234] who used Coulomb's and Neumann's potentials to propose his *electrokinetic potential energy* U:

$$U = Kqq'(1 - \mathbf{v}\cdot\mathbf{v}'/c^2) \tag{4.39}$$

Using Lagrange's method:

$$\mathbf{F} = d(\partial U/\partial \mathbf{v})/dt - \partial U/\partial \mathbf{r}$$

we get Clausius' electrodynamics:

$$\mathbf{F} = Kqq'r^{-3}[\mathbf{r} - \mathbf{r}(\mathbf{v}\cdot\mathbf{v}'/\mathbf{c}^2) + \mathbf{v}'(\mathbf{r}\cdot\mathbf{v})/c^2 - \mathbf{v}'(\mathbf{r}\cdot\mathbf{v}')/c^2 - r^2\mathbf{a}'/c^2] \tag{4.40}$$

$$\mathbf{F} = Kqq'r^{-3}\mathbf{r} + Kqq'/c^2[-\mathbf{r}(\mathbf{v}\cdot\mathbf{v}')/r^3 - d(\mathbf{v}'/r)/dt]$$

$$\mathbf{F} = -q\nabla\phi + q(\mathbf{v}\cdot Kq'\mathbf{v}'/c^2)(-\mathbf{r}/r^3) - qd[Kq'\mathbf{v}'/(rc^2)]/dt$$

$$\mathbf{F} = -q\nabla\phi + q(\mathbf{v}\cdot Kq'\mathbf{v}'/c^2)\nabla(1/r) - qd[\mathbf{A}]/dt$$

$$\mathbf{F} = -q\nabla\phi + q\nabla[\mathbf{v}\cdot Kq'\mathbf{v}'/(rc^2)] - qd\mathbf{A}/dt$$

$$\mathbf{F} = -q\nabla\phi + q\nabla[\mathbf{v}\cdot\mathbf{A}] - qd\mathbf{A}/dt \tag{4.41}$$

$$\mathbf{E} = -\nabla\phi - d\mathbf{A}/dt + \nabla(\mathbf{v}\cdot\mathbf{A}) \tag{4.42}$$

If we define Neumann's potential χ by the following equation:

$$\chi = \mathbf{v}\cdot\mathbf{A} \tag{4.43}$$

then, eq. (4.42) becomes:

$$\mathbf{E} = -\nabla(\phi - \chi) - d\mathbf{A}/dt \tag{4.44}$$

or $$\mathbf{E} = -\nabla\psi - d\mathbf{A}/dt \tag{4.45}$$

where $$\psi = \phi - \chi \tag{4.46}$$

is Clausius' potential. The following development shows the deduction of Maxwell-Lorentz's force from Clausius' field, eq. (4.42), or Marinov's force given by (4.38). Using the convective operator we have:

$$\mathbf{E} = -\nabla\phi - \partial\mathbf{A}/\partial t - (\mathbf{v}\cdot\nabla)\mathbf{A} + \nabla(\mathbf{v}\cdot\mathbf{A}) \tag{4.46*}$$

Vector theorem:

$$\nabla(\mathbf{v}\cdot\mathbf{A}) = (\mathbf{v}\cdot\nabla)\mathbf{A} + (\mathbf{A}\cdot\nabla)\mathbf{v} + \mathbf{v}\text{x}(\nabla\text{x}\mathbf{A}) + \mathbf{A}\text{x}(\nabla\text{x}\mathbf{v})$$

If **v** is function of t only, and not of x, y, z, or constant, then we have:

$$\nabla(\mathbf{v}\cdot\mathbf{A}) = (\mathbf{v}\cdot\nabla)\mathbf{A} + \mathbf{v}\text{x}(\nabla\text{x}\mathbf{A})$$
$$\mathbf{E} = -\nabla\phi - \partial\mathbf{A}/\partial t - \nabla(\mathbf{v}\cdot\mathbf{A}) + \mathbf{v}\text{x}(\nabla\text{x}\mathbf{A}) + \nabla(\mathbf{v}\cdot\mathbf{A})$$
$$\mathbf{E} = -\nabla\phi - \partial\mathbf{A}/\partial t + \mathbf{v}\text{x}(\nabla\text{x}\mathbf{A})\text{; multiplying by q:}$$
$$\mathbf{F} = -q\nabla\phi - q\partial\mathbf{A}/\partial t + q\mathbf{v}\text{x}(\nabla\text{x}\mathbf{A})$$

Q.E.D.

The last equation completes the proof that Clausius' electrodynamic force is identical to Maxwell-Lorentz's force. This equation also proves that Clausius predated Lorentz in rederiving Maxwell's electrodynamic force per unit charge. Maxwell did his derivation in 1873, Clausius in 1877, and Lorentz in 1892. O'Rahilly [56, Vol. I, p. 222], made the following comment about Clausius' electrodynamics and, in consequence, about Maxwell-Lorentz's electrodynamics:

"However defective, this formula of Clausius is the pioneer and model of Liénard-Schwarzschild force-formula which, implicitly or explicitly, is the condensed statement of the form of the electron theory which is almost universally accepted to-day, which in particular is accepted by all followers of Lorentz and Einstein."

Clausius' force, in the form given by eq. (4.45), is very suggestive in relation to Hertz's *hypothesis.* This consists of replacing all the partial derivatives for total derivatives in Maxwell's field equations. The interested reader should consult the book of Phipps [57] titled *Heretical Verities: Mathematical Themes in Physical Description,* and the work of Moon, Spencer *et al.* [58] in relation to Hertz's hypothesis. Here, we will refer very briefly to one of Maxwell's field equations:

$$\nabla \times \mathbf{E} = -\partial \mathbf{B}/\partial t = -\partial(\nabla \times \mathbf{A})/\partial t \qquad (4.47)$$

It is a well-known fact that Maxwell-Lorentz's force cannot be deduced from Maxwell's field equations. The Maxwell-Lorentz force is referred to as a *subsidiary* equation to Maxwell's field equations. Subsidiary means adjunct, auxiliary. In the following study cases, we will use Hertz's hypothesis to investigate the possibility of deducing Maxwell-Lorentz's force from Maxwell-Hertz's field equation. The last equation can be written in the following way:

$$\nabla \times \mathbf{E} = -\nabla \times (\partial \mathbf{A}/\partial t)$$
$$\nabla \times (\mathbf{E} + \partial \mathbf{A}/\partial t) = \mathbf{0}$$

The solution of the last equation is usually expressed in terms of the gradient of Coulomb's *electrostatic* potential. This procedure is rather awkward, because we are trying to solve an *electrodynamic* problem and not an *electrostatic* one. The solution of the last equation should be:

$$\mathbf{E} + \partial \mathbf{A}/\partial t = -\nabla \psi$$

where the potential ψ should be equal to Coulomb's potential plus another potential that at least should be *electrokinetic.* Let us use Neumann's potential along with Coulomb's, as given by eq. (4.46). Solving for $\mathbf{E}$ in the last equation we get:

$$\mathbf{E} = -\nabla \psi - \partial \mathbf{A}/\partial t \qquad (4.48)$$

Now let us consider different cases:

Case 1. Using Hertz's hypothesis in eq. (4.48), and $\psi = \phi - \chi$. We get:

$$\mathbf{E} = -\nabla\psi - d\mathbf{A}/dt \tag{4.49}$$

The last equation is identical to Clausius' ponderomotive force per unit charge given by eq. (4.45), from which we deduced Maxwell-Lorentz's force. However, the total derivative in eq. (4.45) is not a hypothesis but a necessary consequence of Clausius' electrodynamics.

Case 2. In this case we will not use Hertz's hypothesis in eq. (4.48) and make $\psi = \phi$. We obtain:

$$\mathbf{E} = -\nabla\phi - \partial\mathbf{A}/\partial t - (\mathbf{v}\cdot\nabla)\mathbf{A}$$
$$\mathbf{E} = -\nabla\phi - d\mathbf{A}/dt$$

This time Hertz's hypothesis reappears but being only associated to Coulomb's potential.

Case 3. Again we will not use Hertz's hypothesis and make $\psi = \phi - \chi$; we get:

$$\mathbf{E} = -\nabla\phi - \partial\mathbf{A}/\partial t + \mathbf{v}\mathrm{x}(\nabla\,\mathbf{A}) + (\mathbf{v}\cdot\nabla)\mathbf{A} \tag{4.50}$$

This time we obtain Maxwell-Lorentz's force, plus Marinov's [59, p. 217] or *motional-transformer induction,* equal to $(\mathbf{v}\cdot\nabla)\mathbf{A}$. In essence, Marinov's hybrid electrodynamics is Einsteinian electrodynamics.

Wesley's Hybrid Electrodynamics

Another hybrid electrodynamics belongs to Wesley [60]. Wesley introduces a *neomechanics or Kaufmann's mechanics* given by:

$$d[m_o\mathbf{v}(1-v^2/c^2)^{-1/2}]/dt = \mathbf{F}$$

If **F**, in the last equation, is Lorentz' force, then the last equation is *identical* to Einsteinian electrodynamics. Of course **F** is not a Lorentzian force but *Wesley's electrodynamic force,* given by:

$$d[m_o\mathbf{v}(1-v^2/c^2)^{-1/2}]/dt = \mathbf{F} = Kqq'r^{-3}\,[1 - 2\mathbf{v}\cdot\mathbf{v'}/c^2 + 3r^{-2}(\mathbf{r}\cdot\mathbf{v})(\mathbf{r}\cdot\mathbf{v'})/c^2 + \mathbf{r}\cdot(\mathbf{a}-\mathbf{a'})/c^2] \tag{4.51}$$

Wesley eliminated from Weber's electrodynamics all the terms proportional to v'^2/c^2 and v^2/c^2. The left-hand side of eq. (4.51) is still an Einsteinian dynamics. Any future student of Newtonian electrodynamics will find, in Wesley's books, solid landmarks and a gold mine of bibliographic notes to adventure his mind in such fascinating and difficult matters. Wesley's work is a beacon to guide new generations of natural philosophers to keep on searching for the ultimate truth. The scholarship of Wesley reminds us of Prof. H.A. Lorentz. We hope Wesley will come back to his lifetime Newtonianism in electrodynamics. The few dissidents left in this branch of knowledge need Wesley's lighthouse to find the right path in this dark labyrinth of present physics. The interested colleague should seriously study Wesley's [61, p. 289, 1990] *Evidence for Weber-Wesley Electrodynamics;* Wesley's [62, 1991] *Advanced Fundamental Physics,* and Wesley's [63, 1995] *Empirically Correct Electrodynamics.* Another interesting book to study is, Wesley's [64, 1983] *Causal Quantum Theory.*

Phipps' Hybrid Electrodynamics

Phipps [65, chap. 4, 5] revives Hertz's hypothesis and Galilean transformation in a brilliant way in electromagnetics (field equations) and electrodynamics (force interaction). When this author was an undergraduate student, he understood immediately the concept of Newtonian force *invariance:* $F' = m'a' = ma = F$, in classical dynamics. Then when he took the next course in electromagnetics and electrodynamics, he expected to see on the blackboard or in his textbook the following mathematical statements:

If $\mathbf{r'} = \mathbf{r} - \mathbf{v}t$ and $t' = t$

then $\nabla'\mathrm{x}\mathbf{E'} = -\partial\mathbf{B'}/\partial t' = -\partial\mathbf{B}/\partial\mathbf{t} = \nabla\,\mathrm{x}\mathbf{E}$

and $\mathbf{E'} = -\nabla'\phi - \partial\mathbf{A'}/\partial t' + \mathbf{v'}\mathrm{x}(\nabla'\mathrm{x}\mathbf{A'}) =$

$\quad - \nabla\phi - \partial\mathbf{A}/\partial t + \mathbf{v}\mathrm{x}(\nabla\mathrm{x}\mathbf{A}) = \mathbf{E}$

As all of us know, our expectations came to a bitter end. All of us were trained in the *magic* of Einstein-Minkowski's *covariance.* Phipps has produced the *magic* of electromagnetic and electrodynamic *invariance* in the proper sense this author expected in his undergraduate years. The reason this author is an advocate of Newtonian relativistic electrodynamics is that these types of electrodynamic forces, depending on relative distance, relative velocities and relative accelerations, assure *ipso facto* dynamic invariance. The outcome of Trouton-Noble's experiment could have been predicted in an exact way if physicists had used some of the German Newtonian relativistic electrodynamics.

There is something hard to understand. Why is there this obsession to anticipate theoretically the mathematical structure of physical laws in different moving laboratories? This *mathematical structure,* for sure, is inside a fictitious laboratory which moves in the virtual realms of our minds, instead of moving in the external reality. At the end of the stagnant 20th century, we should have been convinced that we did not have the proper electromagnetic nor the proper electrodynamic theories to predicate how the mathematical structure of physical laws, inside a virtual laboratory, is going to be. It is too premature to establish *space-time-coordinate* (STC) transformation in order to look for invariance or covariance of field equations of doubtful electromagnetic theories. On the other hand, we cannot use the *same* STC transformation for material particles and for immaterial electromagnetic waves.

Unfortunately, Phipps' *Neo-Hertzian Electromagnetism* and electrodynamics retain the variation of mass with velocity. Phipps' theory is *fundamentally* different from Einstein's electrodynamics. In his book Phipps [66, p. 193] writes:

> "In this section we have probed a proposed electrodynamic force law and its associated equation of motion in manifestly invariant and noninvariant forms, through examination of a few special cases. No particular problems seem to arise, the observable predictions being identical to those of Einstein's special theory - although the formalism is quite different. The coincidence of predictions is to be expected, since we acknowledged at the outset our acceptance of Einstein's single-worldline (One-body) physics, and have departed from his ideas only in respect to metric, structural, or worldline-relational features of physical description. It may be, though, that further investigation will turn up unacceptable consequences of the present formulas. If so, these will furnish "clues" to the next stage of advance. Meanwhile, it seems to me of value to both electrodynamics

> and kinematics, as in the next chapter, to exploit and extend the gains made thus far in a direction orthogonal to Einstein's course . . . and (I submit) far sounder in its bearings."

If "*. . . further investigation will turn up unacceptable consequences of the present formulas,*" we could not blame Phipps for this failure but the electromagnetics and electrodynamics he used.

4.8 Deduction of Hertz's "Hypothesis."

The fact that we have been unable to deduce Maxwell-Clausius-Lorentz's force law from Maxwell's field equation $\nabla \times \mathbf{E} = -\partial \mathbf{B}/\partial t$ has suggested to some physicists that this field equation is incomplete. In 1928, Mason and Weaver [67, p. 254-260] made a critical analysis of this field equation, including ontological arguments; they wrote:

> "To a large school of physicists the state of the "field" at a given point in space has a definite and describable reality (a reality aided by certain mechanical conceptions of the aether) apart from the nature of that which is producing the "field." That is to say, a given value of B at a certain point is taken to be descriptive of some condition obtaining at that point, and the emphasis is so thrown over onto the importance of this condition and its description by means of B that one is not to be concerned particularly with what has produced the value of B. One who adopts this point of view will say that the last equation above states that the curl of the E' vector is given by l/c times the negative rate of change of the B vector, and will not feel it necessary to qualify this statement at all; in particular, he will not feel it necessary to distinguish between two cases, in which the negative rates of change of B are equal, the cause of the change being quite different in the two cases . . . It is very easy to let the notation carry the burden of the argument, to neglect this discussion, and to hold that the value of curl E' is related to the rate of change of B in every case in the way stated by the last equation. It is important to point out, however, that by so doing one may be overlooking something of fundamental physi-

cal significance, and it is desirable to insist upon the potential importance of keeping in mind the auxiliary nature of the vectors E and B, and the necessity of always going back to the physical case."

If Mason and Weaver are right, then Maxwell's field equation, $\nabla x\mathbf{E} = -\partial\mathbf{B}/\partial t$, is lacking an important term. In 1938, O'Rahilly [68, Vol. 2, p, 583] did not see this possibility of having two sources which *cause* the variation of **B**. He, consequently, strongly criticized the concepts of Mason and Weaver. Hence, if Marinov would have shown experimentally, without the slightest doubt, the reality of his field **M**, the motional-transformer induction, given by:

$$\mathbf{M} = (\mathbf{v}\cdot\nabla)\mathbf{A} \qquad (4.52)$$

then, we could correct Maxwell's field equation, given by eq. (4.47). We will call **M**, the Marinov field equation as given by (4.52). The new Maxwell-Marinov's field equation should read as follows:

$$\nabla x\mathbf{E} = -\partial\mathbf{B}/\partial t - \nabla x\mathbf{M} \qquad (4.53a)$$

or

$$\nabla x\mathbf{E}^* = -\partial\mathbf{B}/\partial t \qquad (4.53b)$$

where

$$\mathbf{E}^* = \mathbf{E} + \mathbf{M} \qquad (4.53c)$$

Now, we plan to demonstrate the following thesis: to deduce, from the last equation, Hertz's hypothesis and Maxwell-Clausius-Lorentz's electrodynamic force. The following sequence of equations demonstrates the thesis.

$$\nabla x\mathbf{E} = -\partial(\nabla x\mathbf{A})/\partial t - \nabla x[(\mathbf{v}\cdot\nabla)\mathbf{A}]$$
$$\nabla x\mathbf{E} = -\nabla x[\partial\mathbf{A}/\partial t] - \nabla x[(\mathbf{v}\cdot\nabla)\mathbf{A}]$$
$$\nabla x\{\mathbf{E}+[\partial\mathbf{A}/\partial t+(\mathbf{v}\cdot\nabla)\mathbf{A}]\} = \mathbf{0}$$

using the convective operator on the square bracket, we get:

$$\nabla x\{\mathbf{E}+d\mathbf{A}/dt\} = \mathbf{0}$$

integrating we get:

$$\mathbf{E}+d\mathbf{A}/dt = -\nabla\psi$$

where ψ is Clausius' potential given by eq. (4.46), or

$$\mathbf{E} = -\nabla\psi - d\mathbf{A}/dt.$$

Q.E.D.

The last equation is identical to Clausius' force per unit charge from which we have deduced Maxwell-Lorentz's force. Phipps [69, p. 129, 130] does very similar work but using the magnetic induction **B** instead of the magnetic vector potential **A**. In Phipps' treatment of the same problem, he does not need Neumann's potential $\chi = \mathbf{v}\cdot\mathbf{A}$. To demonstrate that eq. (4.46*) constitute a necessary and sufficient condition of Clausius' force law, let us use the convective operator on this equation and express Clausius' potential ψ as $\phi - \chi$:

$$\mathbf{E} = -\nabla\phi + \nabla\chi - \partial\mathbf{A}/\partial t - (\mathbf{v}\cdot\nabla)\mathbf{A}.$$

Now, using the curl operator in the last equation, we get:

$$\nabla x\mathbf{E} = -\nabla x(\partial\mathbf{A}/\partial t) - \nabla x[(\mathbf{v}\cdot\nabla)\mathbf{A}]$$

using eq. (4.52), Marinov's force law, we get:

$$\nabla x\mathbf{E} = -\partial(\nabla x\mathbf{A})/\partial t - \nabla x[\mathbf{M}]$$

$$\nabla x\mathbf{E} = -\partial\mathbf{B}/\partial t - \nabla x\mathbf{M}$$

Q.E.D.

In summary, we still want to remain faithful to authentic Newtonian Relativistic Electrodynamics in which the mass of the particles is constant. We should maintain this position until we design experiments with proper interpretations which will falsify the existence of Gauss-Weber's force, and until better experiments decide if Kaufmann's effect is *caused* by a variation of mass with velocity and not by a Newtonian Relativistic Electrodynamics. It is very interesting to note the following relationship between the electrodynamic forces of Neumann (**N**), Marinov (**M**), and Grassmann (**G**):

$$\mathbf{N} = \mathbf{M} + \mathbf{G}$$

The last equation will help some dissidents understand Marinov's discovery.

4.9 Eddington's Model of the Neutron.

In 1933, A. Eddington [70] proposed a model for the new elementary particle, called neutron, which was discovered in the previous year by Chadwick. In this section, we plan to revive this model which describes the neutron as a miniature hydrogen atom.

Theoretical conclusions, based on Newtonian relativistic electro-dynamics (NRED) and experimental results, show that the phenomenon of *cold fusion* is probably due to a new quantum transition in the hydrogen atoms. Using experimental values of magnetic moments of proton, electron and neutron, and the NRED for a circular electronic orbit in the hydrogen atom, we obtain a quadratic equation for the orbit's radius. One solution provides the Bohr radius of normal hydrogen. The other solution gives a radius 18,796 times smaller than Bohr's radius. This latter solution points to the existence of a miniature hydrogen atom or neutron. The energy associated to a quantum transition between the hydrogen ground level, to the new sub-ground level, is of the order of 0.26 Mev, which corresponds to gamma emission.

This new quantum transition can be induced in hydrogen absorbed by the crystalline structure of metals. Decreasing the lattice constant by means of impractical extremely high pressure, or by decreasing the temperature of the metal below the liquid nitrogen temperature, we should expect the emission of neutrons and gamma rays. This phenomenon has been observed experimentally with hydrogen isotopes, but not with normal hydrogen at temperatures higher than the liquid nitrogen temperature.

Present electrodynamics, i.e., Maxwell-Lorentz-Einstein's (MLE) electrodynamics, has been critically analyzed in the last decades. Particularly, the Grassmann force, contained in Lorentz's force, has not passed serious experimental tests in the hands of Graneau, and others.

On the other hand, Grassmann's force has not passed theoretical tests in the hands of O'Rahilly, Wesley and others who have shown that using Grassmann's law, it is possible to design electrokinetic systems which would lift themselves from the ground. The so-called failure of Newton's dynamics in classical electrodynamics was due to the lack of knowledge of all the electrical forces in the interaction of two elementary charged particles in motion. This can be shown by using General Relativity Theory (GRT), as this author did in 1982 [24]. However, it is not necessary to resort to GRT to show the incompleteness or deficiency of MLE electrodynamics. We only need to resort to Newtonian Relativistic Electro-dynamics (NRED). Since 1835, when Gauss created the first NRED, the world has been exposed to sporadic improvements of this forgotten NRED. The names Weber,

Riemann, Ritz, O'Rahilly, Brown, Feynman, Marinov, Wesley, Moon, Spencer and others will forever be associated to NRED. To overcome the present stagnation of MLE's electrodynamics, we have the option of seriously considering NRED.

The NRED contains force terms that depend on relative distances, relative velocities and relative accelerations of the interacting charged particles. The relative motion of the observer has nothing to do with the electrodynamic forces between the interacting particles. Einstein's present MLE electrodynamics is highly subjective. NRED returns objectivity to Natural Philosophy. The mathematical description of the motion of a propagating electromagnetic wave requires a field theory. On the other hand, the mathematical description of the motion of two interacting charged elementary particles requires an electrodynamic theory. If we do not understand the substantial difference, in the ontological sense, between electromagnetic and electrodynamic phenomena, we will fall into the mistake of believing that the same law of velocity composition is applicable to particles and to the propagation of electromagnetic waves. This hidden act of faith led Einstein's mind to extend Lorentz's transformations of propagating electromagnetic waves (transport of immaterial energy) to the motion of electric material particles. After Fizeau, everyone knew that the velocity of light does not obey Galilean transformations of material particles in motion. Everyone knows the price we have paid in the 20th century: contraction of distances, dilation of time, variation of transversal mass due to change of velocity, and the existence in textbooks of longitudinal mass which no one uses. The fusion of space and time, in the Minkowski world, has brought much turmoil in the understanding of cold fusion and, why not say it: a very prolonged half century of confusion in the understanding of extremely expensive hot fusion.

Grassmann's force, in present MLE's electrodynamics, violates Newton's principle of action and reaction. In homopolar induction and Faraday's induction, there is very seldom mentioned electrodynamic action and reaction. The use of NRED, in the long controversy of homopolar induction, brings to mind a clear understanding of action and reaction between magnets and copper conductors in rotation. In NRED, the concept of a magnetic field, with its magnetic lines and tubes, is totally irrelevant. Over the years, physicists and electrical engineers have hypostatized the magnetic lines which were invented as a mere mnemonic rule. These ideas are very important for everyone working in the design of homopolar electric generators.

In Atomistics or old quantum physics, everyone blamed Newtonian dynamics for being unable to explain atomic phenomena. The blame should have fallen on the deficient knowledge of the electric forces, on the interaction of nuclear protons and atomic electrons. Present quantum mechanics was born with the name *atomic elec-*

trodynamics, and we definitively need a *nuclear electrodynamics.* What we really need is one good electrodynamics to account for macroscopic, atomic and nuclear electric interactions, because these seemingly different physical phenomena are only a matter of geometrical and kinematical scaling. Present quantum theory is waiting for the right *Hamiltonian*, which must come from a better electrodynamics. NRED is probably the electrodynamics that will bring the new nuclear electrodynamic theory to the point where we will finally understand the strong nuclear forces. At that point, in the near future, physicists will realize that quantum nuclear chemistry has been with us for almost three quarters of a century, waiting, as we said before, for the correct *Hamiltonian.* Chemists of cold fusion must understand that the virulent and deplorable reaction of physicists is based on the old saying, *"Nobody wants to be told that he has been barking up the wrong tree."* Cold fusion has been discovered experimentally by chemists, though physicists could have predicted it theoretically had they dared to study the history of their own science. In 1991, E. F. Mallove [71, p. XV], in his excellent review of cold fusion, wrote:

> "After reviewing mounting evidence from cold fusion experiments, I am persuaded that it provides a compelling indication that a new kind of nuclear process is at work. I would say that the evidence is overwhelmingly compelling that cold fusion is a real, new nuclear process capable of significant excess power generation."

The main purpose of this section is to present the physical foundations of this new nuclear process named *cold fusion,* based on NRED. We only hope that the young generation of physicists and chemists will take these ideas and further them to see if they are correct or incorrect.

4.9.1 THE NEUTRON: A MINIATURE HYDROGEN ATOM "A LA BOHR"

In 1987, Ne'eman and Kirsh [72, p. 119], speaking about the four basic forces, wrote:

> "It is interesting that in his later years Eddington had some original ideas that were not always accepted in theoretical physics. He was so taken with the model which postulated that the nucleus was made of electrons and protons, that he refused to abandon it, even after the discovery of the

> neutron. He insisted on regarding the neutron as a bound state of a proton and electron, and developed a complicated theory to support this notion."

Eddington's concept that the neutron is a miniature hydrogen atom is absolutely absurd in the context of present Quantum Physics (QP). In 1988, Davies [73, p. 63], in his book *The forces of nature,* wrote:

> ". . . there are good reasons why this explanation is unacceptable. Quantum theory shows that the lowest electrically bound state of the proton-electron system is the ground state of the hydrogen atom . . . [the electron] It would move with a speed of 99.97% of the speed of light. It is hard to imagine such entities smashing around inside the nucleus without leaving some sort of trace."

Davies is right if QP is correct. However, QP is based on MLE electrodynamics if we pay attention to the Hamiltonian function it uses. Also, we have to notice that present QP is a linearized particular case of the more general Hamilton-Jacobi-Bohm nonlinear equation, which contains the quantum collective potential as we proved in chapter 2.

If we only use Coulomb's force or potential in Bohr's model, we necessarily find a linear equation for the orbit's radius. On the other hand, if we use NRED, we end up with a quadratic equation for the orbit's radius in Bohr's model as we will see. Davies tries to imagine the elusive electron going around the proton. However, QP forbids imagining any *thing* (reality) at a nuclear or atomic quantum level. We will see that the electron, indeed, leaves some sort of trace in Eddington's model. This trace has to do with the magnetic moment of neutrons, but let us stop and see another objection. Davies [73, p. 63] points to a strong objection to Eddington's model of the neutron when he writes:

> "There is another reason why the neutron cannot be a composite of electron and proton. It is found that the neutron shares with the latter two particles the property of possessing intrinsic spins ½ (see Section 2.7). The only ways that the electron and proton can combine their half units of spin is to give either one unit or zero, neither of which matches that of the neutron."

This time Davies is not right. It is true that the neutron, electron and proton have intrinsic spins equal to ½. However, Davies has forgotten the orbital angular momentum of the electron in Eddington's model. In the normal hydrogen atom, this orbital angular momentum is *constant* and equal to an integral multiple of h. If we maintain this constancy of the orbital angular momentum of the electron, then, in the miniature hydrogen atom, the total angular momentum would not have the value which corresponds to the established one. The constancy of the orbital angular momentum in QP is by axiomatic definition. It is a consequence, in the old quantum theory of Bohr, to use the *electrostatic* force of Coulomb. However, when we use NRED, the orbital angular momentum of the electron is not constant due to the fact of a non-vanishing transversal component of the electrodynamic force:

$$mr^2 d\theta/dt = h(1-C/r) \tag{4.54}$$

where C is a constant equal to $\alpha Ke^2/mc^2$; α is an adjustable coefficient and $K = 1/4\pi\varepsilon_o$. Eq. (4.54) is formally identical with the expression of the orbital angular momentum of a massive particle in General Relativity Theory, when we study the perihelic precession of planet Mercury (Møller [74], 1960, page 349; 1974, page 494). C has a very small numerical value which can be neglected in the study of the normal hydrogen atom. It can be shown that C/r is approximately equal to 0.5 when r is about 20,000 times smaller than Bohr's radius. Therefore, this time the orbital angular momentum of the electron, in Eddington's model, is equal to the intrinsic spin-angular momentum of the same. Now, we have three spins of the same value, ½ in the miniature hydrogen atom or neutron. Then, the total angular momentum of the neutron can be either 1/2 or 3/2. The next objection to Eddington's model will allow us to exclude the value 3/2 from the neutron. In the context of present QP, the

following objection to Eddington's model is, perhaps, the strongest one. In 1971, Anderson [75, p.104] wrote:

> "The intrinsic spin-angular momentum and the associated magnetic moments of the elementary particles have provided the strongest arguments against the existence of electrons in the nucleus. The electron, having a magnetic moment roughly 1000 times greater than any known nuclear moment, could hardly remain undetected within the nucleus."

Consciously or unconsciously, most physicists forget to include the orbital magnetic moment of the electron in their arguments against the concept that the neutron can be a miniature hydrogen atom. Of all possible configurations of rotation and orbiting directions of the electron around the proton in the neutron, we are left with two possibilities when experimental values are used to discern among these configurations.

Table 4-I describes the possible angular momenta and magnetic moments in the neutron. The following nomenclature will be used.

s_p: proton spin
s_e: electron spin
l : electron orbital angular momentum
s_n: neutron spin
μ_p: proton magnetic moment
μ_e: electron magnetic moment
μl : electron orbital magnetic moment
μ_n : neutron magnetic moment

Table 4-I. Configuration of spins and magnetic moments in the neutron.

Config.	s_p	s_e	l	s_n	μ_p	μ_e	μ_l	μ_n
1	up	dn*	up	up	up	up	dn	dn
2	up	up	dn	dn	up	dn	up	up

* dn = down

Using the experimental values of the magnetic moments of the proton, electron and neutron, we calculated the orbital magnetic moment of both configurations by means of the following equation:

$$m_l = \mu_e \pm (\mu_p + \mu_n) \qquad (4.55)$$

The results are:

$\mu_1 = 924.64 \times 10^{-26}$ (J/T)
$\mu_1 = 929.36 \times 10^{-26}$ (J/T)

These two values, obviously, reproduce the experimental value of the neutron magnetic moment. The two configurations represent the neutron and anti-neutron in which the total spin and total magnetic moment are parallel and anti-parallel, respectively. From Table 4-I, we see that the total spin of the neutron in both configurations is equal to ½.

The trace left by electrons inside the nucleus, due to the large value of its magnetic moment, is precisely to leave no trace because of the existence of its orbital magnetic moment in the neutron. Eddington's model of the neutron, as a miniature hydrogen atom, has enormous possibilities in many fields of knowledge.

4.9.2 ON A NEW QUANTUM TRANSITION IN HYDROGEN ATOMS

In what follows, we use Newtonian dynamics in which the mass of the moving particle is constant. We use, also, NRED in an electrodynamic model of a fixed proton around which an electron moves in a circular orbit. The mathematics is straightforward, leaving the following solution of a quadratic equation for the radius of the circular orbit. The quadratic character, of this equation, was totally unexpected, and logically speaking is a consequence of using the NRED. The radius is given by the following equation:

$$r = 2r_o/[1 \pm (1-4c^2/Q)^{1/2}] \qquad (4.56)$$

where $r_o = e^2/(4\pi\varepsilon_o mc^2)$, and $Q = 4\mu_1^2/(e^2 r_o^2)$. Using experimental values, we get the two solutions of eq. (4.56). They are r_B, Bohr's radius, and r_E, Eddington's radius:

$r_B = 5.2807 \times 10^{-11}$ m
$r_E = 2.8095 \times 10^{-15}$ m

The value of r_B corresponds to the ground state of the normal hydrogen atom, while the value of r_E corresponds to the *sub-ground* state of the same atom of hydrogen or Eddington's level. A quantum transition from Bohr's energy state to Eddington's energy state is given by:

$$h\nu = E_B - E_E$$

Using the proper numeric values, we get the energy of the emitted photon of this new quantum transition equal to 0.255 Mev. This value corresponds to the emission of a gamma ray which transforms a normal hydrogen atom into a neutron.

4.9.3 CONCLUSIONS AND QUERIES

We have shown the sources from which were logically drawn strong objections to Eddington's conception of the neutron. We have shown these objections are not valid in the context of Newtonian Relativistic Electro-dynamics. In doing so, we discovered a new quantum transition which explains the emission of gamma rays and neutrons in the Pons-Fleischmann discovery of cold fusion. We hope these preliminary concepts will be taken by young chemists and young physicists to unveil new truths of the microcosm, and to show that it is not that complicated after all. To induce this new quantum transition of absorbed normal hydrogen in the crystalline structure of a metal, the temperature should descend below the liquid nitrogen temperature, or perhaps down to the liquid helium temperature.

In this last section, we want to speculate about the future of the new science of cold fusion. This, rightly, deserves the name of *Quantum Nuclear Chemistry.* To develop this new chemistry, we have to abandon present MLE electrodynamics. Once in possession of NRED, the next step is to apply it to elliptic orbits in the same way Sommerfeld used Special Relativity Theory (MLE's electrodynamics). A treatment like this provides an intuitive insight of the quantum reality, in addition to the analytical one. Then young chemists should re-do the work of Linus Pauling and E. Bright Wilson by using the new Hamiltonian in Schrödinger's equation. The subject matter is completely developed in the Pauling-Wilson textbook [76, p. 327]. In chapter XII, Section 42 entitled *The hydrogen molecule ion,* it shows how to treat the quantum physical-chemistry of the deuteron. Naturally the new Hamiltonian will be required.

A Nuclear Theory will emerge from this Quantum Nuclear Chemistry. There will then be understanding that centimetric electrodynamics, beta decay and the strong nuclear forces are different manifestations of *one* and the same electrodynamics, i.e., NRED, which only depends on cosmic scale. Einstein perhaps never realized that he formally created the long lost unified field theory. The solution of his field equations, transferred by analogy to electrodynamics and modifying the ontological background, will eventually produce the NRED that can, anyway, be deduced from the Natural Philosophy of Newton.

In 1994, Randall L. Mills [77] published a theory on subground states of the electron in hydrogen atoms. Dr. Mills, in a private communication, sent to this author another paper in which his theory is amply described. Though both theories referred to subground states in hydrogen atoms, they differ in the foundations and the distance of the electron to the proton. Mills' theory opens a vast field of research, and deserves our most serious considerations. On the other hand, the technological implications of cold fusion or quantum nuclear chemistry are multiple. The production of heavy water in laboratories or industry is one of them. Engineers should be reminded of the thought of a biologist: *we can only control what we understand.*

4.10 The Convective Operator.

This theme is mathematical. We will consider two cases. One case is in respect to a point scalar function ψ. The other case is in respect to a vector function. To be specific, let us choose the magnetic vector **A.**

Case 1. Let $\psi = \psi\,[x(t),y(t),z(t),t]$. In this case it is straightforward to conclude that:

$$d\psi/dt = \partial\psi/\partial t + (\mathbf{v}\cdot\nabla)\psi$$

Case 2. Let $\mathbf{A} = \Sigma\, \mathbf{i}_\mu A_\mu\ ;\ \mu = 1,2,3$

$$A_\mu = A_\mu[\{x(t) - x'(t)\},\{y(t) - y'(t)\},\{z(t) - z'(t)\}, t]$$

Now the coordinates x',y',z' belong to the source particle. The component A_μ cannot be identified with a one-point scalar function as ψ in case 1. The component A_μ is a two-point scalar function. Therefore, after some mathematics, the conclusion is that the convective operator, when it is applied to a vector function, should read as follows:

$$d\mathbf{A}/dt = \partial\mathbf{A}/\partial t + [(\mathbf{v} - \mathbf{v}')\cdot\nabla]\mathbf{A}$$

The presence of **v'**, in the last equation, will introduce a force-term directly proportional to v'^2/c^2 in the use of Hertz's hypothesis. All of us use the convective operator with $\mathbf{v}' = \mathbf{0}$. From a mathematical point of view, we can make $\mathbf{v}' = \mathbf{0}$, if we so desire, and there is no objection in doing so. However, from a physical point of view, it is absolutely objectionable because we reduce our electrodynamic study to a quasi-electrostatic one.

Nevertheless, velocity squared force-terms, at laboratory level, are so minute that no serious harm has been done to different electrodynamics. Besides, we are talking about *probable* velocity square force-terms.

However, the last equation has a more important consequence in electrodynamic induction, particularly in the analysis of the first page of Einstein's paper of 1905 [78, p.37], titled *On the Electrodynamics of Moving Bodies*. Einstein provides a qualitative discussion of the *relative motion* in observing electromagnetic induction between a magnet and a closed circuit. Einstein's admiration of this phenomenon of electromagnetic induction is that whether the magnet is in motion and the closed circuit is at rest in the laboratory, or the magnet is at rest and the closed circuit is in motion. The electromotive force is of the same magnitude in the closed circuit in both cases. The line integral of the previous equation provides the same magnitude of the electromotive force if the relative velocity in the previous equation is the same. Thus, here we have a theoretical explanation of the relativity of electromagnetic induction.

4.11 Some Notes on Unipolar Induction

Einstein [78, p. 55] explained away the phenomenon of *unipolar induction* in the following simple terms in the context of SRT:

> "Moreover, questions as to the "seat" of electrodynamic electromotive forces (unipolar machines) now have no point."

Indeed, in the context of SRT, we have no point, no place to discuss any phenomenon dealing with rotational motions as is the case of unipolar rotating induction. However, the "seat" of the unipolar induction took more than 80 years to be determined. In 1987, F.J. Müller [79, pp. 156-169] designed a very clever experiment to localize the source of the induced electromotive force in the circuit of a monopolar induction device. In 1991, J.P. Wesley [80, p. 239] explained theoretically Müller's results using Weber's electrodynamics. Recently, P. Mann [81] published his doctoral thesis *Determination and Verification of the Electrodynamic Postulates*. Mann used *Spencer-Gauss* [82] Electrodynamics to analyze the phenomenon of monopolar induction, and he compared these results with other results using the classical Lorentz's force, and also using Weber's electrodynamics. The thesis is very interesting because it contains many publications of Dr. Domina Spencer and her colleagues on the subject of the original and modified versions of Gauss' Electrodynamics. Prof. Domina Spencer has been, for more than 40 years, the Connecticut champion defending the Newtonian Gauss Electrodynamics. She reminds us of the perseverance of Mme. Curie. Thus, we have reason enough to say that Spencer-Gauss electrodynamics contains one force-term proportional to [½ **U** (**r**•**U**)], where the relative velocity of the test particle, with respect to the source particle, is **U** = **v** - **v'**. The force- term **U** (**r**•**U**) is also contained in Ritz's Electrodynamics, and also in the Newtonian Relativistic Electrodynamics, but not in Weber's Electrodynamics. The factor ½ is vital in explaining Eddington's model of the neutron. It would be very interesting to compare Mann's results with Wesley's theoretical work using Weber's Electrodynamics. Another comparison to be done is Mann's theoretical results with Müller's experimental results.

Conclusions

In the Introduction of this chapter, we learned that Einstein made the accusation that Newton's theoretical method was *deficient* because it was incapable of deducing the mathematical structure of the forces by logical and formal considerations. This accusation is false in respect to the *visible* motion of celestial bodies. However, we took Einstein's accusation as a serious *challenge* to Newton's classical conception of the universe. We decided to deduce, theoretically, a Newtonian relativistic electrodynamics pertaining to the interaction of *invisible moving electrical elementary particles.* On the other hand, we decided to obtain, theoretically, a Newtonian relativistic gravitodynamics about the interaction of moving visible celestial bodies without observing them. Not without self-esteem, we have to admit we successfully fulfilled both goals.

A most interesting theoretical result was the identity of the Parametrized Newtonian Relativistic Electrodynamics with the one we obtained in chapter 3. This was based on an eclectic inspection of a compendium of electrokinetics and electrodynamics. The foundation of the new electrodynamics is in addition to other axioms of Newton's dynamical methodology. Particularly important is the axiom on the mathematical form of the electrodynamic force-terms. An extensive analysis is made in section 4.3 about all the axioms of the new Newtonian dynamical methodology.

In section 4.5, we presented the results of the electrodynamic action on static charges by the equivalent Ampère electric current associated to permanent magnets. We used a modification of Millikan's Apparatus. We decided to leave, in section 4.5.3, the conclusions of this experiment. In section 4.5.4, we have the critical comments of J.P. Wesley of 1991 concerning this modified Millikan's Apparatus experiment.

The Parametrized Newtonian Relativistic Electrodynamics contains Maxwell's generalization of Ampère's force law. It is almost imperative to submit Maxwell's electrokinetics so as to test experimentally the four force-terms it contains. Doing this experimental work, dissidents may discontinue the unfounded discussions of deciding between Grassmann's force and Ampère's force. In section 4.6, we corrected the historical misunderstanding between the official and dissident physicists concerning Lorentz's force. J.C. Maxwell was the first to deduce the so-called "Lorentz force" 19 years before Lorentz. In section 4.7, we criticized three hybrid electrodynamics: Marinov's, Wesley's and Phipps' hybrid electrodynamics in which they have the relativistic variation of mass with velocity.

In section 4.8, we proved that Hertz's hypothesis is not necessary when deducing the Maxwell-Lorentz's force from electromagnetic field equations if we use Marinov's field equation, curl(**v** . grad)**A**. We left the analysis of the field equation, curl(**H**) and Hertz's hypothesis to the reader.

In section 4.9, we critically analyzed the objections of Eddington's model of the neutron as a miniature hydrogen atom. After we successfully rejected the objections, we proceeded to study the hydrogen atom "a la Bohr" using the new Newtonian Relativistic Electrodynamics. To our surprise, we found two radii corresponding to Bohr's ground state and the other to Eddington's ground state. We decided to leave the conclusions of Eddington's model of the neutron at the end of section 4.9.3. The explanatory note in section 4.10 on the convective operator is extremely interesting. It offers a classical theoretical explanation of the relativity of electromagnetic induction.

In relation to Einstein, we defended Newton's dynamical methodology from an unfair, erroneous and unrelenting attack by Einstein. It is interesting to note that Einstein initiates his attack by resorting to the fourth ontological principle, the principle of cause and effect. Einstein accused Newton of violating the causation principle by not assigning a reaction force to the centrifugal force in a noninertial reference system. Every learned physicist knows that Newton's action-reaction principle is not valid in accelerated reference systems. Hence, ontologically and physically Newton's dynamics is not valid in accelerated reference systems. Even Einstein knew this when he imagined a dialogue with Newton about the principle of equivalence. We will expand on this subject in chapter 6. To be fair, ignorance is not an excuse so we find Einstein guilty of raising false testimony against Newton's dynamical methodology.

References

1. I. Newton, *Mathematical Principles of Natural Philosophy and His System of the World* (Univ. of California Press, Berkeley, CA, 1960)
1. A. Einstein, *Ideas and Opinions* (Crown Publishers, Inc., NY, 1982) [On the Method of Theoretical Physics]
2. Ref. 2
4. R.S. Westfall, *Never at Rest* (Cambridge University Press, Cambridge, New York, New Rochelle, Melbourne, Sydney, 1980)
5. Ref. 4
6. Ref. 2
7. A. Einstein, *Relativity, the Special and General Theory* (Mathuen, London, 1920)
8. J.C. Curé, Galilean Electrodynamics, **2**, no. 3, 43-47 (1991) [The Perihelic Rotation of Mercury by Newton's Original Method]
9. Ref. 2, p. 276
8. V. Bush, J. Math. Phys., **5**, 129 (1926) [The Force between Moving Charges]
9. J.C. Maxwell, *Matter and Motion* (Dover Publications, Inc.) [from 1877]
10. M. Born, *Einstein's Theory of Relativity* (Dover Publications, Inc., New York, 1965)
13. Ref. 2
14. A.K.T. Assis, *Weber's Electrodynamics* (Kluwer Academic Publishers, Dordrecht, Boston, London, 1994)
15. A. Einstein, *Sidelights on Relativity* (Dover Publications, Inc., NY, 1983)
16. J.C. Maxwell in *The Scientific Papers of James Clerk Maxwell* (W.D. Niven, Dover Publications Inc., NY, 1966)
17. Ref. 1

18a. In Collection: *The Principle of Relativity* (Dover Publications, Inc., New York, [1923])

18b. A. Einstein, *The Meaning of Relativity* (Princeton University Press, Princeton, NJ, 1922)

19. H. Thirring, Phys. ZS., **19**, 33 (1918)
20. J. Lense and H. Thirring, Phys. ZS., **19**, 156 (1918)

21. A. Einstein, Preuss. Akad. der Wiss., Sitzber. pt. 2, pp. 831-839 (1915)
22. K. Schwarzschild, Preuss. Akad. der Wiss., Sitzber., pp. 189-196 (1916)
23. G.B. Brown, *Retarded Action at a Distance* (Cortney Publications, 95/115 Windmill Road, Luton Bedfordshire, Great Britain, 1982)
24. J.C. Curé, Phys. Lett. B, **116**, 158 (1982)
25. D.W. Sciama, *The Physical Foundations of General Relativity* (Doubleday & Company, Garden City, New York, 1969)
26. D.W. Sciama, *The Unity of the Universe* (Faber, London, 1959)
27. J.C. Maxwell, *A Treatise on Electricity and Magnetism* (Dover Publications Inc., NY, 1954)
28. A. O'Rahilly, *Electromagnetic Theory* (Dover Publications, Inc., NY, 1965)
29. Ref. 14
30. J.B Tate, Master's Thesis (University of Houston, Library, 1968)
31. W.F. Edwards, Proceedings of the Utah Academy of Science, Arts, and Letters, **51**, part.2, pp. 1-7 (1974)
32. W.F. Edwards, C.S. Kenyon, and D.K. Lemon, Physical Review D, **14**, 922-938 (1976)
33. J.P. Wesley, *Advanced Fundamental Physics* (Benjamin Wesley - Publisher, Weiherdammstrasse 24, 7712 Blumberg, West Germany, 1991)
34. Ref. 24
35. Ref. 30, 31, 32
36. Ref. 33
37. R. Sansbury, Rev. Sci. Instrum., **56**, no.3, 415-417 (1985)
38. Wesley, [Empirically Correct Electrodynamics] Private communication (1995)
39. D.K. Lemon, W.F. Edwards, and C.S. Kenyon, Phys. Lett., A **162**, 105-114 (1992)
40. S. Marinov, *The Thorny Way of Truth* (Int. Publishers East-West, Morellenfeldgasse16, 8010 Graz, Austria, 1990), pp. 36-39; *Proceedings of the International Symposium on New Energy* (Ed. M. Albertson, Ph.D. and M. Shaw, M.S., Rocky Mountain Research Institute, Room 203, Weber Building, Colorado State University, Fort Collins, CO 80523, Dr. M.L. Albertson, 303/482-3731), pp. 299-320
41. W.G.V. Rosser, Am. J. Phys., **30**, 509-511 (1992) [Second-Order Electric Field to a Conduction Current]
42. Ref. 14
43. Ref. 14
44. G. Bonnet, Phys. Lett., **82** A, 465 (1981)

45. C.G. Adler, Am. J. Phys., **55** (8), 739-743 (1987)
46. L.B. Okun, Physics Today, 31-36, June 1989
47. Ref. 45, 46
48. P. Graneau and N. Graneau, *Newton versus Einstein* (Carlton Press, Inc., New York, 1993)
49. Ref. 40
50. E. Whittaker, *A History of the Theories of Aether and Electricity* (Thomas Nelson and Sons Ltd., London, NY, 1951)
51. Ref. 27
52. Ref. 50
53. S. Marinov, *Eppur Si Muove* (Int. Publishers East-West, Morellenfeldgasse16, 8010 Graz, Austria, 1977 and 1981), p. 46
54. S. Marinov, Deutsche Physik, **4**, no. 13, 45-52 (1995)
55. Ref. 28
56. Ref. 28
57. T.E. Phipps, Jr. *Heretical Verities: Mathematical Themes in Physical Description.* Classic Non-Fiction, Urbana, IL, U.S.A.
58. P. Moon, D.E. Spencer, A.S. Mirchandaney, U.Y. Shama, and P.J. Mann (Private communication on «The Gaussian Electrodynamic Force», 1994)
59. Ref. 40. p. 37
60. Ref. 38
61. J.P. Wesley, Proceedings of the Conference on *Foundations of Mathematics & Physics,* Perugia, Italy, 1989 (Ed. U. Bartocci and J.P. Wesley, Benjamin Wesley - Publisher, 7712, Blumberg, Germany, 1990), pp. 289-343
62. Ref. 33
63. Ref. 38
64. J. P. Wesley, *Causal Quantum Theory* (Benjamin Wesley - Publisher, 7712, Blumberg, Germany, 1983)
65. Ref. 57
66. Ref. 57
67. M. Mason and W. Weaver, *The Electromagnetic Field* (Dover Publication, Inc., New York, 1929)
68. Ref. 28
69. Ref. 57

70. Eddington, Arthur, *New Pathways in Science* (Ann Arbor Paperback, The University of Michigan Press, (p. 31, 1959) See also Slater, Noel B. *The Development & Meaning of Eddington's Fundamental Theory* (Cambridge at the University Press)
71. E.F. Mallove, *Fire from Ice* (John Wiley & Sons, Inc.,NY,NY, 1991)
72. Y. Nee'man and Y. Kirsh, *The Particle Hunters* (Cambridge University Press, Cambridge, New York, 1987)
73. P.C.W. Davies, *The Forces of Nature* (Cambridge University Press, Cambridge, New York, 1988)
74. C. Møller, *The Theory of Relativity* (Oxford at Clarendon Press. Glasgow, New York, Toronto, Melbourne, 1960 &1974)
75. E.E. Anderson, *Modern Physics and Quantum Mechanics* (W.B. Saunders, Philadelphia, London, Toronto, 1971)
76. L. Pauling, and E.B. Wilson, *Introduction to Quantum Mechanics.* MacGraw-Hill Book Company, Inc., New York, London.
77. R.L. Mills, Fusion Technology, **25**, no. 1, 103-119 (Jan., 1994)
78 A. Einstein, in *The principle of Relativity: A collection of Original Memoirs* (Dover Publications, Inc., first published in 1923) [On the Electrodynamics of Moving Bodies.]
79 F.J. Müller, in *Progress in Space-Time Physics 1987* (Edited by J.P. Wesley, and Published by Benjamin Wesley - Publisher, Weiherdammstrasse 24, 7712 Blumberg, West Germany, 1987)
80 Ref. 33
81 P. Mann, *Determination and Verification of the Electrodynamic Postulates* (Doctoral Thesis, Approved by Professor Domina E. Spencer, Department of Mathematics, University of Connecticut, 1999)

CHAPTER 5

ON THE IDENTITY OF THE COSMIC ETHER

TABLE OF CONTENTS

"According to the general theory of relativity, space without ether is unthinkable."

A. Einstein

Introduction.

In this chapter, we will search for an ontological interpretation of the luminiferous or cosmic ether as formally understood by Einstein in 1920. We will then go back to the same ontological problem, but this time analyze it from a classical point of view. These two interpretations lead to the conclusion that the nature of the ether is pure energy. Michelson-Morley's experiment is analyzed with this new conception of the ether, concluding that scientists and philosophers of the last century overlooked the energetic ether associated with planet earth. It is also shown that the postulate of relativity is incompatible with an ontological identification of the ether, and that the speed of light is faster in interstellar space compared with the terrestrial determination of this speed. At the end of this chapter, we will present the *eikonal* equation to analyze the light deflection in the vicinity of the sun, showing that this deflection can be explained in 3-Dimensional space as a simple phenomenon of light refraction in the energetic medium around the sun.

5.1. Einstein's Resuscitated Ether.

In 1988 L. Kostro [1], in an extraordinary exhaustive bibliographic research concerning the intellectual evolution of the concept of ether in Einstein's mind, wrote that *"Einstein's conception of the ether is today almost unknown by physicists and philosophers and sometimes also by historians of Relativity Theory."* Kostro's work seems, also, to unveil the mystery surrounding the active or passive role of the null result of Michelson-Morley's experiment in the genesis of the Special Theory of Relativity.

Kostro distinguishes three stages in Einstein's conception of the ether. The first one lasted until 1899, the second until 1916 and the third until his death. Before 1899, Einstein was a firm believer "in the existence of a stationary elastic ether which transports light at different velocities." In 1894 or 1895, he wrote his first concepts about the ether titled *"On the Investigation of the State of the Ether in the Magnetic Field."* We should notice that the young Albert was, at most, 16 years of age. At the end of this chapter, we will propose an experiment to investigate the state of the ether in a magnetic field. He also conceived an experiment similar to Michelson-Morley's design. Kostro [2] has found a very significant document in which Einstein declares:

"While I was thinking of this problem in my student years, I came to know the strange result of Michelson's experiment. Soon I came to the conclusion that our idea about the motion of the earth with respect to the ether is incorrect, if we admit Michelson's result as a fact. This was the first path which led me to the special theory of relativity."

This statement of Einstein seems to put an end to the long debate about the role played by Michelson-Morley's experiment in the genesis of the Restricted Theory of Relativity. Einstein's belief in a stationary ether comes to an end in 1899 when, in a letter to his future wife, Mileva Maric, he expressed his doubts about the existence of the ether. In [1] we read:

"The introduction of the word 'ether' in the theories of electricity conducts to the idea of a medium about the motion of which we speak without the possibility, as I think, to attribute any physical sense to such speech."

Einstein could not have been more correct in his criticism. At that time and until today, no one knew anything about the very nature of that strange metaphysical entity called ether. In the second stage, we see Einstein denying the existence of the ether. As it is well known, Einstein, in 1905, called the ether "superfluous." He forced himself to reject the idea of an ether. This rejection was not based on total ignorance of physicists concerning the nature of the ether, this ether which was supposed to pervade the entire universe. Einstein's rejection was purely on kinematical grounds. H.A. Lorentz had identified the ether with absolute space, and absolute space, according to Newton, was immovable. This *immobility* of the ether was in total contradiction with a consequence of Einstein's principle of relativity, to which there is no privileged reference system in the universe.

The third and last stage in Einstein's conception of the ether began in 1916, and it lasted for the rest of his life. As Kostro [1] has discovered: *"On 6th June 1916, Lorentz wrote a long letter to Einstein in which he tried to convince Einstein of this."* Einstein answered Lorentz's letter on June 17, 1916:

"I agree with you that the general relativity theory is nearer to an ether hypothesis than is the special relativity theory. However, this

> new ether theory would not violate the principle of relativity, because the state $g_{\mu\nu}$ = Aether would not be that of a rigid body in an independent state of motion, but its state of motion would be a function of position determined via the material processes."

Einstein has brought us very close to the identity of the ether when he wrote $g_{\mu\nu}$ = Aether. In 1920, upon the request of the journal *Nature* [1], Einstein wrote a long essay out of which only a three- page summary was published in 1921. The original paper is in the Morgan Library in New York. In 1988, Kostro [1] quoted part of this manuscript concerning the new and last conception of the ether in Einstein's mind:

> "Therefore, in 1905, I was of the opinion that it is no longer allowed to speak about the ether in physics. This opinion, however, was too radical, as we will see later when we consider the general relativity theory. . . But it is not allowed to attribute to this medium . . . a state of motion in any point. Further, it is, by no means, homogeneous and its state has not autonomous existence but it depends on the matter which gets up the field. Since, in the new theory, the metrical facts cannot be separated from the 'properly' physical ones, therefore the notion of 'space' and the notion of 'ether' fuse together."

This last quotation deserves two brief comments. First, when Einstein says, "But it is not allowed to attribute to this medium . . . a state of motion in any point," we can interpret this lack of motion of the ether as being at *absolute rest.* But certainly this was not Einstein's intention when he was talking about this subject. Einstein, to be consistent with his principle of relativity must face an incredible task to convince himself that nothing should be predicated about the kinematical attributes of the ether. Second, Einstein reaffirmed Lorentz's and Newton's identification of the ether with space. Nevertheless, we must understand that there is an abyss between Einstein and Lorentz-Newton's conception of space.

Finally, we come to an extraordinary lecture Einstein delivered on May 5, 1920, at the University of Leyden [4]. The bottom line of this lecture is Einstein's declaration that, *"According to the general theory of relativity, space without ether is unthinkable."* It is in the second part of the Leyden lecture which Einstein introduced his new conception of the ether as contained in his General Relativity Theory. This author elaborated, in 1987, more extensively on this lecture [5]. Here we wish to emphasize the point in which Einstein truly grazes the essence of the ether. He wrote:

> "There can be no space nor any part of space without gravitational potentials; for these confer upon space its metrical qualities, without which it cannot be imagined at all. The existence of the gravitational field is inseparably bound up with the existence of space. On the other hand a part of space may very well be imagined without an electromagnetic field; thus in contrast with the gravitational field, the electromagnetic field seems to be only secondarily linked to the ether, the formal nature of the electromagnetic field being as yet in no way determined by that of gravitational ether. From the present state of theory it looks as if the electromagnetic field, as opposed to the gravitational field, rests upon an entirely new formal motif, as though nature might just as well have endowed the gravitational ether with fields of quite another type, for example, with fields of a scalar potential, instead of fields of the electromagnetic type."

In the previously quoted "Morgan Manuscript," we can read Einstein declaring: *"And so one can say that the ether has resuscitated anew in the general relativity theory, though under a more sublimated form."* The "sublimated form" and subtle essence of the new Einsteinian ether is hidden in the following statement:

> "There can be no space nor any part of space without gravitational potentials."

It is here where Einstein gave the clue to the essence of the cosmic ether. But unfortunately, an ontological identification of the relativistic ether is going to be almost impossible in the context of his theory because of the principle of relativity. For this reason, Einstein is left only with a formal identification.

5.2. Einstein's Ether is a Metaphysical Entity.

For the very first time, Einstein revived the ether in his answer to Lorentz in 1916. He realized he faced an almost insurmountable problem. He wrote to Lorentz that *"However, this new ether theory would not violate the principle of relativity, because the state* $g_{\mu\nu}$ *= Aether would not be that of a rigid body in an independent* *state of motion, but its state of motion would be a function of position determined via the material processes."* Here Einstein is speaking of a *state of motion* of his new ether, which is a *function of position* determined by material objects. Later on, Einstein will insist that we abstain from speaking of the kinematics of his relativistic ether. Nevertheless, he is giving another clue to reinterpret his new ether in the way Lorentz would have liked it. In another letter that Einstein wrote to Lorentz in 1919 [1], we read:

> "It would have been more right if I had limited myself, in my previously published papers, to lay emphasis only on the nonexistence of any velocity of the ether instead of the defense of the total nonexistence of the ether."

The "nonexistence of any velocity of the ether" is in flagrant contradiction with "its state of motion which would be a function of position determined via the material process." This contradiction is rooted in the same *cause* Einstein had in 1905. This cause is the *immobility of absolute space or ether.* The *immobility* of Lorentz-Newton's ether has to be avoided at any cost to save the principle of relativity. It is very clear in what follows that this relativistic ether had to be expelled, once again, from experimental physics, relegating it to the foggy domain which lies even beyond metaphysics. Kostro [1] quotes Einstein in 1920, saying:

> ". . .a more exact reflection shows that this denial of the existence of the ether is not demanded by the restricted principle of relativity. We can assume the existence of an ether; but we must abstain from ascribing a definitive state of motion to it, i.e., we must divest it by abstraction of the last mechanical characteristic which Lorentz left it . . . As regards the mechanical nature of Lorentz's ether, one might say of it with a touch of humor, that immobility was the only mechanical property which Lorentz left it."

In the last three quotations, we see Einstein forbidding any statement about the kinematics of the new relativistic ether. In 1988 and 1989, Kostro elaborated on Einstein's identification of the new ether with physical spacetime, saying that this relativistic ether *"constitutes an ultra-referential fundamental reality."* In plain language, Kostro [3] reduced the new ether to an ultra-metaphysical entity in which nothing kinematical can be predicated about it. But it is better to read Kostro [1] textually:

> "Einstein, when he identifies ether with the physical space, he makes a very clear distinction between the physical space as such which is one and unique and the reference spaces the number of which is infinite.
>
> "According to him, the ether cannot be identified with anyone of the reference spaces because it would mean that one of them is favoured or privileged with respect to the others. This contradicts the principle of relativity. According to Einstein, only the physical space as such constitutes the relativistic ether because it is, by no means, a reference space. Such a net distinction between the physical space as such and the reference spaces do not exist in the pre-Einsteinian physics. The absolute space of the pre-Einsteinian physics, connected or identified with the ether in the old sense, constituted also itself a reference space. It was one of the reference spaces. It was the privileged reference space at absolute rest.
>
> "In the Relativity Theory the notion of motion can be applied only with respect to the reference spaces because only they move or are at rest with respect to each other."

According to Einstein, we are presented with an interpretation of an entity called *physical space* which cannot move nor cannot be at rest. What kind of an entity is this "physical space" in which we are forbidden to even think about its state of motion? This relativistic ether is as absurd as the classical ether which is *not* dragged at all by the moving planet Earth (Bradley's stellar aberration), but it is dragged completely by the moving planet Earth of the Michelson-Morley's experiment. This new concept forces Relativity Theory to transcend into a metaphysical agnosticism. The divorce of the *physical space, the total field, relativistic ether, the components of the metric tensor* and reference spaces are an ingenious discursive exposition to avoid a devastating contradiction between the new ether and the principle of relativity. This contention is laudable from a logical point of view so as to conform with the rational requirements of *coherence* in any theory, but in so doing, we lose any possibility to categorize the Einsteinian ether. Once again, the principle of relativity has killed the ether.

5.3. A Classical Identification of Einstein's Ether.

By now we must conclude that any concept of an immovable ether is incompatible with the principle of relativity. Therefore, if we again want to resuscitate the concept of ether, we must abandon, forever, the principle of relativity. This does not mean we must discard everything from Relativity Theory, particularly Einstein's field equations from GRT, but the conceptual background must suffer profound changes in many respects. Today, we have enough experimental evidence on the invalidity of the constancy of the speed of light and the existence of a privileged reference system at absolute rest [6, 7, 8, 9]. On the other hand, Einstein's relativistic interpretation of Maxwell-Lorentz's electrodynamics has recently been seriously questioned on an experimental basis [10,11,12]. All this evidence is obviously pointing to a serious revision of 20th century physics.

The root of all these problems is not in the Relativity Theory, but in the incapacity of Classical Physics, at the end of the 19th century, to interpret correctly the outcome of experiments concerning the propagation of electromagnetic waves. One of these problems was the total ignorance of the essence of the cosmic ether. In 1892, H.A. Lorentz in an almost desperate letter to Lord Rayleigh, [13], wrote:

> "I am totally at loss how to solve the contradiction and yet I believe that if Fresnel's wave theory is abandoned, we should have no aberration theory at all . . . Can there be some point in the theory of Mr. Michelson's experiment which has as yet been overseen?"

What was overlooked was the *nature* of the luminiferous ether. Even today, and in spite of the new relativistic ether, we are still at a loss concerning the *essence* of the cosmic ether. Let us go back to Einstein's Leyden lecture when he said,

> "There can be no space nor any part of space without gravitational potentials."

This time let us try to find the very nature of the cosmic ether. Instead of considering gravitational potentials, we might as well consider gravitational potential energy, or better yet, density of gravitational potential energy. If we only derive from the metric tensor the component g_{00} in the weak approximation, then in Newtonian language, we are only considering a gravitostatic field generated by a material body. Under these conditions, we can rephrase Einstein's statement to read:

> There can be no space nor any part of space without a certain density of gravitostatic potential energy.

In fact, around any material body in the universe, we can assign a density of gravitostatic potential energy ρ_G, given by:

$$\rho_G = GM^2/(8\pi r^4) \tag{5.1}$$

We see that according to eq. (5.1), this density of energy decreases very fast in relationship to the distance from the material body. *Thus, universal matter creates a cosmic ocean of energy in which it immerses itself,* or in other words, *the universal ocean of energy is the primordial source of matter.* Here we face a particular identification of the cosmic ether. *The essence of the cosmic ether is gravitostatic potential energy.* Einstein, in his Leyden's lecture, inquired about the future of his resuscitated ether in the following terms:

> "As to the part which the new ether is to play in the physics of the future we are not yet clear. We know that it determines the metrical relations in the space-time continuum, e.g., the configurative possibilities of solid bodies as well as the gravitational fields; but we do not know whether it has an essential share in the structure of the electrical elementary particles constituting matter. Nor do we know whether it is only in the proximity of ponderable masses that its structure differs essentially from that of the Lorentzian ether; whether the geometry of spaces of cosmic extent is approximately Euclidean."

Now that we have identified the nature of the cosmic ether, we are in a position to answer one of Einstein's queries. According to eq. (5.1) and eq. (5.2), it is only in the neighborhood of material celestial bodies that this energetic cosmic ether "differs essentially from that of the Lorentz ether." In intergalactic space, a uniform density of cosmic gravitostatic potential energy must exist. As we enter a galaxy, this density of energy increases due to the proximity of the corresponding galactic matter. Here, inside a galaxy, we should have an interstellar density of energy ρ^*. Finally, if we approach a planet, this density of energy will increase to a value ρ, given by

$$\rho = \rho^* + GM^2/(8\pi r^4) \tag{5.2}$$

Any celestial body can be pictured as being surrounded by an atmosphere of gravitostatic potential energy. In particular, planet earth should superimpose its gravitational atmosphere of energy onto the interstellar energy, and consequently, planet earth will *carry* its own atmosphere of energy while it moves in space. In another section, we will translate this intuitive picture into the language of mathematical physics to explain Michelson-Morley's almost null experimental result. We will do this in the context of the energetic conception of the cosmic ether. The above intuitive picture is our answer to Einstein's concern about Lorentz's ether. Now, if an electromagnetic experiment is conducted in a terrestrial laboratory, then eq. (5.2) should be generalized to the following equation:

$$\rho = \rho^* + GM^2/(8\pi r^4) + \tfrac{1}{2}\mathbf{E}\cdot\mathbf{D} + \tfrac{1}{2}\mathbf{B}\cdot\mathbf{H} \tag{5.3}$$

In eq. (5.3) we see three forms of energy: gravitational, electric and magnetic. But should not energy be one and the same entity? J.C. Maxwell [14] in one of his scientific papers wrote:

> "In speaking of the energy of the field . . . I wish to be understood literally. All energy is the same as mechanical energy, whether it exists in the form of motion or that of elasticity, or in any other form . . . . On our theory [the energy] it resides in the electromagnetic field, in the space surrounding the electrified and magnetic bodies, as well as those bodies themselves, and in two different forms, which may be described without an hypothesis as magnetic polarization and electric polarization, or, according to a very probable hypothesis, as the motion and the strain of one and the same medium."

For Maxwell, energy is one and the same entity no matter what its source is. Once energy is released from any source: chemical, physical or whatever, *its being is one and the same.* This Maxwellian identification of all forms of energy constitutes the ontological generalization of Einstein's cosmic ether. Thus, the *"onto"* of the ether is *energy.* Maxwell, again, in his *Treatise* [15; Vol.II, Ch. XX] wrote:

> "According to the theory of undulation, there is a material medium which fills the space between the two bodies, and it is by the action

of contiguous parts of this medium that the energy is passed on, from one portion to the next, till it reaches to the illuminated body.

"The luminiferous medium is therefore, during the passage of light through it, a receptacle of energy. In the undulatory theory, as developed by Huygens, Fresnel, Young, Green, &c., this energy is supposed to be partly potential and partly kinetic. The potential energy is supposed to be due to the distortion of the elementary portions. We must therefore regard the medium as elastic. The kinetic energy is supposed to be due to the vibratory motion of the medium. We must therefore regard the medium as having a finite density."

Now, the only concept we must replace in the previous quotation is "material medium." Instead we will replace it with "immaterial energetic medium," according to our energetic conception of the ether. Through Einstein and Maxwell, we have decided on Newton's speculations concerning the nature of the agent which causes gravity. In 1692, Newton wrote to Bishop Bentley that *"Gravity must be caused by an agent acting constantly according to certain laws; but whether this agent be material or immaterial, I have left to the consideration of my readers."* Einstein's considerations were that the gravitational agent is the relativistic immaterial metric tensor caused by the matter-energy tensor. But this was a formal answer to Newton's query. Our answer is an essential one. Thus, the nature of Descartes's "plenum," Newton's "sensorium Dei" and Bohm's "unbroken wholeness of the entire universe," *is plain energy.*

According to Maxwell, *"The luminiferous medium is therefore, during the passage of light through it, a receptacle of energy."* According to our interpretation, the luminiferous medium is not a material medium that is a receptacle of the propagating energy, but pure energy in itself. An electromagnetic wave, then, must be brought into existence by a modification of the local density of energy which propagates subsequently. In this respect, an electromagnetic wave should be the propagation of a density wave of energy. In the past, an argument against the existence of the ether medium was the absence of a density wave in electromagnetic theory. Nevertheless, Maxwell's theory contains all the elements to deduce the propagation of an energy density wave. In what follows, we will see the existence of D'Alambert's equation for the density of electromagnetic energy. To do this, let us use eq. (5.3) to make this metaphysical and seemingly impractical speculation more respectable. We will translate it into the language of mathematics. Let us prove that an electromagnetic wave is, indeed, the propagation of energy density in the cosmic energetic medium. In a terrestrial laboratory, the first two terms, on the right-hand side of eq. (5.3), can be considered constants. Taking the partial derivative with respect to time of eq. (5.3), and using Maxwell's equation in "vacuum," we immediately obtain the continuity equation in terms of Poynting's vector **ExH**:

$$\partial\rho/\partial t = -\nabla\bullet(\mathbf{E}\mathrm{x}\mathbf{H}) \qquad (5.4)$$

Taking again the partial derivative of eq. (5.4), with respect to time, we get:

$$\partial^2\rho/\partial t^2 = c^2\, \nabla\bullet\{\mathbf{D}\mathrm{x}(\nabla\mathrm{x}\mathbf{E}) + \mathbf{B}\mathrm{x}(\nabla\mathrm{x}\mathbf{H})\} \qquad (5.5)$$

Now taking the gradient of eq. (5.3), we obtain:

$$\nabla\rho = \mathbf{D}\mathrm{x}(\nabla\mathrm{x}\mathbf{E}) + \mathbf{B}\mathrm{x}(\nabla\mathrm{x}\mathbf{H}) + \mathbf{T} \qquad (5.6)$$

with $$\mathbf{T} = (\mathbf{D}\bullet\nabla)\mathbf{E} + (\mathbf{B}\bullet\nabla)\mathbf{H} \qquad (5.7)$$

Now taking the divergence of eq. (5.6), we get:

$$\nabla^2\rho = \nabla\bullet\{\mathbf{D}\mathrm{x}(\nabla\mathrm{x}\mathbf{E}) + \mathbf{B}\mathrm{x}(\nabla\mathrm{x}\mathbf{H})\} + \nabla\bullet\mathbf{T} \qquad (5.8)$$

Finally, combining eq.s (5.5) and (5.8), we get an inhomogeneous D'Alambert equation for the energy density associated to an electromagnetic wave:

$$\nabla^2\rho - c^{-2}\partial^2\rho/\partial t^2 = \nabla\bullet\mathbf{T} \qquad (5.9)$$

where **T** can be related to Maxwell's stress tensor. If the electromagnetic wave, generated at the surface of the earth, propagates into space, then the second term, in eq. (5.3) can no longer be considered constant and the mathematics of this other problem becomes more complicated.

If Maxwell's generalization of the concept of energy is correct, and if an electromagnetic wave is indeed an energy density wave as shown by eq. (5.9), then before us, we have the possibility of developing an *ergodynamics of the ether* instead of the old elastic theory of the ether. This was the hope Maxwell had in his *Treatise* [15] when he wrote:

> "The mathematical expressions for electrodynamic action led, in the mind of Gauss, to the conviction that a theory of the propagation of electric action in time would be found to be the very keystone of electrodynamics. Now we are unable to conceive of propagation in time, except either as the flight of a material substance through space or as the propagation of a condition of motion or stress in a medium already existing in space . . . In fact, whenever energy is transmitted from one body to another in time, there must be a medium or substance in which the energy exists after it leaves one body and before it reaches the other . . . If we admit this medium as a hypothesis, I think it ought to occupy a prominent place in our investigations and that we ought to endeavour to construct a mental representation of all the details of its action; and this has been my constant aim in this treatise."

The conception of the ether as pure energy has far reaching possibilities which can be only grasped intuitively at this stage. Again, if Einstein is correct in conceiving matter as condensed energy, then the spinning elementary particles would be nothing but vortices of energy. In the present, as far as this author is acquainted, these ideas are being developed by T.G. Barnes and P. Tewari [16]. Perhaps in the 21st century, we will rid ourselves of the concept of positive and negative charges, and we will develop an *Ergodynamics of moving vortices* as Descartes conceived long ago. Recently, Bergman and Wesley, Rado, and Ginzberg, have advanced the concept that an electron is a rotating *toroid*. We will come back to this important subject in chapter 7.

The offensive principle of Complementarity will be dissolved in this cosmic ocean of energy if one day, we will decide to run experiments in order to show that a moving elementary corpuscle does not suffer an ontological metamorphosis. That is to say, it will not stop being what it *is* to become what it *is not.* On that day, perhaps, we will realize that what is really diffracted is the *action potential wave* generated by the moving electrons immersed in the energetic ether. The electrons themselves are dispersed by the diffracted *action potential.* The secret of this possibility lies in the Hamilton-Jacobi equation extended by Madelung and Bohm. From this extended equation, we see that Schröedinger's equation is just a particular case, and that the Uncertainty Principle only has a restricted validity. The Quantum Hidden Potential is still hidden in our ontological ignorance of new electrodynamic forces which are contained in disguise in Einstein's field equations of GRT. All these possibilities are there in the cosmic ocean of energy, but to see them, we have to ontologically reinterpret Einstein's new conception of the ether, identified with the components of his metric tensor, and be prepared to modify our conceptions about nature. We saw these changes in chapter 2. Also will see these changes in chapters 6 of this book.

Being that energy is the one and same entity in the entire universe, there is no essential difference between gravitational energy, electromagnetic energy, weak nuclear energy and strong nuclear energy. If energy is the line integral of force, then we will discover that only one category of force in the universe exists: electrodynamic, which manifests itself in different intensities according to the metric scale of the dynamic phenomena we observe. In 1938, O'Rahilly [17], writing about Ritz's Electrodynamics, said:

> "We are tempted to add some cognate remarks (18a); but we omit any detailed discussion or proof, as the subject is beyond our scope. It is easy to see the possibility, on Ritz's theory, of explaining gravitational attraction as residual statistical forces between groups of moving charges; no such possibility is available from the Lorentz-Lienard theory. These forces must be due to terms of a high order and the forces will be small relatively to the first order forces familiar in electromagnetics."

According to Relativity Theory, do we not have a strong clue to the unification of gravitodynamic fields and electrodynamic fields in the common speed of propagation which is equal to c? What about longitudinal waves propagating in the cosmic ocean of energy? Could it be possible that only transversal electromagnetic waves propagate in the cosmic energetic ether at the speed of light? If this is so, then we terrestrials will be limited to sending only *electromagnetic smoke signals* into space, and we will forever remain isolated in this vast universe from potential communications with probable existing advanced galactic civilizations. This is a terrifying thought of cerebral inferiority. However, *In Electronics and Wireless World,* February 1990, in an editorial note entitled *Maxwell was half right,* we read about electrical signals traveling at superluminal velocities in spite of the relativistic barrier:

> "It comes without surprise to find that the experiments of Alexis Guy [Obolensky] and Harold went unreported even though they were conducted over 11 years ago. Recently the Obolensky-Pappas experiments demonstrated electrical signals propagating in the range 2c and over 100c under special conditions.
>
> "No theory is in hand to explain these velocities . . .The weakness of the observed superluminal signals and the special conditions for their propagation support the above ideas. Therefore, research is needed to investigate techniques for effective transmission of unitary signals, to investigate optimum media for their propagation.
>
> "If unitary waves exist, they are certainly used by advanced civilizations. Most likely, the development of unitary waves by a civilization is the minimum qualification to join the next to the human group of civilization."

A.G. Obolensky and P.T. Pappas [18], reported in December 1988, the experimental evidence of electrical signals traveling faster than the speed of light. The editorial note above informs us that "No theory is in hand to explain these velocities." This is not altogether true. Landau and Lifshitz in their book, *The Classical Theory of Fields,* when performing a Fourier resolution of the electrostatic field as a superposition of plane waves, ended up with *longitudinal waves.* But the theoretical existence of these waves was known since the 19th century, even before the advent of Maxwell's Electromagnetic Theory.

Whittaker [19], in Chapter V, entitled *The Aether as an Elastic Solid,* described the works of Stokes, Navier, Cauchy, Green, MacCullagh, Neumann and Boussinesq in relation to the reflection of optical waves and their propagation in anisotropic crystals. Of particular interest is the work of Green read to the Cambridge Philosophical Society in December 1837, and published the following year in the Transactions of the Society.

Now, we must ask a pragmatic question concerning the identification of the ether which we have presented in this chapter. How can we use this energetic ether to develop an electromagnetic field theory which, in addition to deducing the presence of transversal electromagnetic waves, will show the existence of longitudinal electrical waves as well? In other words, is it possible to develop an ergodynamic theory? To attempt this task, it would be equivalent to reinventing the wheel. This work was done in the 19th century, and the ironic episode was that everyone who embarked on this analytical enterprise tried their best to eliminate from the equations the presence of longitudinal waves propagating in the old elastic ether. The starting point of Green was to write the *density of potential energy* due to a state of stress of an isotropic medium which is strained. Then, he introduced this energy density, in the well-known variational equation of generalized dynamics, to obtain the equation of motion in the D'Alambertian form. Whittaker comments on Green's work in the following terms:

> "The result found by Green was that if the vibration of the ethereal molecules is executed at right angles to the plane of incidence, the intensity of the reflected light obeys Fresnel's sine law, but the inertia ρ to vary from one medium to another . . . and that the optical difference between media are due to the different densities of ether within them.
>
> "It now remained for Green to discuss that case in which the incident light is polarised at right angles to the plane of incidence, so that the motion of the ethereal particles is parallel to the intersection of the plane of incidence with the front of the wave. In this case it is impossible to satisfy all the six boundary conditions without assuming that longitudinal vibrations are generated by the act of reflection...
>
> "The work of Green proved a stimulus not only to McCullogh, but to Cauchy, who now (1839) published yet a third theory of reflection. This appears to have owed its origin to a remark of Green's; that the longitudinal wave might be avoided in either of two ways- namely, by supposing its velocity to be indefinitely great or indefinitely small. Green curtly dismissed the latter alternative and adopted the former, on the ground that the equilibrium of the medium would be unstable if its compressibility were negative...
>
> "The energy carried away by the longitudinal waves is infinitesimal, as might be expected, since no work is required in order to generate an irrotational displacement."

Let us not believe that these works of the 19th century can be used today in their original form. Much work is left to be done to adapt them to our present knowledge. In 1963, Dirac [20] wrote:

> "One of these ideas is to introduce something corresponding to the luminiferous ether, which was so popular among physicists of the 19th century. We said early that physics does not evolve backwards. When we talk about reintroducing the ether, we do not mean to go back to the picture of the ether one had in the 19th century, but we do mean to introduce a new picture of the ether that will conform to our present ideas of quantum theory."

These were the concepts B.G. Wallace [21] had in mind, in 1973, when he published *The Unified Quantum Electrodynamic Ether.* In 1987, L. Kostro [2] made a similar but highly formal approach on the same subject using Einstein's relativistic ether. The concept of ether, as energy presented here, opens the technological possibility of drawing energy from this inexhaustible cosmic ocean of energy, and channeling it into the circuits of special electrodynamic devices as has been reported lately [22]. In conclusion, our identification of the cosmic ether, as pure energy, brings us back to the Lorentzian stationary or immovable ether in interplanetary, interstellar and intergalactic spaces. The density of this energetic ether is not uniform or homogenous in the universe as Lorentz conceived it. Its density increases as we approach celestial bodies. This time we can predicate its kinematical status saying that the ether, associated to earth for example, is completely carried at the same absolute velocity of our planet. We can also say that light, according to Fresnel-Fizeau's dragging coefficient, propagates in a terrestrial laboratory, not with respect to the stationary cosmic energy nor with respect to the moving energy "atmosphere" of earth, but with respect to the ergocenter of these two realms of energy. Let us later translate these *metaphysical principles of natural philosophy* into the language of mathematics to make them formally more respectable.

5.4 On Michelson-Morley's Experiment.

It was the metaphysical incapacity of Classical Physics to identify the essence of the cosmic ether which forced the ingenious creation of Relativity Theory. It was also the incredible error of Classical Physics to have applied the *same* Galilean transformations of the *composition of velocities of material corpuscles,* to the composition of velocities of electromagnetic waves, when every physicist *knew* the law of velocity composition was established by Fresnel and experimentally verified by Fizeau. Classical physicists did not see the essential difference between the motion of a material object and the motion or propagation of an electromagnetic wave. The difference in the following two laws of composition of velocities is extremely evident. Galilean composition of velocities for material particles: $v = v' + ut$, in which v' is the velocity of a particle in reference system S', moving rectilinearly with constant velocity u with respect to an inertial system S. Fresnel and Fizeau's composition of velocities for electromagnetic waves is:

$$v = v' + (1 - 1/n^2)\, ut$$

where $v' = c/n$ is the velocity of light propagating in a medium of index of refraction n. The parenthesis, in the last equation, represents the Fresnel dragging coefficient. It is hard to believe that classical physicists of the 19th century expected, or wanted to have, *one* law of velocity composition for the propagation of two ontologically different entities such as particles and waves. This conception would have been considered absolutely insane. However, after 1905, we have learned to live with this insanity. Classical physicists did not know that the Fresnel dragging coefficient depended on the gravitostatic energy density of planet earth, as we will now prove. If a metaphysical conception of an ethereal cosmic medium does not modify any physical equation, then we have to agree with O'Rahilly's criticism in [17]:

> "The purpose is the important one of distinguishing between the discourse of physicists and the quantitative formulation of physics. Once we establish this distinction, we shall have acquired a technique for getting rid of such sterile discussions as those concerning the aether, the field, lines of force, dimensions, etc. We have a criterion for separating what pertains to genuine scientific physics from what pertains to the discourse - often a farrago of philosophy, paradox and imagination - in which physicists so often indulge when writing semi-philosophical or popular books and even when writing textbooks. The question to be asked concerning any hypothesis is this: Does it in any way influence the measure-numbers which are to be tested by the man in the laboratory. If it does not, it may be good or bad philosophy; but it is not physics. Accordingly we have no difficulty in deciding that the hypothesis of an "aether," whatever it is supposed

to mean, does not nowadays pertain to the science of physics; for its denial does not make any alteration in any formula."

Let us remedy this defect. An evident analogical conjecture, about the speed of light, is to say that it is inversely proportional to the square root of the energy density of the medium. In interplanetary space we must have:

$$c^* = K/\sqrt{\rho^*} \quad (5.10)$$

This equation follows the work of Green and McCullogh as mentioned in [19]. Inside a terrestrial laboratory we have:

$$c = K/\sqrt{(\rho^* + \rho_G)} \quad (5.11)$$

From the last two equations, we can immediately deduce Fresnel's hypothesis which says that the density of ether ρ, is proportional to the square of the index of refraction $n = c^*/c > 1$:

$$n^2 = 1 + \rho_G/\rho^* = (\rho^* + \rho_G)/\rho^* = \rho/\rho^* \quad (5.12)$$

An immediate consequence of these elementary equations is that the magnitude of the speed of light in interplanetary space must be slightly greater than on earth. Now if we call $\phi = 1 - 1/n^2$ the Fresnel dragging coefficient, the expected shifts of the fringes in the interference pattern of the Michelson-Morley's experiment is given by:

$$s = [2L\beta^2/\lambda]f(\phi,\beta) \quad (5.13)$$

$$f(\phi,\beta) = 2(1/A - 1/\sqrt{B})/\beta^2 \quad (5.14)$$

$$A = 1-\beta^2(\phi-1)^2 \quad (5.15)$$

$$B = \{\sqrt{[1-\phi^2\beta^2(1-\cos^2\psi)]}+\phi\beta\cos\psi\}^2 - \beta^2 \quad (5.16)$$

$$\cos\psi = \beta/\sqrt{(1+2\,\phi\,\beta^2-\phi^2\beta^2)} \quad (5.17)$$

The square bracket in eq. (5.13), represents the classical shift multiplied, this time, by the ether factor $f(\phi,\beta)$. Eq. (5.13) shows the modification of a physical law which was caused by our philosophical speculations concerning the essence of the ether.

The first interesting consequence of this factor f is its numeric value when the gravitostatic energy field of the earth is ignored. In this case, making the gravitostatic energy density of the earth equal to zero, i.e., $\rho_G = 0$, then n = 1 in eq. (5.12), and therefore, Fresnel's dragging coefficient ϕ becomes zero. With $\phi = 0$ the ether coefficient f becomes unity: $f(\phi,\beta) = 1$.

Now, the numeric value of the classical shift was expected to be equal to 0.4. The experimental data, published by Michelson and Morley in 1887, showed a small shift in the interference pattern. This small shift oscillates as the Earth rotates and translates in space. The result was not a total *null result* as it is believed today. Michelson and Morley [23] wrote in their historical paper: "*The actual displacement was certainly less than the twentieth part of this, and probably less than the fortieth part.*" Let us assume that the displacement was one-fortieth part of the expected 0.4 shift. With a cosmic energy density ρ^* equal, at least, to 1.076×10^{10} (J/m^3), we get exactly a shift equal to 0.4/40 = 0.01. In this way, we satisfactorily explain the shocking outcome of the Michelson-Morley experiment, and at the same time, we have some estimation of the magnitude of the *cosmic density of energy*. However, this estimation is not based on an almost null shift in the interference pattern of other repetitions of Michelson-Morley's experiments. In chapter 6, we will make a better estimation of the cosmic density of energy based on the starlight deflection by the energy field of the sun.

If Michelson-Morley's experiment is repeated inside a space craft, far from planet Earth, the predicted shift will be identical to the classical one. The reason is that Fresnel's dragging coefficient $f(\phi,\beta)$ will be equal to unity, because the gravitostatic density of energy, associated to the matter of the space craft, is absolutely negligible with respect to the cosmic density of energy. The conception of performing this space experiment is more than a century old. In fact, Whittaker [19], commenting on a paper written by Michelson in 1897, wrote:

> "Michelson concluded that if there were no choice but between the theories of Fresnel and Stokes, it would be necessary to adopt the latter, and to suppose that the earth's influence on the ether extends many thousand kilometers above its surface."

This "earth's influence on the ether," in our interpretation, is the earth's gravitostatic energy density, which at a distance of six terrestrial radii from the center of our planet (38,000 km), has decreased more than 1,000 times with respect to the value on the surface of the earth. At this altitude, Frenel's dragging coefficient $\phi = (1 - 1/n^2) = 0.009$. This last numeric value brings the ether coefficient almost equal to unity. These last considerations open the possibility of sending electromagnetic pulses from one stationary satellite to another in order to observe periodic diurnal variations of the time-of-flight of the pulses to indirectly test the classical shift. Michelson's intuition brought his mind very close to discovering the *energetic essence* of his "beloved ether." In 1973, Dorothy Michelson-Livingston [13], Michelson's daughter, wrote:

> "Had Michelson lived to see the astronauts fly to the moon, he would no doubt have urged that an interferometer be taken along to test his theory in outer space."

This proposed experiment of Michelson's daughter is long overdue. Are we afraid to discover the ontological truth about the luminiferous ether? Newton did not hypostatize absolute space when he identified it with the cosmic ether. The absolute space always has had the invisible essence of energy. In 1909, Sir Oliver Lodge, in his book *The Ether of Space,* wrote: "*The existence of a continuous space medium, for instance, is probably regarded by most educated people as a more or less fanciful hypothesis, a figment of the scientific imagination - a mode of collating and welding together a certain number of observed facts, but not in any physical sense a reality, as water or air are realities.*" Today, as truly natural philosophers, we must ignore all these "educated people" who know so many mathematical theorems but do not understand anything about reality, about ontology.

In 1983, T. Theocharis [24] published an excellent interpretation of Maxwell's ether, identifying it with an earth-generated electromagnetic field which is carried by our planet. Theocharis wrote:

> "As we have already said, the SRT was accepted on the basis of theoretical grounds alone, long before a genuine experimental test, as indicated here, could be possibly carried out. Consequently, hardly anybody could realize the crucial significance of the test in suggestion when the time came - the opening of the space age. Thus, a genuine test of the SRT is now more than 20 years overdue."

On our theoretical grounds, we can predict that the SRT is not correct in interplanetary or interstellar space. To verify this prediction, we need to repeat Michelson-Morley's experiment in interplanetary space. Which nation will carry out this test? In the meantime, we should study B.G. Wallace's analysis of radar signals in interplanetary space [25]. In 1984, Wallace wrote:

> "When the radar calculations were based on Einstein's second postulate, the 1961 observed-computed residuals ranged to over 3 msec of the expected error of 10 μsec from the best possible general relativity fit the Lincoln Lab could generate[4], a variation range of over 30,000%. My analysis of the published 1961 data showed a component that seemed to be relativistic in the Galilean sense, and we called for a complete objective analysis."

Wallace has also proposed a *dynamic ether* but devoid of an ontological identity. Hatch [26] has proposed a similar unontological ether, very descriptive but not too quantitative. In the end, did Einstein determine the essence of his metric ether? The following quotations contain the answer to this question. The quotations are taken from Kostro [4], and we will add short comments.

Q1 "Since according to a consequent field theory also the ponderable matter i.e. the elementary particles, which do constitute the latter, have to be regarded as "fields" of particular kind or as particular "states of space" thereafter one can change the opinions of today's physicists in such a way that in the notion of the ether all objects of physics can be embraced." [27]

Here, we do not find a definite identity of Einstein's ether. However, we observe that the ether, whatever its essence, is the Being of all entities. Matter or elementary particles are conceived by Einstein as "states of space" or "fields" of a particular kind. There is no identification of the very essence of the Einsteinian ether. Here is the embryonic singularity of the field that will eventually give birth to the mass of particles.

Q2 "According to the views here presented, the axiomatic foundation of physics appears as follows: The real is conceived as a four-dimensional continuum with a unitary structure of a definite kind (metric and direction). The laws are differential equations which the structure mentioned satisfies, namely, the fields which appear as gravitation and electromagnetism. The material particles are positions of high density without singularity . . ." [28]

Let us ignore the so-called "real" formality of the mental construct of 4-Dimensional continuum. Once the singularity in the ether is actualized, it then becomes a material particle without singularity. This conception is well expressed by Einstein.

Q3 "The strange conclusion to which we have come is this - that now it appears that space will have to be regarded as a primary thing and that matter is derived from it, so to speak, as a secondary result. Space is now turning around and eating up matter. We have always regarded matter as a primary thing and space as secondary result. Space is now having its revenge, so to speak, and is eating up matter. But that is still a pious wish." [29]

Here we can ask a misleading question. What was first, matter or "space" (ether according to Einstein)? The answer should be: *neither of them.* The primordial eternal entity that has always been is the *action-potential*, as we saw in chapter 2.

Q4 "It is interesting to note that the problem of space, ether, and field was the subject of Einstein's lecture delivered at the 2nd World Power Conference (Berlin 16-25 June 1930) devoted to the resources of energy. It means that Einstein was convinced that physical space possessing the properties of an ether and conceived as the total field possessing energy and able to generate elementary particles, is the most fundamental resource of energy." [30]

This reference should be thoroughly studied by all patent officers in the world before they reject patent projects based on the Zero-Point Energy of Vacuum (Ether). Ignorant scientists should not be allowed to become patent officers. Einstein is a good example of a very knowledgeable patent officer in Bern.

Q5 "Physical space and the ether are only different terms for the same thing; fields are physical states of space." [31]

Here we see that Einstein felt that ether was a synonym for physical space, also a synonym of field. A physical field, like the gravitational field, is a physical state of space. One gets extremely desperate, and therefore, very frustrated, when reading Einstein's writings about the ultimate nature of the ether. If we agree that the most outstanding attribute of any field is its energy content, then we certainly should have expected Einstein to say that the ultimate essence of any field is *pure energy*. That the essence of the cosmic or universal space is *pure energy*, and that the *Being* of all entities is *pure energy*.

Q6 "The neutral as well as the electrical particle is a portion of space." [32]

Beyond the year 2,000, we will learn that electrons and protons are rotating *toroids* of condensed energy, of condensed ether, of condensed "space," of condensed fields. We learned, in chapter 4, that the neutron is a miniature hydrogen atom, as Eddington conceived it. To understand this model, we need a totally new Electrodynamics which was presented in chapter 4. The conclusion in this chapter, up to here, is that the essence of the classical ether is pure energy. Occasionally, we will refer to this classical conception of the ether as the Primordial Energy Theory.

5.5. Derivation of the Eikonal Equation.

From a physical point of view, the eikonal (iconal) equation transforms the physics of waves (D'Alambert's wave equation) into the physics of rays (mutilated Hamilton-Jacobi's equation). In other words, the eikonal equation transforms physical optics into geometrical optics. From a mathematical point of view, the eikonal equation is a nonlinear first-order partial differential equation, very similar to Hamilton-Jacobi's equation. The latter is given by:

$$\partial S/\partial t + (1/2m)[(\partial S/\partial x)^2 + (\partial S/\partial y)^2 + (\partial S/\partial z)^2] + U = 0 \qquad (5.18)$$

where S is the momentum-energy potential or action potential (in textbooks of Calculus of Variations S is called the "generating function of the canonical transformation,") and U is the potential energy of the system acting on a particle of mass m. Taking into account that:

$$\partial S/\partial t = - E \qquad (5.19)$$

where E is the total energy of the system, eq. (5.18) becomes:

$$(\partial S/\partial x)^2 + (\partial S/\partial y)^2 + (\partial S/\partial z)^2 = 2m(E - U) \qquad (5.20)$$

In chapter 2, we have dealt more extensively with eq. (5.20). Our present thesis is to deduce the eikonal equation given by:

$$(\partial\phi/\partial x)^2 + (\partial\phi/\partial y)^2 + (\partial\phi/\partial z)^2 = n^2k^2 \qquad (5.21)$$

where ϕ represents the space phase of an electromagnetic wave; n is the index of refraction of the medium in which the electromagnetic wave propagates, and k is the wave number, given by $k = \omega/c = 2\pi/(Tc) = 2\pi/\lambda$. As we can see, there is a remarkable similarity between eqs. (5.21) and (5.20). One of these equations deals with the motion of particles, and the other equation deals with rays that correspond to the trajectory of packages of electromagnetic energy. This similarity anticipates the erroneous duality of particles and waves as we have seen in chapter 2. D'Alambert's equation, for one component ξ of the electric or magnetic field, is given by:

$$\nabla^2\xi - (1/v^2)\, \partial^2\xi/\partial t^2 = 0 \qquad (5.22)$$

Let us make the following substitution in eq. (5.22):

$$\xi = f(x,y,z)e^{-i\omega t} \qquad (5.23)$$

The result is:

$$\nabla^2 f + (\omega^2/v^2)\partial^2 f/\partial t^2 = 0 \qquad (5.24)$$

The index of refraction n and the speed v, are given by:

$$n = c/v, \text{ from which } v = c/n \qquad (5.25)$$

The wave number k is given by:

$$k = \omega/c \qquad (5.26)$$

Introducing eqs. (5.25) and (5.26) in eq. (5.24) we get:

$$\nabla^2 f + n^2 k^2 = 0 \qquad (5.27)$$

For a plane wave, the solution of the previous equation is:

$$f = A\exp[i\, n(\mathbf{k}\cdot\mathbf{r})] \qquad (5.28)$$

or $$f = A\exp[i\, n(k_x x + k_y y + k_z z)]$$

For any plane electromagnetic wave, the amplitude and the phase remain constant in space and time. In a plane wave, all the points of a wave front are coherent, i. e., all the points have the same phase. Now, when the surface of the wave-front is not a plane, the analysis of the propagation of the wave is rather complicated. However, an approximation can be made when the curvature of the wave-front is small. This type of approximation is acceptable when the wave length is very small, or equivalently, when the wave number $k = 2\pi/\lambda$ is very large. We will use this condition of k to neglect a Laplacian term in what follows. Now, let us try a solution of eq. (5.27) of the form:

$$f = A\exp[i\, \phi(x,y,z)] \qquad (5.29)$$

here ϕ is called eikonal . The eikonal differs little from the phase of a plane wave. Finally, let us introduce another function σ through the following substitution:

$$\phi = k\sigma(x,y,z) \qquad (5.30)$$

where σ is a quasi-linear function of the space coordinates. Introducing eq. (5.30) in eq. (5.29) we get:

$$f = A\exp[i\, k\sigma(x,y,z)] \qquad (5.31)$$

Introducing the last equation in eq. (5.27) and simplifying the equation by $k^2 e^{ik\sigma}$, we get:

$$(\nabla\sigma)^2 - (i/k)\nabla^2\sigma = n^2 \qquad (5.32)$$

The magnitude of the function σ is independent from k. Therefore, for very large values of k, we can neglect the second term on the left-hand side of eq. (5.32). Now we have:

$$(\nabla\sigma)^2 = n^2 \tag{5.33}$$

Introducing eq. (5.30) in the previous equation, we get:

$$(\nabla\phi)^2 = n^2k^2 \tag{5.34}$$

The last equation is called the *eikonal equation.* Before we continue with the last two equations, we must mention the nature of the neglected Laplacian term of σ. This term represents a collective quantum potential as we have clearly seen in chapter 2. In relation to the eikonal equation, we did not establish any restrictions to the index of refraction n. In consequence, n can be any function of the space coordinates. If ϕ is a solution of the eikonal equation, then:

$$\phi = k\sigma(x,y,z) = \text{const.} \tag{5.35}$$

The last equation represents a surface of equal phase. Now, the propagation of an electromagnetic wave is in the direction perpendicular to the surface of constant ϕ or constant σ. In other words, the wave propagates in the direction of $\nabla\phi$ or $\nabla\sigma$. Thus, we have transformed the physics of waves, given by D'Alambert's equation (5.22), into the physics of geometrical rays given by the eikonal equation (5.33) or (5.34). The gravitostatic energy density ρ_G, around a spherical celestial body of Mass M and radius R, is given by:

$$\rho_G = GM^2/(8\pi r^4) \tag{5.36}$$

where $r \geq R$ is measured from the center of the celestial body. Introducing eq. (5.36) in eq. (5.12) we get:

$$n^2 = 1 + GM^2/(8\pi\rho^* r^4) \tag{5.37}$$

We see from the last equation, the index of refraction offers spherical symmetry $n = n(r)$. Finally, let us write the eikonal equation (5.33) in spherical coordinates:

$$(\partial\sigma/\partial r)^2 + (\partial\sigma/\partial\theta)^2/r^2 + (\partial\sigma/\partial\phi)^2/(r^2\sin^2\theta) = 1 + a^2/r^4$$

(5.38)

where

$$a^2 = GM^2/(8\pi\rho^*)$$

It would be interesting to solve the starlight deflection problem using eq. (5.38). However, eq. (5.38) is too much tool to solve this problem. In chapter 6, we will solve this problem using Snell's law of refraction in the energy field of the sun.

5.6. Experimental proposals.

The purpose of the following experimental proposals is to determine the stellar energy density ρ^* from experimental or observational data.

Energy Sandwich Experiment.

To test the reality of the energetic essence of the cosmic ether at laboratory level, we propose to pass one of the rays of a Michelson interferometer back and forth through the gap of an electromagnet. The whole set up should be in a vacuum chamber. The magnetic energy density will increase the total density of energy, consequently decreasing the speed of light. The index of refraction in the magnetic gap will change, introducing an interference shift. Preliminary calculations indicate that magnetic pulses of 40 Teslas will show the predicted shift. Recently this author found that Albert Einstein, when he was fifteen or sixteen years of age, proposed this conception more than a century ago. This reference to Einstein is found in the excellent book by Gerald Holton titled *Thematic Origins of Scientific Thought* [33]. Holton writes: "*I am grateful to Professor Pelseneer for a copy of the six-page essay.*" Einstein sent this essay to his uncle, Caesar Koch, in 1894 or 1895. Holton describes Einstein's proposal in the following terms:

> "For this purpose he suggested sending a lightbeam into a magnetic field as a probe. Any effects on the measurable speed or wavelength of such a beam would reveal the 'elastic deformation' of the ether or field."

It is obvious, at that particular time, that young Albert had not heard about the expected shift of the interference fringes in the famous Michelson-Morley experiment. The author's proposal is a combination of these two experiments.

Starlight Deflection by the Sun.

In chapter 6, we will analyze, in detail, the starlight deflection by the gravitational field of the sun. This will be a classical analysis based on the familiar phenomenon of refraction of the starlight ray by the gravitostatic energetic solar field. Some elementary Euclidean theorems of classical geometry will be required. The mathematical expressions, which we will deduce for the angle of total deflection, will allow us to search for the proper value of the interstellar energy density. In this way, we will be able to match the astronomical observations of the solar deflections.

Michelson-Morley's experiment in interplanetary space.

Let us have an intuitive-imaginative mental picture of the cosmic ocean of energy. Let us start by having a technicolor allegory of this universal medium of energy, and assign a light blue or bluish color to the interplanetary cosmic energy. At the same time, let us assign a yellow color to the earth's gravitostatic energy of our planet. Now we can see, in our imagination, that our planet is surrounded by a green energetic atmosphere. As we recede from the surface of the earth, the green color begins to fade and gradually disappears as it becomes a pure bluish color.

When we have a man-mission to Mars, the astronauts should take a Michelson-Morley interferometer. Halfway between the earth and Mars, they will have ideal conditions to verify the prediction of the19th century. The mass of the spacecraft is so insignificant that its gravitostatic energy is negligible. This means that the spacecraft is not surrounded by any weak green color. Halfway between the two planets the spacecraft becomes a cosmic "submarine" totally immersed in the bluish ocean of cosmic energy. In this way, the astronauts will be able to measure the *absolute* velocity of the spacecraft.

Conclusions

The most outstanding aspect of this chapter is Einstein's acceptance of the existence of the cosmic ether in 1920. However, for more than 60 years textbooks of physics emphasized, instead, Einstein's rejection of the cosmic ether in 1905.

Einstein's new ether is its identification with the four-dimensional space-time components of the metric tensor. These components represent the geometrodynamic potentials in GRT. This revival of Einsteinian ether is highly formal, if not to say, highly mathematical. This new ether of Einstein is so metaphysical that its creator said; "*We can assume the existence of an ether; but we must abstain from ascribing a definitive state of motion to it, i.e., we must divest it by abstraction of the last mechanical characteristic which Lorentz left it.*" A statement like this cannot belong to Natural Philosophy, but to meta-mathematics.

We finally removed a serious objection against the existence of a cosmic ether. This objection consisted in denouncing the absence of a wave equation for the density of ether in physics. We proved first that the essence of the cosmic ether is pure energy which pervades the entire universe. We then proved that a D'Alambert equation, for the density of the energetic ether, does indeed exist. This firms the existence of the so called *Zero-Point Energy of Vacuum*, and identifies the being of ether as energy.

We explained the outcome of Michelson-Morley's experiment by using the dragging coefficient of Fresnel. Nevertheless, the numeric value of the interstellar energy density was much smaller than the one determined using the photon deflection by the energy field of the sun, as it is done in chapter 6. We left to the young generation the problem of taking more terms of higher order in Fresnel's dragging coefficient, to see if the interstellar energy density is increased after solving the problem created by Michelson-Morley's experiment. Finally, we proposed some experiments which might verify that the speed of light is inversely proportional to the square root of the energy density.

Going back to Einstein's rejection of the ether in 1905 in his SRT, and his revival of a strange metaphysical ether in 1920 in his GRT, we find Einstein guilty of promoting the nonexistence of a real ether. His obstinate attitude of rejecting the existence of the cosmic ether caused him to pay a high price. He was unjustly rejected from the quantum mechanics interpretation of the Copenhagen circles. In chapter 2, we took the defense of Einstein when he declared that quantum mechanics was an incomplete theory.

References

1. L. Kostro, *Proceedings of The British Society for the Philosophy of Science*, September 1988 [Outline of the History of Einstein's Relativistic Ether Conception]

2. L. Kostro, *Quantum Uncertainties - Recent and Future Experiments and Interpretations* (Ed. W.M. Honing, D.W. Kraft and E. Panarella, Plenum Press, NY, London, 1987) Series B: Physics Vol. 162, p. 435-448 [Einstein's Conception of the ether and its-up-to-date Applications in the Relativistic Wave Mechanics]
3. L. Kostro, Proceedings of the Conference on Foundations of Mathematics & Physics, Perugia, Italy, 1989 (Edited by U. Bartocci and J.P. Wesley - Benjamin Wesley Publisher, 7712 Blumberg, Germany, 1990) [Albert Einstein and the Theory of the Ether]
4. A. Einstein, *Sidelights on Relativity* (Dover Publications, Inc.,NY, 1983)
5. J.C. Curé, J. Class. Phys., **6**, part I, 1-30 (1987) [Einstein and the luminiferous ether],
6. E.K. Conklin, Nature, **222**, 971 (1969); P.S. Henry, Nature, **231**, 516, (1971)
7. S. Marinov, Gen. Rel. Grav., **12**, 57 (1980)
8. S. Marinov, *The Thorny Way of Truth,* Part II (East-West, Graz, Austria, 1984)
9. W. Silvertooth and S.F. Jacobs, Applied Optics, **22**, 1274 (1983)
10. P.T. Pappas, Il Nuovo Cimento, **76B**, no. 2, 189-196 (1983); P.T. Pappas and P.G. Mossydes, Phys. Lett., vol.**111A**, no. 4, 193 (1985); P.G. Mossydes and P.T. Pappas, J. Appl. Phys., vol.59, no. 1, 19 (1986)
11. P. Graneau, IEEE Trans. Magnetics, vol. Mag. **20**, no. 2, 444 (1984)
12. J.C. Curé, Phys. Lett, vol.**116B**, no.9, 158 (1982)
13. D. Michelson-Livingstone, *The Master of Light* (Charles Scribner's Son, NY, 1973)
14. J.C. Maxwell in *The Scientific Papers of James Clerk Maxwell* (W.D. Niven, Dover Publications Inc., NY, 1966)
15. J.C. Maxwell, *A Treatise on Electricity and Magnetism* (Dover Publications Inc., NY, 1954)
16. T.G. Barnes, *Space Medium, The Key to Unified Physics* (Geo/Space Research Foundation, P.O. Box 13560, El Paso, Texas 79913, 1986); P. Tewari, *Beyond Matter* (Print Well Publications, Aligarh, India, 1984);
17. A. O'Rahilly, *Electromagnetic Theory* (Dover Publications, Inc., NY, 1965)
18. P.T. Pappas and G. Obolensky, Electronics & Wireless World, 1162 (December 1988)
19. E. Whittaker, *A History of the Theories of Aether and Electricity* (Thomas Nelson and Sons Ltd., London, NY, 1951)

20. P.A.M. Dirac, Sci. Am., 45 (May 1963)
21. B.G. Wallace, Found. Phys., **3**, no. 3, 381-388 (September 1973)
22. P. Tewari, Magnets in your Future, **2**, December 1987; The Journal of Borderland Research, USA, Vol. XLV, no. 5, (September-October 1989)
23. A.A. Michelson and E.W. Morley, Am. J. Sci., 3rd series, **34**, 333-341 (1887)
24. T. Theocharis, Lett. Novo Cimento, **36**, no. 11, 325-332 (1983)
25. B.G. Wallace, Spec. Sci. Tech., 9, no. 1, 9-17 (1984)
26. R.R. Hatch, *Escape from Einstein* (Published by the KNEAT KOMPANY, 1142 Lakme Ave., Wilmington, CA 90744, 1992)
27. Ref. 3, p. 160
28. Ref. 3, p. 160
29. Ref. 3, p. 161
30. Ref. 3, p. 151
31. A. Einstein, *Ideas and Opinions* (Crown Publishers, Inc., NY, 1982) p. 281
32. A. Einstein and N. Rosen, Phys. Rev., Vol. **48**, Jul. 1, pp. 73-76 (1935) [The Particle Problem in the General Theory of Relativity],
33. Gerald Holton, *Thematic Origins of Scientific Thought* (Harvard University Press. Cambridge, Massachusetts and London, England, 1988)

CHAPTER 6

NEWTONIAN RELATIVISTIC GRAVITODYNAMICS

TABLE OF CONTENTS

"A new scientific truth does not triumph by convincing its opponents and making them see the light, but rather because its opponents eventually die, and a new generation grows up that is familiar with it."

Max Planck

Introduction

In this chapter, we present two serious accusations which Einstein made against Newton's *Principia.* The first one, which is in relation to the anomalous motion of planet Mercury, will be presented in section 1. The most incredible historical fact is that the theoretical solution of this so-called *anomalous motion of planet Mercury* is contained in the *Principia* of 1687. In 1979, after Prof. R.H. Dicke read the content of section 1, he pleasantly articulated his dismay to this author. Many years have passed since this epistolary encounter so it can be only paraphrased today: *It is incredible that Newton's solution escaped the attention of generations of physicists before and after Einstein.*

The second accusation is presented in section 3. This accusation is about the incapacity of classical mechanics to explain the equality of the inertial mass with the gravitational mass. Both accusations are proven wrong. In these two sections, Einstein is found guilty. In section 2, we take the defense of GRT by proving that if the sun *is* oblate GRT will not be proven wrong. On the contrary, GRT proves to be very useful in determining the angular momentum of the sun even if the sun is oblate. In section 4, we analyze and criticize Mach's concept of mass as a ratio of accelerations. This *ratio of acceleration, acquired by interacting bodies,* is what Einstein will later call Mach's Requirement. In section 5, we draw conclusions from the identity of inertial and gravitational mass which we established in section 3. In this section, we claim that Einstein's Principle of equivalence is a classical corollary of Newton's mechanics. In section 6, we present different statements of Mach's Principle, and we proceed to interpret one of them as the *cosmic collective potential energy*. In section 7, we offer a classical explanation of starlight deflection using geometrical optics in the solar energetic envelope.

In section 8, we analyze the so-called *Doppler cosmological red shift*. In section 9, we analyze and speculate about the possibility that *gravitational forces* may be statistical residues of *electrodynamic forces* between electric dipoles. In section 10, we compare Einstein's field equations with Hamilton-Jacobi-Bohm's equation. Finally, in section 11, we advance the existence of nonlinear electrodynamic field equations.

6.1 Newton's Explanation of the Anomalous Motion of Planet Mercury.

This section is entirely based on a paper published by the author in Galilean Electrodynamics, in 1991. The purpose of this section is to show that the dynamical solution of the so-called "anomalous motion" of planet Mercury, or excess perihelic motion of the planets, appeared for the first time in Newton's *Principia* in 1687. This was long before the actual astronomical phenomenon was discovered by Leverrier in 1859. This historical fact invalidates Einstein's assertion that "Classical Mechanics is powerless" to explain this astrodynamic phenomenon. It is also shown that this Newtonian solution of the excess perihelic motion of the planets provides two additional gravitational terms: one repulsive and the other attractive.

In 1915, Einstein [1] solved the problem of the direct excess of precessional motion of the perihelion of planet Mercury, discovered by Leverrier [2] in 1859. The following year he formally published [3] his General Relativity Theory (GRT). Einstein [4], in 1917, was very outspoken when he accused Newtonian dynamics or Classical Mechanics of being *powerless* to explain this astronomical phenomenon. He writes:

> "[Relativity theory] has already explained a result of observation in astronomy against which Classical Mechanics is powerless."
>
> "This effect can be explained by means of classical mechanics on the assumption of hypotheses which have little probability, and which were devised solely for this purpose."

After Leverrier's astronomical discovery of the perihelic rotation, equal to 38 "/ century, and the secular variation of the eccentricity of the orbit of planet Mercury, different classical solutions of the "anomalous motion" of this planet were published prior to Einstein's solution. In 1749, Clairaut, mentioned by Moulton [5], advanced the hypothesis that the "old" Newtonian gravitational force should be substituted by $A/r^2+B/r^3$ in order to solve the mathematical equivalent problem of lunar perigee. Clairaut ignored the fact that this was not a hypothesis in Newtonian dynamics, but a proven thesis of Proposition XLIV, Book I in the *Principia* [6]. Tisserand, mentioned by Sciama [7], in 1872, and Levy, mentioned by O'Rahilly [8], in 1890, published solutions to this problem using authentic Newtonian gravitodynamics based on analogies with 19th century electrodynamics. In 1884, Hall [9] advanced the hypothesis that the "old" Newtonian gravitational force should be changed to GMm/r^{2+e} in order to explain the unusual excess motion of

Mercury's perihelion. But again, this is not a hypothesis in classical dynamics but simply a logical consequence of Proposition XLV, Book I in the *Principia*. In 1897, Newcomb [10] hypothesized an oblate sun caused by solar rotation. While this hypothesis can explain an excess perihelic rotation of planet Mercury equal to 41.6 "/century, it introduces other secular variations in Mercury's orbit which have not been observed astronomically. Newcomb's hypothesis was revived by Dicke, in 1964, as we will see in the following pages.

In 1898, Paul Gerber [11] solved the problem of the excess perihelic motion of planet Mercury by introducing a finite speed of propagation of gravitodynamic interactions equal to the speed of light. The same equation, derived by Gerber, was found by Einstein years later, but he made no reference to Gerber's work. K. Demys [12], in 1985, made the following remark:

> "The famous physical theorist Ernst Mach had singled out Gerber's work for special mention already in the 4th edition of Mach's classic book on mechanics, and again in the 5th edition, in 1904, on page 201. Einstein, whom Mach admittedly deeply influenced, was a keen student of Mach's *Mechanik*, citing it importantly (for instance on page 769 of the *Annalen der Physik* in 1916). As a careful reader of Mach, he miraculously escaped noticing the explicit mention of Gerber's breakthrough by Mach in two successive editions both of which appeared before 1905 and well before 1916, when Einstein announced his gravitational views seven years after Gerber's death."

In 1908, Walter Ritz, mentioned by O'Rahilly [8], again solved the problem of the excess perihelic motion of planet Mercury developing another Newtonian gravitodynamics, by analogy, with his own Galilean Electrodynamics. After Einstein's work of 1915, G.B. Brown [13], in 1955, published again another classical solution to the "anomalous" motion of planet Mercury using a Newtonian gravitodynamics, by analogy, with an induced electrodynamics of his own and arriving at the same equation of Gerber. Finally, Dicke [14], in 1964, and Dicke and Goldenberg [15], in 1967, revived the solar oblateness of Newcomb in order to only explain a small fraction of the 43 "/century of Mercury's excess perihelic rotation. The published data of the solar oblateness, in 1967, raised an unusual reaction from both relativists and astrophysicists. Two things were very clear in those years: GRT was inadequate to explain the excess perihelic motion of the planets, and very little was known about the physics of the solar core. In 1975, Hill and Stebbins [16] published new measurements of solar oblateness, concluding that their measurements removed a serious consequence derived from Dicke's and Goldenberg's refutation of Einstein's GRT. Nevertheless, Hill *et al.* [17], in 1982, published new solar oblateness measurements indicating that this was a real effect of the rotation of the sun. If physicists keep on treating the sun as an absurd point-like particle, then GRT will be proven totally powerless in explaining the excess perihelic rotation of the planets. In 1982, some newspapers published declarations of some scientists who said that Einstein's GRT was in error, but these scientists were in total historical ignorance. The so-called error had been removed from GRT in 1918 which was 64 years before.

In the past few decades, a series of papers have been published on the possibility that a real deviation exists from the original inverse square law of Newton's gravitational force [18 - 26]. Different authors have pointed out that the present data, associated with laboratory and geophysical measurements, cannot ban a deviation from the r^{-2} gravitational law on an astrophysical large scale. This deviation can be analyzed in three different mathematical ways: (1) The exponent of the original Newtonian gravitational law is not 2 but (2+e). This is the case of Hall's assumption. (2) The universal gravitational *constant* is not constant, but depends on the distance of separation of the interacting bodies. This is the case of the most recent publications on this subject. This approach adds an extra exponential short-range gravitational term to Newton's gravitational law. (3) The recent gravitational measurements are indicating the existence of new gravitational terms. This analytical approach will take us from Newtonian *gravitostatics* to Newtonian *gravitodynamics*. When Ampere published his electrodynamics, physicists never claimed that the electrostatic constant of Coulomb's force law was not constant, viz., they never asserted that Ampere's electrodynamics violated Poisson's equation, but that they were in the presence of new electrodynamic terms in addition to Coulomb's electrostatic term. In GRT, neither Einstein nor relativists have indicated that the universal gravitational constant is not constant, in spite of the fact that relativistic gravitational force has many more terms than the single Newton gravitostatic term. The trend to consider the gravitational constant G, a pseudo-constant which depends on the distance of the interacting bodies, is acceptable from a mathematical point of view, but not from the point of view of Natural Philosophy. Sooner or later, new and more precise astronomical measurements will indicate that the gravitational pseudo-constant depends, also, on the velocity and acceleration of the planets and probably on the time of the observations. The latter dependency has already been advanced in the context of Brans-Dicke's theory [27]. Considering theoretical estimations and geophysical and planetary orbit data, the time variation of this gravitational pseudo-constant has been assigned an approximate value of the order of 10^{-10} years.

In 1987 Stacy *et al.* [28], in an extensive review article, declared that:

"Kepler's laws of planetary motion are as free of doubt or qualifying conditions now as they were three centuries ago when Isaac Newton used them to derive the inverse square law of gravitational force."

This is indeed an exaggerated statement. Leverrier showed, in 1859, that Kepler's first law was incorrect. Leverrier's correction, in mathematical language, is given by

$$r = p/[1+e\cos(k\theta-w)] \qquad (6.1)$$

The parameter k in eq. (6.1), slightly less than unity, expresses the excess perihelic rotation of the planets. Kepler's first law corresponds to k=1. Leverrier's correction gave birth to extensive theoretical and astronomical research up to the present time.

Eq. (6.1) can be deduced with the help of GRT. Einstein's GRT, as well as other Newtonian gravitodynamics based on analogies with electrodynamics of the Weber or Ritz type, show, on theoretical grounds, that Kepler's second law is also incorrect. The correction is given by:

$$r^2(d\theta/dt) = h(1- K/r) \qquad (6.2)$$

Kepler's second law corresponds to K=0. Unfortunately, no experimental or astronomical verification of eq. (6.2) is found in scientific literature. Nevertheless, very interesting consequences can be derived from eq. (6.2) in gravitodynamics as well as in electrodynamics. An extensive account on secular variations of orbital parameters is offered by Whittaker[29]. Of particular interest are the variations of the mean motion of celestial bodies in relation to Kepler's third law.

6.1.1 NEWTONIAN DYNAMICAL METHODOLOGY

Newton's dynamical methodology is clearly established in the preface of his *Principia*. J.C. Maxwell [30] describes it in the following terms:

"The process of dynamical reasoning consists in deducing from successive configurations of the heavenly bodies, as observed by astronomers, their velocities and their accelerations, and in this way determining

> the direction and the relative magnitude of the force which acts on them.
>
> "Kepler had already prepared the way for Newton's investigation by deducing from careful study of the observations of Tycho Brahe the three laws of planetary motion which bear his name."

The lack of precision in Brahe's measurements, performed in the second half of the 16th century, undoubtedly had a repercussion in the number of gravitational terms determined by Newton in his gravitational law. If Newton had had Leverrier's data at his disposal, he most certainly would have determined more than one gravitational term in his gravitational force law. It is incredible that physicists and astronomers never used this Newtonian dynamical methodology immediately after Leverrier published his data on the "anomalous motion" of planet Mercury. Einstein never would have accused Newtonian dynamics of being "powerless" to explain Mercury's "anomalous motion" if he had taken the time to classically explain this astrodynamical problem. Einstein [31, p. 300] was well acquainted with Newton's dynamical methodology. He says:

> "Classical Mechanics is only a general scheme: it becomes a theory only by explicit indication of the force laws (d) as was done so very successfully by Newton for celestial mechanics. From the point of view of the aim of the greatest logical simplicity of the foundations, this theoretical method is deficient in so far as the laws cannot be obtained by logical and formal considerations, so that their choice is *a priori,* to a large extent arbitrary."

Einstein is mistaken in the second sentence of the above quotation. Newton did not arbitrarily obtain his gravitational force law. Newton obtained it by logical and formal considerations, applying the axiomatic structure of his theory of dynamics to factual or empirical data. By no means are we entitled to accuse Newton of having indicated the mathematical structure of his gravitational force law simply *a priori*; on the contrary, he obtained his gravitational force law *a posteriori*. GRT never would have been considered a physical theory if Einstein [32] had not adjusted his constant with the help of Newton's universal gravitational constant G, via Poisson's equation. This fact seems to have been forgotten when physicists claim that GRT has no need to adjust any parameter in the solution of Mercury's excess perihelic rotation.

Let us now update the classical gravitational force law while strictly following Newton's dynamical methodology. To do this, we use Newton's second axiom, in plane polar coordinates, along with the improved measurements of Leverrier, represented by eq. (6.1) and Kepler's second law given by eq. (6.2) with K=0. Using the differential equation of the orbit given by Binet's equation, we get the radial component of the force acting on the planet

$$F = - k^2 h^2 p^{-1} m/r^2 - (1-k^2) h^2 m/r^3 \qquad (6.3)$$

where p is the *semi-latus-rectum*; h is the specific orbital angular momentum of the planet, and k is a co-efficient which is slightly less than unity which, consequently, conforms the direct excess motion of the precessional motion of the planet's perihelion. Now let us make

$$k = 1-Q \qquad (6.4)$$

where $Q \ll 1$. Introducing eq. (6.4) in eq. (6.3), and making the well known substitution $GM=h^2p^{-1}$, where G is the universal gravitational constant, we get

$$F = - GMm/r^2 + 2QGMm/r^2 - 2Qh^2m/r^3 \qquad (6.5)$$

Now if the differential equation of the orbit is written as

$$d^2u/d\theta^2 + u = GM/h^2 + \Sigma\, C_n u^n \qquad (6.6)$$

with $u=r^{-1}$, then the excess perihelic rotation is given by the author's solution:

$$\Omega = (\pi/T)\, \Sigma\, nC_n\, p^{1-n} \qquad (6.7)$$

where T is the period of revolution of the planet. The last equation, for the particular form of the force given by eq. (6.5), becomes

$$\Omega = 2\pi Q/T \qquad (6.8)$$

Using the approximate value of 43 "/century for Mercury's excess perihelic rotation in eq. (6.8), we get $Q=8\text{x}10^{-8}$. This result allows us to disregard the term $2QGMm/r^2$

when compared with GMm/r^2 in eq. (6.5) for this planetary problem.. Eq. (6.5) becomes

$$F = - GMm/r^2 - Lm/r^3, \text{ where} \quad (6.9)$$

$$L = 2Qh^2 \quad (6.10)$$

From eq. (6.10), eq. (6.8), and $GM=h^2p^{-1}$, we get

$$L = \Omega pTGM/\pi \quad (6.11)$$

In Table 6-I, we indicate the numeric values of L for different planets along with the excess perihelic rotation of the same and their percentage errors as determined from astronomical data. The high percentage errors of Venus and earth are probably responsible for the deviations in L. In 1972, I.I. Shapiro, (as quoted by Misner, Thorne and Wheeler [33]), after analyzing thousands of observations, assigned an uncertainty factor of 9.3% to the calculated value of Mars' excess perihelic rotation and only 1% for Mercury.

Table 6-I. Astronomical excess perihelic rotations of the planets and Leverrier's constant L.

Planet	Ω ("/century)*	Error %	$Lx10^{24}$ (ISU)
Mercury	42.6 + 0.94	2.2	1.1635
Icarus**	9.8 + 0.8	8.2	1.1471
Venus	8.4 + 4.8	57.1	1.1445
Earth	4.6 + 2.7	58.7	1.4080
Mars	1.5 + 0.04	2.7	1.3045

* W.M. Blanco and S.W. Mc Cuskey, *Basic Physics of the Solar System* (Addison Wesley, Readings, Mass., 1961)

** S. Weinberg, *Gravitation and Cosmology* (John Wiley & Sons, NY, NY, 1972)

Table 6-II Newtonian, astronomical and Einsteinian excess perihelic rotations of different planets

Excess perihelic rotations in ("/century)

Planet	Newtonian	Astronomical	Einsteinian*
Mercury	42.56	42.56	42.86
Icarus	9.95	9.8	10.02
Venus	8.54	8.4	8.60
Earth	3.80	4.6	3.83
Mars	1.34	1.5	1.35

*L. Motz and A. Duveen Essentials of astronomy (Wadsworth Publishing Company Inc., Belmont, CA, 1966)

In what follows, we will make the assumption that L is a planetary constant and refer to it as Leverrier's constant. To calculate the value of this new constant, we will use eq. (6.10) and the astronomical excess perihelic rotation of planet Mercury. Leverrier's constant becomes 1.1635×10^{24} (m^4/s^2). Table 6-II lists the Newtonian calculated excess perihelic rotations of various planets, together with the corresponding astronomical values and Einstein's results.

In examining Table 6-II, we see the Newtonian figures correspond more closely to the astronomically determined excess perihelic rotations than the Einsteinian figures do. In fact, Newton's mean square deviation is 0.188 while Einstein's is 0.207, thus, demonstrating that "Classical Mechanics" can hardly be considered "powerless" in explaining the "anomalous motion" of Mercury and the other planets. As the figures of Table 6-II show, Newtonian dynamics is not only capable of explaining the excess perihelic rotations of the planets, but it renders superior results without the alleged necessity of framing "hypotheses which have little probability and which were devised solely for the purpose."

6.1.2 PROPOSITION XLIV, THEOREM XIV

In the *Principia*, Book I, under the heading *Proposition XLIV, Theorem XIV*, Newton proves the following thesis:

"The difference of the forces, by which two bodies may be made to move equally, one in a fixed, the other in the same orbit revolving, varies inversely as the cube of their common altitudes."

Newton then proceeds to analyze the motion of two bodies of equal mass in two separate but identical ellipses. He assumes that one ellipse is fixed (with respect to an inertial reference system), while the other revolves around a center of force located in one of the foci of the ellipse. Newton's purpose, in this Proposition, is to investigate how the force, acting on the body that moves in the fixed ellipse, is modified when the whole elliptic orbit revolves around the focus. It is interesting to notice that this Proposition XLIV is under the heading titled *The motion of bodies in movable orbits; and the motion of the apsides*. The line of apsides connects the perihelion and the aphelion of the elliptic orbit, and, therefore, is equivalent to referring to the rotation of the apsides or the rotation of the perihelion. Newton's theoretical conclusion in modern notation reads as follows:

$$\text{Difference in the forces} = F - (-C_1/r^2) = -C_2/r^3$$

Where C_1 and C_2 are constants. Eq. (6.9) is formally identical to the previous equation derived by Newton in Corollary II of Proposition XLIV. Perhaps, the reason why this Newtonian solution of the excess perihelic rotation of the planets escaped the attention of generations of physicists, for more than three hundred years, was the geometrical techniques used by Newton. Proposition XLIV extends Newton's own inverse square gravitational law, and constitutes the dynamical solution of the perihelic rotation of the planets as is clearly shown in Table 6-II.

Newton was well acquainted with the gravitational perturbations of the planets, among themselves, as is clearly established in the Scholium of Proposition XIV, Theorem XIV, in Book III, in the *Principia*. There he writes:

"Since the planets near the sun (viz., Mercury, Venus, the earth, and Mars) are so small that they can act with but little force upon one another, therefore their aphelions and nodes must be fixed, except so far as they are disturbed by the actions of Jupiter and Saturn, and other higher bodies. And hence we may find, by the theory of gravity, that their aphelions move forward a little, with respect of the fixed stars, and that as the 3/2th power of their several distances from the sun. So that if the aphelion of Mars, in the space of a hundred years, is carried forwards 33' 20", with respect to the fixed stars, the

> aphelions of the earth, of Venus, and of Mercury, will in a hundred years be carried forwards 17' 40", 10' 53", and 4' 16", respectively. But these motions are so inconsiderable, that we have neglected them in this Proposition."

Let us finish this section with a curious observation. To what higher planets was Newton referring to in the above quotation, in 1687, when he wrote: "... by the actions of Jupiter and Saturn and other higher bodies?" Uranus was discovered by William Herschel in 1781!

6.1.3 GRAVITODYNAMICS AND GEOMETRODYNAMICS

An interesting comparison can be made between Newtonian gravitodynamics and Einsteinian geometrodynamics in relation to the excess perihelic rotation. The Einsteinian excess perihelic rotation is given by

$$\Omega' = 6\pi GM/(Tpc^2) \quad \text{(rad/s)} \tag{6.12}$$

where c is the speed of light. Eq. (6.12) and eq. (6.11) can both be written in the same form with the help of Kepler's third law:

$$GM=4\pi^2a^3/T^2$$

$$\Omega = Ca^{-5/2}(1-e^2)^{-1} \tag{6.13}$$

where "a" is the semi-major axis of the elliptic orbit. The constant C in Newton's gravitodynamics is equal to $5.0515\text{x}10^{13}$, while in Einstein's geometrodynamics $C=5.0982\text{x}10^{13}$, being both numeric values expressed in the ISU. This small discrepancy explains the agreement in the numerical results of the excess perihelic rotation shown in Table 6-II.

With respect to geometrodynamics, Møller [34] gives the following expression for the gravitational relativistic force:

$$F = -m\alpha c^2/(2r^2)$$

where $\alpha=2GM/c^2$; $m=m_o(1-v^2/c^2-\alpha/r)^{-1/2}$, and v is the velocity of the planet. Using the binomial expansion in m's definition, the previous equation becomes:

$$F = -GMm_o/r^2 - \tfrac{1}{2}(v^2/c^2)GMm_o/r^2 - (GM/c^2)^2 m_o/r^3 - 3(GM)^3 m_o/(2c^4r^4) \quad (6.14)$$

Now eq. (6.11) can be written as:

$$\Omega pT = L\pi/(GM) = A \quad (6.15)$$

where A is a constant independent of the planets. The constancy of the product ΩpT could have been discovered empirically many years ago. This constant represents another planetary law. Using eq. (6.10) and eq. (6.15) in eq. (6.5), we get:

$$F = -GMm/r^2 + (Ap^{-1}/\pi)GMm/r^2 - (A/\pi)GMm/r^3 \quad (6.16)$$

Now comparing eq. (6.14) and eq. (6.16), we observe that relativistic geometrodynamics does not contain a *repulsive* gravitational force in the planetary system. Both Newtonian gravitodynamics and Einsteinian geometrodynamics contain a small attractive gravitational term inversely proportional to the cube of the distance. Nevertheless, the relativistic inverse-cube term is 5.9 times smaller than the Newtonian term. It is this classical term which so adequately reproduces the astronomical excess perihelic rotations of the planets in the Newtonian solution. Finally, according to our Newtonian interpretation, the so-called "fifth" and "sixth" forces are essentially gravitational forces.

6.2 Angular Momentum of the Sun.

Expensive space missions have been proposed to determine the intrinsic angular momentum (IAM) of the sun using General Relativity Theory (GRT). Here we will present an inexpensive method with which to calculate the solar IAM. To accomplish this task, we use the astrometric determination of the excess perihelic rotation of planet Mercury in combination with different gravitodynamic theories. We show that Einstein's solution of 1915, and Schwarzschild's solution of 1916 of the perihelic rotation of planet Mercury are useless in determining the IAM of the sun. We also show the 1918 solution of Lense and Thirring, used to solve the same problem of planet Mercury, provides a realistic method with which to calculate the IAM of the sun. We also discuss theories which introduce the concept of gravitodynamic induction **B***. In this type of gravitodynamics, we include Einstein's proposal which was published in 1912 in an obscure journal of medicine. This proposal, written in German, was immediately translated to Russian. The intention of this author, when writing this section, is not to defend or to attack any one person, but to attack, unmercifully, any dishonest falsehood or fraud in science, and, therefore, to defend what the author honestly believes is the truth in Natural Philosophy. The sun contributes a very small percentage to the total angular momentum of the solar system. It is obvious that we cannot send a probe to the interior of the sun in order to measure its IAM. It has been proposed that some expensive space missions measure the solar IAM by observing the precession of gyroscopes in spacecrafts orbiting close to the sun. Not knowing the radius of the solar core, nor the angular velocity of the core of the sun, we can only estimate. It would be important to know the magnitude of the IAM of the sun in order to test some models of the sun core. In 1975, Hass and Ross [35], commented on the solar IAM.

> "Models of the formation of the solar system have always been plagued with the fact that the Sun appears to be deficient in angular momentum relative to the planets. It is known that stars of spectral type earlier than the Sun rotate rapidly. The theory of stellar interiors indicates that the Sun may have a remnant rapidly rotating core (Roxburgh 1964) which might increase its angular momentum by as much as a factor of 80 over the presently accepted value. A direct measurement of the angular momentum of the Sun would be very valuable."

Mathematical physicists have always had a *mania* for reducing natural *things* to point-like entities. Thus, the sun was reduced to a point in 1915 when Einstein [1], approximately, solved the problem of the excess perihelic motion of planet Mercury. The following year Schwarzschild [36], without introducing any approximation, solved Einstein's field equations of GRT in order to explain again the anomalous motion of planet Mercury. Schwarzschild demonstrated that a nonrotating spherical body is gravitationally equivalent to a point-like particle in GRT. However, point-like bodies *cannot* have physical or real angular momentum.

To determine the solar IAM, in an inexpensive way, we need two tools. One is a good gravitodynamic theory. The other tool is a reliable astrometrical measurement of some astronomical phenomenon observed in the solar system. Since 1859, when Leverrier [2] published the astrometrical determination of the perihelic rotation (PR) of planet Mercury, we have had the real data of an astronomical phenomenon. If the gravitodynamic theory is able to *deduce* the PR of any planet in terms of the IAM of the sun, then we can determine the magnitude of the intrinsic angular momentum of the sun. Hence, any gravitodynamics which has more terms beyond Newton's *gravitostatic* term, in principle, should be able to determine the IAM of the sun if the extra terms depend on the angular velocity of the source massive body.

To honor historical precedence, we must insist that Einstein was not the first scientist to solve, in 1915, the so-called *anomalous motion of planet Mercury*. In 1917, Einstein [5, p. 102] made the following accusation against Newton's classical mechanics:

> "[Relativity theory] has already explained a result of observation in astronomy against which Classical Mechanics is powerless. "

This is not true! Isaac Newton theoretically established the theory to solve the problem of the "anomalous motion of planet Mercury," or of any other planet or satellite, 172 years before the real astronomical problem was discovered by Leverrier. The theoretical solution of this problem is contained in Newton's *Principia*, Book I, Proposition 44, Theorem 14. Thus, Classical Mechanics [6] is very powerful in solving the excess perihelic rotation of the planets. It is unbelievable that Newton's solution of the PR of the planets escaped the attention of generations of physicists before and after Einstein. What is more incredible is that the most outstanding critic of the *Principia* never studied Proposition 44, Theorem 14 of Newton's *Principia*. For classical solutions of the perihelic rotation of the planets before Einstein's solution of the same, the interested reader should consult reference [6].

The following list of explanations regarding the excess perihelic rotation of planets or satellites does not pretend to be complete. Nevertheless, the list offers many antecedents with which to judge Einstein's unacceptable accusations opposing Newton's Dynamics.

In 1687, Isaac Newton [6] anticipated the theoretical solution to a problem which was to be discovered almost two centuries later. The radial force proposed by Newton was:

$$F = - A/r^2 - B/r^3 \tag{6.17}$$

In 1991, 304 years later, this author [37] published the solution of the anomalous motion of planet Mercury using Newton's original extended gravitational law given by eq. (6.17). This paper was published after 13 years of unsuccessful attempts.

In 1749, Clairaut [38] proposed exactly eq. (6.17) as a hypothesis to solve the perigee rotation of our Moon. Obviously, Clairaut, like Einstein, did not adequately study the *Principia*.

In1884, Hall [39] advanced what he thought was an hypothesis given by

$$F = - A/r^{(2+\delta)} \tag{6.18}$$

However, the last equation represents a simple logical corollary in Newton's treatment of the perihelic motion of planets. In 1897, Newcomb [40] made an assumption that the sun is oblate because of its rotation. This hypothesis is very probable. While this hypothesis can explain an excess perihelic rotation of planet Mercury equal to 41.6 ("/ century), it introduces other secular variations in Mercury's orbit which have not been observed astronomically. Newcomb's hypothesis was revived by Dicke [41] in 1964. For more than a decade, the oblateness of the sun united relativists and astrophysicists against Dicke's concepts and his somewhat few advocates. Even relativists, in those days, admitted that if the sun is oblate, then GRT should be considered inadequate. This belief is *totally* incorrect. The reason why this statement is erroneous is because the sun, in the minds of mathematicians, is still a point-like particle. Later on, we will prove that GRT is perfectly capable of explaining the excess perihelic rotation of the planets even in the real case of an oblate sun.

There have been many other attempts to explain the excess of perihelic rotation of the planets based on gravitodynamics, established by analogy, with different electrodynamics. In 1898, Gerber published an equation giving the excess perihelic rotation of planet Mercury. In 1915, Einstein again deduced Gerber's equation of the excess perihelic motion of Mercury using GRT. Einstein *never* acknowledged the work of Gerber.

In 1994, A. Assis [42] presented an excellent bibliographic review involving the excess perihelic rotation of the planets. Assis emphasized the application of Weber's gravitodynamics, established by analogy, with Weber's Electrodynamics. Assis's bibliography on the excess perihelic motion of planets includes up to the 1990s decade. There are three other gravitodynamics, established by analogy with electrodynamics, that Assis did not mention. One is the work of M. Lévy in 1890 in which Lévy proposed a gravitokinetic potential which is a linear combination of Weber's and Riemann's gravitokinetic potentials. This reference appears in the second volume of *Electromagnetic Theory,* written by A. O'Rahilly [43] in 1938. In this same reference, O'Rahilly shows, on page 544, the mathematical expression of the excess of the perihelic rotation of Mercury obtained by Ritz in 1908. Ritz's gravitodynamics was established, by analogy, with Ritz's Electrodynamics. Finally, we should mention the work done by G. Burniston

Brown [44], in 1955 and 1982, pertaining to the excess of perihelic rotation using his own gravitodynamics, obtained by analogy, with Brown's electrodynamics established by empirico-logical induction.

Until 1982, the analogy from electrodynamics to gravitodynamics was always used. The inverse analogy, i.e., from gravitodynamics to electrodynamics might provide new insights into the electric interaction of two moving electrically charged particles. In 1982, this author [45] published a paper about the probable existence of a new electrodynamic force proportional to the square of the velocity of the source electric charge. This time the analogy was used from GRT, geometrodynamics, to electrodynamics. If this new electrodynamic force is real, then Einstein's GRT will prove that Special Relativity Theory (SRT) is incorrect because it is *incomplete*.

Up until 1918, none of the scientists, who used the so-called gravitodynamics by analogy with different electrodynamics, treated the sun as a material spherical object. As we mentioned before, any of these gravitodynamics containing terms proportional to the square of the velocity of the source particles, or proportional to the product of velocities, are capable of providing the excess perihelic motion of the planets and satellites in terms of the intrinsic angular momentum of the central rotating body. In 1918, Lense and Thirring [46] solved Einstein's GRT field equations up to a first degree of approximation in the case of a rotating solid sphere. In the calculations of the components of the gravitational field intensities, these authors called a group of physical parameters a constant K. Lense and Thirring, more concerned with the mathematics of the problem than the physics of the same, did not notice that K was equal to two times the intrinsic angular momentum of the sun (IAM). Thus, Lense and Thirring lost the opportunity to determine the IAM of the sun as early as 1918. In what follows, we will solve this problem of the solar IAM using Lense-Thirring results. However, it is worthy to mention that in the work of Lense and Thirring there are off-diagonal terms in the matrix representation of the components of the metric tensor which modifies Schwarzschild's metric [47] of 1916. Not until 1960 was this modification introduced formally in Schwarzschild's metric by Shiff [48]. The modified geodesic element ds^2 becomes:

$$ds^2 = c^2[1 - 2GM/(c^2r)]dt^2 - [1 - 2GM/(c^2r)]^{-1}dr^2 - r^2d\phi^2 + (2/5Mb^2\Omega)[4G/(c^2r)]d\Omega dt \qquad (6.19)$$

Obviously, the parentheses in the fourth term of the last equation represent the IAM of the spherical body of radius b, and rotating with angular velocity Ω. As a lateral comment, we should mention that the second term in the previous equation contains, in potency, a mathematical singularity which is responsible for the *conceptual or metaphysical* creation of black holes.

6.2.1 LENSE-THIRRING RESULTS OF 1918

The intrinsic angular momentum J* of a spherical body of mass M, of radius R and rotating with angular velocity ω, is given by

$$J^* = 2MR^2\omega/5 \tag{6.20}$$

It has become a tradition to refer to any aspect of GRT as *geometrodynamics*. This author will use the word *gravitodynamics* to refer to any Newtonian Relativistic Gravitational Theory introduced, by analogy, with different electrodynamic theories. Lense and Thirring introduced a Cartesian Stationary reference system (the axes of reference are *fixed* at *infinity)*. They calculated the geometrodynamic components of the acceleration acting on a planet of mass m moving in the equatorial plane of the central rotating spherical body of radius R, mass M, and rotating with a *clockwise* angular velocity **ω**, parallel to the positive Z-axis. Their results are:

$$a_x = -GMx/r^3 + 2GJ^*/(c^2r^3)[3yzv_z/r^2 + (x^2+y^2-2z^2)v_y/r^2] \tag{6.21a}$$

$$a_y = -GMy/r^3 + 2GJ^*/(c^2r^3)[3zxv_z/r^2 + (x^2+y^2-2z^2)v_x/r^2] \tag{6.21b}$$

$$a_z = -GMz/r^3 + 2GJ^*/(c^2r^3)[xv_y - yv_x]3z/r^2 \tag{6.21c}$$

This author, before being acquainted with Lense-Thirring's work, proposed a gravitational theory in a memoir [49] published in 1976. He proposed to write, with an asterisk, all the terms in Maxwell-Lorentz's electrodynamics. We will refer to this analog gravitational theory as *Asterisk Theory of Gravitation*. The mathematics of this theory represents a particular case of GRT. However, the physics and terminology of this asterisk theory belong to a completely new interpretation of gravitation. The Asterisk-Lorentz's gravitodynamics is given by the following equation, in which we recognize Newton's gravitostatic force, Faraday's induction force and Grassmann's gravitokinetic force:

$$\mathbf{F} = \sigma[\,-q^*\nabla\Phi^* - q^*\partial\mathbf{A}^*/\partial t + q^*\mathbf{v}x\mathbf{B}^*] \tag{6.22}$$

$$\mathbf{B}^* = \nabla x\mathbf{A}^* \tag{6.23}$$

B* is the gravitodynamic induction which equals the curl of the gravitodynamic vector potential **A***. Also we have:

$$\varepsilon_o^* = 1/(4\pi G) \tag{6.24}$$

gravitostatic permittivity of "vacuum", and

$$\mu_o^* = 4\pi G/c^2 \tag{6.25}$$

gravitodynamic permeability of "vacuum", and

$$\sigma = -1 \tag{6.26}$$

to only adequate attractive gravitostatic forces

Q* and q* should be replaced by M and m. Eq. (6.22) becomes:

.

$$\mathbf{F} = -\,GMm\,\mathbf{r}/r^3 + m\partial\mathbf{A}^*/\partial t - m\mathbf{v}x\mathbf{B}^*$$

or

$$\mathbf{F}/m = \text{grad}\,\phi + \partial\mathbf{A}^*/\partial t + \text{curl}(\mathbf{A}x\mathbf{v}) \tag{6.27}$$

After a long vectorial work, Lense-Thirring equations (6.21) can be written as follows:

$$\mathbf{F}_{LT} = -\,GMm\,\mathbf{r}/r^3 - m\mathbf{v}x\mathbf{B}^* \tag{6.28}$$

where **B*** is given by:

$$\mathbf{B}^* = [\mu_o^*/(4\pi r^3)][3(\mathbf{d}\cdot\mathbf{r})\,\mathbf{r}/r^2 - \mathbf{d}] \tag{6.29}$$

where **d** is given by:

$$\mathbf{d} = \tfrac{1}{2}\,\mathbf{J} \tag{6.30}$$

d should be called the *gravitodynamic dipole moment* of the rotating sphere. Equation (6.28) shows that the Lense-Thirring first degree of approximation to Einstein's nonlinear field equations, in geometrodynamics, cannot generate the Faraday's analog induction force $m\partial\mathbf{A}^*/\partial t$. In relativistic circles eq. (6.29) is interpreted as an induced dragging angular velocity of rotation of the local system with respect to a reference system fixed at infinity, which Newton used to call *absolute space*. The whole of GRT can be *reinterpreted* in Newtonian terms if a Newtonian Relativistic Gravitodynamics is used, and the "asterisk" terminology is adopted to describe in 3-Dimensional space all gravitational phenomena. Before we determine the solar IAM, it is intriguing that Einstein himself asked a significant question about $m\partial\mathbf{A}^*/\partial t$, or Faraday's gravitodynamic force. Einstein formulated the question in a paper which he published, in German, in 1912. The same year the paper was translated into Russian. This question is the title of a paper Einstein [50] published in a little known German Journal of Medicine. The question was this: *In Gravitation, is there an Analogous Effect to Electrodynamic Induction?* In other words, in gravitodynamic theories is there gravitodynamic forces proportional to the acceleration of the source bodies? The question was excellent but the so-called experts of that time were not so astute, and Einstein's excellent conception was lost in a little known medical journal! In May 1921, Einstein [51] offered a lecture at Princeton University which was published in 1922 under the title *The Meaning of Relativity*. In his book, the first equation of Einstein's group of three equations contained in his eq. (118), on the right hand side, has identical terms with the same mathematical signs of our eq. (6.27). Thus, Einstein answered his own question of 1912 in an affirmative way. Now we may say that in gravitation there is an effect identical to Faraday's induction phenomenon which is caused by the acceleration of the source material body.

6.2.2 INTRINSIC ANGULAR MOMENTUM OF THE SUN

We will study the motion of planet Mercury assuming its solar orbit is contained in the equatorial plane of the rotating sun. Using the Lense-Thirring result, given by eq.(6.28), we get the acceleration of planet Mercury:

$$\mathbf{a} = -GM\mathbf{r}/r^3 - \mathbf{v}\mathrm{x}\,[\mu_o^*/(4\pi r^3)][3(\mathbf{d}\cdot\mathbf{r})\,\mathbf{r}/r^2 - \mathbf{d}] \tag{6.31}$$

In the previous equation, omitting the term inversely proportional to the fifth power of r, we get:

$$\mathbf{a} = -GM\mathbf{r}/r^3 + \mathbf{v}\mathrm{x}\,[\mu_o^*/(4\pi r^3)]\mathbf{d}$$

or using, in this last equation, eq. (6.25) and eq. (6.30), we have:

$$\mathbf{a} = -GM\mathbf{r}/r^3 + [G/(2c^2r^3)]\mathbf{v}\mathrm{x}\mathbf{J} \tag{6.32}$$

Now we will introduce plane polar coordinates in the equatorial plane of the rotating sun. At this point, we must recall that Lense and Thirring had a sun rotating clockwise $\mathbf{J} = -\mathbf{k}J$. Now we are introducing a sun rotating counter-clockwise $\mathbf{J} = \mathbf{k}J$ to use it in the previous equation in order to determine the radial and transversal components of the acceleration. The prime notation means derivative with respect to time: $r' = dr/dt$.

$$a_r = r'' - r\theta'^2 = -Gm\,\mathbf{r}/r^3 - [G/(2c^2)]\,J\theta'/r^2 \tag{6.33}$$

$$a_\theta = r^{-1}d(r^2\theta')/dt = [G/(2c^2)]\,r'J/r^3 \tag{6.34}$$

or
$$d(r^2\theta') = [G/(2c^2)]Jr^{-2}dr \tag{6.35}$$

The presence of a transversal component of the acceleration, different from zero, anticipates that the orbital angular momentum is not a conserved magnitude. In other words, Kepler's second law is incorrect. Leverrier proved, in 1859, that Kepler's first law was also incorrect. For some enigmatic reason, books on GRT do not emphasize this remarkable violation of the conservation of the orbital angular momentum of planets. On the contrary, some respectful authors like Professor C. Møller, in his otherwise excellent book *The Theory of Relativity* [52], eliminated, in the edition of 1974, the following paragraph which was in the edition of the same book in 1952:

> "However, the left hand side of (18) cannot in general be interpreted as angular momentum, since the notion of a 'radius vector' occurring in the definition of the angular momentum has an unambiguous meaning only in a Euclidean space."

We are repeating what we said in chapter 4, because we think this subject matter is very important for the future of GRT. Eq. (18) of Professor Møller is

$$r^2\theta' = C[1 - 2GM/(c^2r)] \tag{6.36}$$

Now, the integration of our eq. (6.35) is:

$$r^2\theta' = h[1 - \tfrac{1}{2}(j/h)GM/(c^2r)] \qquad (6.37)$$

where j = J/M is the solar IAM per unit mass, and h is the orbital angular momentum per unit mass of the planet. $GM/(c^2r) \approx 2.5\text{x}10^{-8}$ in the case of planet Mercury. The presence of the factor (j/h) in eq. (6.37), and the absence of this factor in eq. (6.36) is due to the fact that eq. (6.37) corresponds to a real spherical sun, while eq. (6.36) corresponds to a point-like sun. The difference in the numeric factor in these last two equations is due to different definitions of constants. The violation of Kepler's second law is very minute. Hence, eq. (6.37) can be approximated to:

$$r^2\theta' = h \qquad (6.38)$$

Introducing this last equation in eq. (6.33), and making the traditional change of variable r=1/u, we get the differential equation of the orbit of the planet:

$$d^2u/dt^2 + u = GM/h^2 + [GM/(2c^2)]\,(j/h)u^2 \qquad (6.39)$$

If the rotating central celestial body is oblate, then Newton's potential is modified by the quadrupole moment potential:

$$\phi = -GM/r - GMR^2\Delta/(3r^3) \qquad (6.40)$$

where Δ is the oblateness, given by $\Delta = (1 - R_p/R)$, in terms of the polar radius R_p and the equatorial radius R of the sun. The oblateness of the rotating body modifies eq. (6.39) and the total potential:

$$d^2u/dt^2 + u = GM/h^2 + [GM/(2c^2)]\,(j/h)u^2 + GMR^2\Delta u^2/h^2 \qquad (6.41)$$

where h is the orbital angular momentum per unit mass of the revolving planet:

$$h^2 = GMa(1 - e^2) \qquad (6.42)$$

where a is the semi-major axis of the elliptical orbit, and e is the eccentricity of the same. The solution of eq.(6.41), obtained by the known method of successive approximations, provides the excess of perihelic rotation Ω of a planet:

$$\Omega = \alpha\pi GM/[Tc^2a(1 - e^2)] + 2\pi R^2\Delta/[Ta^2(1 - e^2)^2] \qquad (6.43)$$

or $$\Omega = \Omega_{LT} + \Omega_{DN} \tag{6.44}$$

$$\Omega_{LT} = \alpha\pi GM /[Tc^2a(1 - e^2)] \tag{6.45}$$
(Ω Lense-Thirring)

$$\Omega_{DN} = 2\pi R^2\Delta/[Ta^2(1 - e^2)^2] \tag{6.46}$$
(Ω Dicke-Newcomb)

$$\alpha = j/h \tag{6.47}$$

Eq. (6.43) is solid proof that GRT will not be proven wrong if the sun is oblate. The most important coefficient in eq. (6.43) is α, from which we can calculate the solar IAM per unit mass. If we introduce observational values for $\Delta=5x10^{-5}$, and for Ω = 42.56 ("/century), and all the standard numeric values of the orbital parameters, we get:

$$42.56 = 42.94\alpha + 4.24 \tag{6.48}$$

Using eq. (6.42) and eq. (6.47) in the last equation, we get the numeric value of the solar IAM per unit mass:

$$j = 2.42x10^{15}\ (m^2/s) \tag{6.49}$$

By 1970, Ostriker [53] had estimated an interval for the solar IAM per unit mass between 10^{13} and 10^{15} (m^2/s). If we calculate the solar j', using the optical observations of the solar spots to determine the period of revolution of the sun in its equatorial belt which is approximately 27 days, we get j'= $5.43x10^{11}$ (m^2/s). These considerations show that R.H. Dicke [54] was right when he claimed that the sun had an inner oblate core rotating much faster than the photosphere of the sun. However, Dicke was *not* right when he claimed that GRT would be a wrong theory if the sun is oblate. Here we have proved that a solution of Einstein's field equations of GRT, obtained only up to a first degree of approximation, is perfectly capable of accounting for the excess perihelic motion of the planets even in the presence of an oblate sun. We also remind the reader that this same solution is obtained in terms of a Newtonian gravitodynamics called "asterisk" theory represented by "Lorentz's" gravitodynamic force, containing Newton's gravitostatic force and "Grassmann's" gravitokinetic force.

The proposition made by Lense and Thirring of combining their first degree approximated solution with Einstein's second degree approximated solution, in the problem of the excess perihelic rotation of the planets, is totally inadmissible. From a mathematical point of view, this combination is illicit for nonlinear equations as Einstein's field equations are. On the other hand, and from a physical point of view, this proposition combines the solution of two completely different problems. In Einstein's case, the sun is an absurd geometrical point. In Lense-Thirring's case the sun is a *real* ball of matter. Lense-Thirring's solution corresponds to a problem of natural philosophy. Einstein's solution corresponds to a fictitious problem of

Table 6-III. Excess perihelic rotation of some planets. The numeric values of this table were calculated with the best value of j corresponding to planet Mercury.

Excess perihelic rotation of planets in ("/century)				
Planet	Lense-Thirring	This Work using (j/h)	Astro metric	Einstein (Sun is a point)
Mercury	42.94	42.46	42.56	42.86
Icarus	10.07	10.05	9.8	10.02
Venus	8.62	8.10	8.4	8.60
Earth	3.95	3.66	4.6	3.83
Mars	1.35	1.23	1.5	1.35

metamathematics. Someone in the near future should solve the Lense-Thirring problem up to a second or higher degree of approximation in order to improve the agreement between Einsteinian geometrodynamic calculations of the excess of perihelic rotation of the planets. Table 6-III summarizes our calculations. Δ was made equal to $5\text{x}10^{-5}$ according to reference [55].

Note that the oblateness of the sun leaves Einstein's mathematical solution with an irreparable error of 9.9% in the case of planet Mercury. Our present physico-mathematical solution proves that Einstein's GRT is an adequate theory even if the sun is oblate. Can we test GRT in the case of Jupiter's satellites? The answer is in the affirmative, but in a totally unexpected way.

6.2.3 EXCESS OF PERIJOVIAN ROTATION OF JUPITER'S SATELLITES

The first part of the answer to the previous question is rather disappointing but extremely interesting in another sense. The Galilean satellites: Io, Europa, Ganymede and Callisto, have eccentricities equal to zero, with the exception of Callisto, as Table 6-IV indicates. They practically move in circular orbits, and therefore, any point in their orbits are perijovian points, indistinguishable from each other. Thus, astrometric observations of excess perijovian rotations would be impossible to make. But this is not the main disappointing point when testing GRT in the motion of Jupiter's satellites. The influence of the quadrupole moment potential is gigantic in the perijovian rotation of the more distant satellites of Jupiter with significant eccentricities. Jupiter's oblateness Δ is equal to 0.066 which is 1320 times larger than the sun's oblateness. This astronomical fact shows that relativistic effects, in the perijovian rotation of Jupiter's satellites, are completely negligible in comparison with the large effect caused by the modification of Newton's gravitostatic potential due to the large quadrupole moment of Jupiter. This author has been unable to find the astrometric determination of the perijovian rotation of Jupiter's distant satellites. Jupiter's IAM per unit mass was calculated with optical astrometrical observations of planet Jupiter. Most likely, Jupiter has a faster rotating inner core made from metallic hydrogen. Astrometric data of the perijovian rotation of the most distant satellites of Jupiter would be very valuable to determine the inner IAM of Jupiter. Before we can use an astrometric determination of perijovian rotations, we must solve the problem of the perturbations on one jovian satellite caused by all the other jovian satellites plus all the planets. This previous work will allow us to determine the excess perijovian rotations of Jupiter's most distant satellites. This is a stupendous problem which must be solved by the next generation. In the case of the jovian satellites, Newcomb in 1897, Dicke in 1964, and Dicke and Goldenberg in 1967 would have been absolutely right about the tremendous influence caused by Jupiter's quadrupole moment on the perijovian rotation of its satellites.

At the beginning of this section, we have the second part of the initial question. This question shows another experimental confirmation of a *new prediction* which we can make with Einstein's GRT. Everyone knows that GRT offers new gravitational forces as well as any good Newtonian relativistic gravitodynamics

(NRG). These forces are *analogous, never identical*, to the Coriolis force of classical mechanics and to the centrifugal force of classical mechanics. GRT offers another gravitodynamic force absolutely unknown in classical mechanics. This is the *axial gravitodynamic force*, parallel to the axis of rotation, and proportional to the square of the angular velocity of a spherical rotating body. The axial force has an unusual characteristic. If a test material body is above the equatorial plane of the rotating massive sphere, the axial force pulls it down to the equatorial plane. On the other hand, if the test material body is below the equatorial plane, the axial force pulls it up to the equatorial plane. In 1918, Lense and Thirring [56] could have made two new predictions based on GRT. Unfortunately, they neglected, at the very beginning of their paper, terms proportional to the square of the angular velocity ω of the rotating solid sphere. They wrote: "*. . . the terms of the centrifugal force, proportional to* ω^2*, are eliminated, and only the Coriolis terms appear.*" This approximation is acceptable for the planets but not for Jupiter's satellites. Today we can make two new gravitational predictions based on GRT or based on our NRG:

P1 The orbit of celestial bodies revolving fast, and close to the central rotating body, will be on the equatorial plane of the rotating central body due to the action of the axial gravitodynamic force. In other words, the angle of the inclination of the revolving test body will be zero with respect to the equatorial plane of the rotating central body.

A second prediction is based on the *wrongly* called centrifugal force of GRT. The essence of this force is authentically gravitational, and its direction is radial, perpendicular to the axis of rotation. We will call this force *pseudo-centrifugal gravitodynamic force*. The second new gravitational prediction of GRT or NRG, is the following one:

P2 The presence of the pseudo-centrifugal gravitodynamic force on a test material body, revolving initially in an elliptical orbit, will deform the orbit from elliptical into circular over a span of time. In other words, the orbit's eccentricity of the revolving test body will be zero.

The unexpected answer to our initial question is that both predictions, P1 and P2, have been verified by the Galilean satellites of planet Jupiter. This was verified before humans began to think. Table 6-IV, clearly shows the astrometric verification of the above predictions. The data was taken from the book *Jupiter* (1976), edited by T. Gehrels [57].

This Table clearly shows an extraordinary verification of GRT by astrometric observations of the Galilean satellites of Jupiter. The perijovian rotation of Jupiter's satellites, calculated by Lense and Thirring, are completely overwhelmed by the enormous perijovian rotation caused by the quadrupole moment of the oblateness of Jupiter. If Einstein [58, p. 103] had known Table 6-III, most likely he would not have written the following arrogant metaphysical statement:

> "Although all of these effects are inaccessible to experiment, because κ is so small, nevertheless they certainly exist according to the general theory of relativity."

Table 6-IV. Astronomical data of Galilean Satellites of Jupiter.

Satellites	Semimajor Axis (10^3 km)	Inclination Angle (°)	Eccentricity
Io	421.6	0.0	0.000
Europa	670.9	0.5	0.000
Ganymede	1070	0.2	0.001
Callisto	1880	0.2	0.01

These insignificant effects only had a noetic existence. A Natural Philosopher begins to accept a physical theory only if some of the theoretical (metaphysical) conclusions are verified in the external world of things, in reality.

6.3 Inertial mass, gravitational mass and the equivalence principle

Einstein, in his book *The Meaning of Relativity* [59, p. 56], raises again false testimony against Newton's classical mechanics. Einstein begins criticizing classical mechanics for showing a deficiency when dealing with reference systems in accelerated motion. Einstein forgot that classical mechanics is only valid with respect to absolute space, or with respect to reference systems moving with constant velocity with respect to absolute space. Einstein also forgot that classical mechanics can perfectly well handle dynamical problems in accelerated reference systems with respect to absolute space. This is done by the introduction of inertial forces at the cost of invalidating the classical principle of action and reaction. What Einstein wanted, indeed, was to extend the principle of relativity to reference spaces in accelerated motion. Here we see the genesis of Einstein's principle of equivalence referring to reference systems K and K' in relative accelerated motion. Then Einstein changes the subject to the ratio of gravitational mass divided by inertial mass, emphasizing the fact that these two *masses* are defined fundamentally (essentially) in two different ways. Then Einstein writes:

> "The equality of these two masses so differently defined, is a fact which is confirmed by experiments of very high accuracy (experiments of Eötvös), and classical mechanics offers no explanation for this equality."

We have written this section 6.3 with the sole purpose of proving that Einstein is wrong in accusing Newton's classical mechanics of offering no explanation for this equality. In what follows, we will prove that classical mechanics is perfectly capable of proving the equality of gravitational mass to inertial mass if a difference ever existed. We will also prove that this difference should never have existed. More importantly, we will demonstrate that the gravitational mass is *identical* to the inertial mass. Why is there this confusion even in Einstein's mind? To answer this question, let us quote Mach [60, p. 172], out of context- "*The embarrassment of the neophyte, which also overcame the great investigators in the face of the great mass of new material presented, alone could have led them to conceive the same fact as two different facts and to formulate it twice.*" - Who, when, where and why was the concept of *gravitational mass* introduced into physics? It seems that scientific history has no record of the scientist who is responsible for introducing this concept in physics. G. B. Brown [61] says:

> "The term *gravitational* or *attractive mass*, which can be symbolized by m_g, appears to be surrounded with serious confusion. The earliest mention of this term seems to be in Thompson and Taits *Treatise on Natural Philosophy*, new edition, where it can be found in the index but not in the text."

The new edition of Thompson and Tait's Treatise was published in 1879. Nevertheless, Sommerfeld [62, p. 312] writes:

"The prerequisite for this is the *identical character of gravitational and inertial mass*, which was expressed in Vol. I, by the equation $m_g = m_i$.

"Only if this is satisfied, is the 'weight' $m_{grav}\, g$ equal to the 'inertial reaction' $m_{inert}\, g$ and only then is the period of oscillation the same for all pendulums of equal length.
In detail the period is $\tau = 2\pi\, [(m_{inert} L)/(m_{grav} g)]^{1/2}$

"Already Newton saw that a profound physical problem was hidden herein and Bessel pursued the problem by making extremely careful measurements on pendulums of different materials[1]."

In footnote l, Sommerfeld writes:

"F. W. Bessel, 'Experiments on the Force, with which the Earth Attracts Different Kinds of Bodies,' Abhandlgen d. Preuss. Akad. 1830; 'Studies on the Length of the Second Pendulum,' loc. cit. 1826 - reprinted in Ostwald's Klassiker Nr. 7."

It seems that as early as 1826, Bessel handled the concepts of inertial and gravitational mass. It also seems that Sommerfeld suggests that Newton was aware of these two types of masses in relation to the period formula of pendulums. At any rate, today many authors establish that Newton performed his experiments with pendulous bodies in order to determine the ratio m_g/m_i to an accuracy of 1 part in 10^3. In the author's opinion, Newton's purpose in running his experiments with pendulous bodies was to answer two questions:

Ql. What is the quantitative relationship between the weight of a body and its mass?

Q2. Does the acceleration of falling bodies, having equal weights in the same locality, depend on their chemical composition?

Newton must have worded the first question as follows: Is the weight of a body proportional to its mass? Newton faced the problem of measuring the mass of bodies for no other reason than he was the first one to introduce the physical quantity "m" in an equation represented by his second "axiom" of motion:

$$d(m\mathbf{v})/dt = \sum \mathbf{F}_k = \mathbf{F} \qquad (6.50)$$

with the additional assumption that m is constant. Let us agree in calling m the "inertial mass" and let us denote it by m_i. Thus Newton's second "axiom" of motion becomes

$$m_i \, \mathbf{a} = \sum \mathbf{F}_k = \mathbf{F} \qquad (6.51)$$

F. Cajori in the Appendix to Newton's *Principia* [63, p. 639] writes:

> "In the use of the concept of mass, as distinguished from weight, Newton has forerunners who perceived the difference between mass and weight more or less clearly."

It fell to Newton to clarify the difference between the concepts of the weight of a body and its mass. This clarification was typically a logico-empirical task performed by Newton.

In order to talk about inertial mass and gravitational mass, one should go beyond empiricism and enter endless discussions on the nature or essence of matter and its philosophical accidents or attributes, one of which is inertia. This attribute, the inertia of a body, once it is quantified is called *mass.* Knowing the aversion Newton felt towards framing hypotheses, it is very doubtful he distinguished any such two "*essentially*" different concepts of mass. On the other hand, Newton very seldom mentioned the word mass in his writings, referring to this property of matter as *body* or *quantity of matter* or simply *matter.* Newton's Definition I and its commentary [64, p. 1] throws light on these questions of mass and weight. In the first part of Definition I, Newton writes that mass is the product of the density of a body times its volume. This definition was rightly and strongly criticized by Mach. Then Newton writes about "quantity of matter":

> "It is this quantity that I mean hereafter everywhere under the name of body or mass . . . for it is proportional to the weight, as I have found by experiments on pendulums, very accurately made, which shall be shown hereafter."

In Newton's Definition I, we find that m =V·d where V is the volume of the body and d is its density. This has been widely rejected. To Mach it was "unfortunate;" to Einstein it was "illusory;" to Sommerfeld it was "a mock definition;" to G. B. Brown it was "relative density" and, therefore, Definition I is "not a circular one." After E. Mach's criticism of Newton's definition of mass, it seems that everyone concentrated on m = V·d in order to attack Newton's concept of mass perhaps with the exception of Sommerfeld [65, p.19] who wrote in a footnote the following concepts:

> "Incidentally we would like to direct the reader's attention to an interesting sentence occurring in Newton's *Mechanics*. At the beginning of this work, under Definition I, Newton says: 'Through very carefully performed experiments with pendula I have verified that mass and weight are proportional'."

Let us follow Sommerfeld's suggestion and focus our attention on that "interesting sentence under Newton's Definition I."

6.3.1 THE WEIGHT OF A BODY IS PROPORTIONAL TO ITS INERTIAL MASS

When Newton, writing about the quantity of matter (mass) of a body in his Definition I, said: "which shall be shown hereafter," he meant that all we have to do is study his Proposition XXIV. Theorem XIX. Book II of his *Principia* [66, p. 303] which says:

> "The quantities of matter in pendulous bodies, whose centres of oscillation are equally distant from the centre of suspension, are in the ratio compounded of the ratio of the weights and the squared ratio of the times of the oscillations in a vacuum."

Following Newton, in his reasoning, is not an easy task. His mathematical techniques are proportions and euclidean geometry. To make things even more difficult, proportions were generally expressed verbally. These circumstances make the study of the *Principia* a little bit cumbersome to our modern taste and practice when using our familiar algebraic notations. Any physics teacher knows how relaively easy it is to prove, using

Newton's principles and the concept of the weight of a body, the *theoretical conclusion* on the period T of a pendulum which oscillates with a small angular amplitude:

$$T = 2\pi (m_i L/w)^{1/2} \tag{6.52}$$

where m_i is the inertial mass necessarily *(logically)* coming from eq. (6.51); L is the length of the pendulum and w is the weight of the pendulum bob which is taken as a *datum* and determined *experimentally* with a scale. When eq. (6.52) is applied to two different pendulums made up of the *same substance,* we can write

$$m_{i1}/m_{i2} = (w_i/w_2)\,(T_1^{\,2}/T_2^{\,2})\,(L_2/L_1) \tag{6.53}$$

If in eq. (6.53) we make $L_1 = L_2$, we finish the proof of Newton's Proposition XXIV. Theorem XIX, Book II. Let us now see how Newton proceeded to prove his Theorem XIX. Book II:

> "For the velocity which a given force can generate in a given matter in a given time is directly as the force and the time, and inversely as the matter. The greater the force and the time is, or the less the matter, the greater the velocity generated. This is manifest from the second Law of Motion. Now if pendulums are of the same length, the motive forces in places equally distant from the perpendicular are as the weights; and therefore if two bodies by oscillating describe equal arcs, and those arcs are divided into equal parts; since the times in which the bodies describe each of the correspondent parts of the arcs are as the times of the whole oscillations, the velocities in the correspondent parts of the oscillations will be to each other directly as the motive force and the whole times of the oscillations, and inversely as the quantities of matter: and therefore the quantities of matter are directly as the forces and the times of the oscillations, and inversely as the velocities. But the velocities are inversely as the times, and therefore the times are directly and the velocities inversely as the square of the times; and therefore the quantities of matter are as the motive forces and the squares of the times, that is, as the weights and the squares of the times. Q.E.D."

After this proof, Newton went on to establish seven corollaries. Corollary V is the verbal expression of eq.(6.53) from which Corollaries I through IV are necessary consequences. Corollary VII is particularly interesting:

> "Cor.VII. And hence appears a method both of comparing bodies one with another, as to the quantity of matter in each; and of comparing the weights of the same body in different places, to know the variation of its gravity. And by experiments made with the greatest accuracy, I have always found the quantity of matter in bodies to be proportional to their weights."

Thus, in eq. (6.53), making $T_1 = T_2$ and $L_1 = L_2$, we get

$$m_{i1}/m_{i2} = w_1/w_2,$$

or

$$m_{i1}/w_1 = m_{i2}/w_2, \text{ or in general} \tag{6.54}$$

$$m_i = kw \tag{6.55}$$

where k is a proportionality constant. One thing should be clear by now: *the concept of mass which appears in all the previous equations corresponds* to "inertial mass," and this is so because *logically* the only way to introduce the term *m* in this discussion is through Newton's second "axiom" of motion where by *convention* the term *m* is called "inertial mass." From eq. (6.52) we obtain

$$w = (4\pi^2 L/T^2)m_i \tag{6.56}$$

Eq.(6.56) emphasizes the *formal fact* that it is a *theoretical conclusion* or theoretical law obtained with the help of Newton's Axioms or «laws» of motion. Eq.(6.56), as a theoretical conclusion, is not enough to establish that «mass is proportional to weight.» Eq.(6.56) has to be tested, checked and verified experimentally in the same locality to find out if $(4\pi^2L/T^2)$ is constant or not. We do this by using bodies of the same substance but of different weights. Only if $(4\pi^2 L/T^2)$ is constant in the same locality, can we say that «mass is directly proportional to weight.» Was eq.(6.56) known to Newton? In [67, p. 408) Newton says:

> "... for a pendulum oscillating seconds in the latitude of *Paris* will be 3 *Paris* feet, and 8 lines ½ in length, as *Mr. Huygens* has observed. And the space which a heavy body describes by falling in one second of time is to half the length of this pendulum as the square of the ratio of the circumference of a circle to its diameter (as Mr. Huygens has also shown), "

In our modern notation the above quotation is represented by:

$$g = \pi^2 L/(T/2)^2 = 4\pi^2 L/T^2 \qquad (6.57)$$

where g is the acceleration of falling bodies. Combining eqs. (6.56) and (6.57) we obtain

$$w = m_i g \qquad (6.58)$$

Eq. (6.58) summarizes theoretical research done by Newton, and the experimental discovery (guided by his theory of classical mechanics) that ($4\pi^2$ L/T^2) is constant in the same locality, and finally, the identification of this constant with the acceleration of falling bodies in that same locality. Thus, eq. (6.58) is a *natural law* established theoretically and tested experimentally. Eq. (6.58) is not a definition unless we are willing to act arbitrarily in order to introduce into physics an ontological ghost. The term m_i in eq. (6.58) is nothing but the "inertial mass" which should simply be called "mass."

It is interesting to note that E. Mach [68, p. 265] makes Newton responsible for *framing the hypothesis* w = mg from which w/w' = m/m' is deduced. If we translate Newton's proof of his Prop. XXIV. Book II to our modern notation, we would see with astonishing clarity Newton's way of reasoning and arriving at w/w' = m/m'. Mach should have known that a hypothesis is not deduced and signed at the end of its proof with Q.E.D., as Newton did, knowing quite well he was proving a *theoretical proposition* and not framing a senseless hypothesis. This is not the first time Mach is caught in wishful interpretations. M. Bunge [69] says:

> "Mach himself acknowledged his ignorance of the history of philosophy. That he was a poor historian of mechanics is shown by the fact that he ignored the whole Middle Ages, and - as shown by Trusdell[3] - that he drew from unreliable sources."

Mach, in [70, p. 240], comments on Newton's verification that "under special circumstances the mass of a body can be determined by its weight," and makes no reference to Newton's Prop. XXIV, but offers very "obscure" arguments to *justify* the equation w = mg. Mach's critique of Newton's mechanics is completely unfounded.

6.3.2 ACCELERATION OF FALLING BODIES OF DIFFERENT CHEMICAL SUBSTANCES

Pendulums in Newton's epoch provided a considerably accurate method of measuring the acceleration of falling bodies. Newton himself performed a series of experiments with pendulous bodies made up of different chemical substances in order to investigate experimentally if $g_A = g_B$, for the case of two bodies of the same weight oscillating with strings of the same length.

$$g_A = 4\pi^2 L/T_A^{\,2} = w_A/m_{iA}, \tag{6.59a}$$

$$g_B = 4\pi^2 L/T_B^{\,2} = w_B/m_{iB} \tag{6.59b}$$

where $w_A = w_B$. Here A and B indicate the different chemical substances of the pendulous bodies. In a final analysis, the experiment consisted of finding out whether $T_A = T_B$ or not. If we assume complete ignorance of Newton's gravitational law, an experiment of this type is completely justified. Later we will analyze this experiment in the light of Newton's gravitational law. Here we shall emphasize the fact that Newton did not perform these kinds of experiments to determine the numerical value of the ratio m_{grav}/m_{inert}, though many authors claim that Newton determined this ratio to an accuracy of 1 part in 10^3. Let us read what Newton wrote under Proposition VI. Theorem VI. Book III [71, p. 411].

> ". . . it has been, now for a long time, observed by others that all sorts of heavy bodies (allowance being made for the inequality of retardation which they suffer from a small power of resistance in the air) descend to the earth *from equal heights* in equal time; and that equality of times we may distinguish to a great accuracy, by the help of pendulums. I tried experiments with gold, silver, lead, glass, sand, common salt, wood, water, and wheat. I provided two wooden boxes, round

and equal: I filled the one with wood, and suspended an equal weight of gold (as exactly as I could) in the centre of oscillation of the other. The boxes, hanging by equal threads of 11 feet, made a couple of pendulums perfectly equal in weight and figure, and equally receiving the resistance of the air. And, placing the one by the other, I observed them to play together forwards and backwards, for a long time, with equal vibrations. And therefore the quantity of matter in the gold (by Cor.I and VI, Prop. XXIV. Book II) was to the quantity of matter in the wood as the action of the motive force (or *vis motrix*) upon all the gold to the action of the one to the weight of the other: and the like happened in the other bodies. By these experiments, in bodies of the same weight, I could manifestly have discovered a difference of matter less than the thousandth part of the whole, had any such been."

If we use eq. (6.53), replacing the subindices 1 and 2 by A and B, respectively, and then we make $L_A = L_B$ and $w_A = w_B$ we get:

$$m_{iA}/m_{iB} = T_A^2/T_B^2$$

Newton measured the discrepancy in equality between T_A and T_B, and logically, this result or difference, according to the above proportion, corresponds to m_{iA}/m_{iB}. If we write

$$m_{iA}/m_{iB} = T_A^2/T_B^2 = 1 \pm 10^{-3}$$

$$m_{iA} = m_{iB} \pm 10^{-3}\, m_{iB}$$

$$m_{iA} - m_{iB} = \pm 10^{-3}\, m_{iB}$$

Newton's "difference of matter" between bodies A and B is $(m_{iA} - m_{iB})$, and Newton's "less than the thousandth part of the whole" is $(\pm 10^{-3}\, m_{iB})$. To say that Newton determined the equality between the gravitational mass and the inertial mass with a precision of one part in one thousand, is to act on bad faith or on plain ignorance. This is one more deplorable *misinterpretation* or false testimony raised against Newton which has unreasonably propagated for more than 300 hundred years. In the equation:

$$m_{iA}/m_{iB} = 1\pm10^{-3}$$

we have only *inertial masses*. There was never any need to introduce the unreasonable concept of gravitational mass. The logic is impeccable in deducing eq. (6.58) in which only the inertial mass appears. To say that the weight of a body is equal to its mass times the acceleration of gravity is an *assumption* made by Newton, as so clearly was written by Mach, is to publicly declare the unforgivable ignorance of the *Principia.* This ignorance was revealed by the so-called "greater critic of Newton." But to say, what nearly everyone says, that eq. (6.58) defines the gravitational mass, is an act of irrationality in physics which must be eradicated as soon as possible. This *must* be done before the next generation takes over.

In conclusion, we may say that Newton "found" or "verified," according to Sommerfeld, "the quantity of matter in bodies to be proportional to their weights." The verification Newton performed with pendulous bodies was the verification of his own theoretical conclusion shown in Proposition XXIV. Book II. In this proposition, Newton explicitly says: "*This is manifest from the second Law of Motion,*" and this implies that he introduced in his theoretical proof the universally accepted concept of "inertial mass." Therefore, Newton's experimental verification can be expressed as: - The inertial mass of a body is proportional to its weight, or as we say today, the weight of the body is proportional to its mass. -

Being that this statement is a theoretical conclusion, verified experimentally, we should refer to it as a *natural law* and by no means as a definition. Therefore, to say that "the gravitational mass" is defined as the ratio of the weight of a body to the acceleration of gravity as M. Born proposed [72, p. 44], among many others, is to give no credit to the founder of Natural, Experimental or Scientific Philosophy.

At this point, we will draw no conclusions in respect to the equality of the acceleration of falling bodies of different substances (chemical composition), for, we first need another critical analysis of the origin of Newton's gravitational law. Finally, in a rigorous interpretation, the alleged accuracy of 1 part in 10^3 of the whole in Newton's experiments on pendulous bodies has nothing to do, whatsoever, with the accuracy in the experimental determination of the ratio $m_{grav} = W/g$. We can *interpret* Newton's experiments, with pendulous bodies, as leading to the determination of m_{grav}/m_{inert}. But this interpretation, which Newton never had in mind, takes us into an obscure metaphysical labyrinth created by Mach and distorts Newton's objectives. The most important conclusion of this section is the following:

The weight of a body is directly proportional to the inertial mass: $W = gm_i$

Finally, the arguments presented in this section do not pretend to be a historical account expressed in modern notation. This is meant to be a rigorous interpretation of Newton's experiments with pendulous bodies from a logico-empirical point of view. This is done with the sincere hope of reducing to a minimum the amount of confusion, misinterpretations and nonsense found in relation to Newton's Natural Philosophy.

6.3.3 NEWTON'S DYNAMICAL METHODOLOGY

In this section, we present the theoretico-empirical derivation of Newton's gravitational law from Kepler's astronomical laws and Newton's "axioms" of classical mechanics. If we behave rationally, as Newton did, we will conclude that the gravitational law only contains inertial masses. In high school, the student is left with the impression that Newton, when "struck by a falling apple," was illuminated or had a vision in which he saw $F=GM\,m/r^2$. The pupil was given the understanding that there was no process of reasoning at all. When that graduate finishes his or her college training, the high school impact has only been reinforced.

There are only a few authors who present, in their textbooks, a logico-empirical derivation of Newton's gravitational law. This deprives their students of the most spectacular application of Newton's axioms of classical mechanics to planetary motion using Kepler's laws. Our students miss, perhaps, the first opportunity to see a "dynamic-reasoning" treatment when solving a mechanical problem using an established theory and empirical facts.

M. Born [73, p. 60] presents a derivation of Newton's gravitational law for circular orbits using elementary mathematical techniques quite adaptable for high school level. Unfortunately, M. Born, in the author's opinion, did not point out the *logical fact* that the physical quantity m appearing in Newton's $\mathbf{F}=m\mathbf{a}$ is necessarily the *same m* in his deduced Newton's planetary force-law. The author [74, p. 45], using Born's derivation but slightly modified, presents this important point which is missing in textbooks of classical mechanics.

J.C. Maxwell [75, Chap. VIII] derives Newton's gravitational law from elliptical orbits. Maxwell does not distinguish between gravitational and inertial mass, and this is consistent with G. B. Brown's suspicion that the earliest mention of gravitational mass, in the English language, seems to be in Thompson-Tait's *Treatise on Natural Philosophy* in

1879. This was two years after Maxwell wrote the preface to his *Matter and Motion* [76, p. 105], where we can read:

> "The most instructive example of the method of dynamical reasoning is that by which Newton determined the law of the force with which the heavenly bodies act on each other.
>
> "The process of dynamical reasoning consists in deducing from the successive configurations of the heavenly bodies, as observed by astronomers, their velocities and their accelerations, and in this way determining the direction and the relative magnitude of the force which acts on them.
>
> "Kepler had already prepared the way for Newton's investigation by deducing from a careful study of the observations of Tycho Brahe the three laws of planetary motion which bear his name."

A most simple and short derivation of Newton's inverse law is presented by P. G. Bergmann in *The Riddle of Gravitation* [77, p. 216] for the case of planetary circular orbits. Bergmann says:

> "This appendix shows that, from this fact alone (Kepler's third law of planetary motion), it follows that the acceleration of a planet toward the sun must be inversely proportional to the square of its distance from the sun."

When Bergmann says "must be," obviously he means "necessarily is" or "logically is." The only *fact* he uses in his derivation is an *empirical fact* represented by Kepler's third law. This fact is introduced in Newton's theory of classical mechanics and is worked out *logically.* The result is a logico-empirical one in which the term m is *logically* the same m that appears in $F = ma$.

Once the logico-empirical result $F = GMm/r^2$ is obtained for the planetary problem, we must ask, what is the cause of gravitational attraction, or ask, why heavy bodies fall down upon our planet. If we answer this last question by saying that heavy bodies fall down upon our planet because of gravitation, we have actually said that heavy bodies fall down because they are heavy. The word gravitation derives from the latin word *gravitas* which means heavy or *weight*.

To speculate on the cause of gravitational attraction is to try to look for a causal explanation to account for the observed *behavior* (motion) of bodies. Some people think that if we discover the *essence* or the *very nature* of matter, we should be able to *explain,* among other observed empirical facts, the riddle of gravitation. Max Jammer put it this way: "Does matter do what it does because it is what it is?"

Now if we ask what matter is, we might be given the answer that matter *is* that it *is*. This answer might be all right for traditional philosophers but not for natural philosophers. Newton was a natural philosopher, and if we had asked him what the cause of gravitational attraction was, he would have answered publically in this way:

> ". . . the cause of gravity is what I do not pretend to know, and therefore would take more time to consider of it."

This answer is taken from a letter Newton wrote to Bentley [78, p. 633]. But Newton, as we all are, was chained by the tyranny of words. In Newton, it is more understandable, because he was trying to divorce his Natural or Experimental Philosophy from Mother or Traditional Philosophy. But in doing so, he carried with his Natural Philosophy the vocabulary of the Mother Philosophy; otherwise, he would have been unable to write a single word about his discoveries in experimental philosophy. In 1958, M. G. Evans [79] commented that Newton never attempted any further explanation of gravity. Actually, Newton is very specific in his *Principia.* He left for posterity the probe to discover the nature of gravity. Newton, in his attempt to separate Natural Philosophy from traditional philosophy behaved as a positivist. M. G. Evans commented in this respect: "*The difficulty which Newton encountered in his search for the ultimate cause of gravity, shaped up his classic remark on method - hypotheses non fingo.*" Then Evans quoted Newton:

> "But hitherto I have not been able to discover the cause of those properties of gravity from the phenomena; for whatever is not deduced from the phenomena is to be called an hypothesis; and hypotheses, whether metaphysical or physical, whether of occult qualities or mechanical have no place in experimental philosophy[21]."

Newton, indeed, was a positivist in these matters of "experimental philosophy." If we accept that Natural Philosophy is just an attempt to apprehend the seemingly unreachable essence of Nature, then it appears there is nothing wrong in framing metaphysical or mathematical hypotheses to create a theoretical structure in physics, that is, as long as we are constantly aware of the probable *feigned* character of our assumptions. The ignorance of this Berkelian lesson has produced painful controversies in the history of physics and an unjust definition of philosophy which is shared by too many physicists today. Take for example the one offered by a classmate of W. Heisenberg [80, p. 30]:

> "Philosophy is the systemic misuse of nomenclature specially invented for the purpose."

It seems, then, that it is important to be aware of these probable metaphysical or mathematical ghosts, otherwise, we might waste a lifetime trying to detect experimentally an entity which never existed in reality but only in the noetic realm of our minds. If we become extremists in defending and practicing Empiricism, we shall find ourselves unable to talk or write about our work in quantum physics. In the 20th century, all of us were trained in "western quantum mysticism," with electronic detectors. The author Capra is right in his book, *The Tao of Physics* [81], on the *method of modern physics*. This quantum "tao" is a *silent path* full of mathematical landmarks having no physical reality. No wonder Einstein wrote [81, p. 301]:

> "We now realize, with special clarity, how much in error are those theorists who believe that theory comes inductively from experience. Even the great Newton could not free himself from this error (*'Hypotheses non fingo'**)."

Since we expect that physics is not only metaphysics, nor is it only mathematics, nor is it only the gathering of experimental data, we should attempt to be moderate metaphysicists, mathematicians, empiricist and that *special* something "else." Perhaps, then, we shall become humble physicists who trans-rationally know we are only approaching asymptotically the *essence of the universe.*

6.3.4 ACCELERATION OF A PLANET

Let us follow Newton's objective of experimental philosophy:

> "...for the whole burden of philosophy seems to consist in this - from the phenomena of motions to investigate the forces of nature, and then from these forces to demonstrate the other phenomena."

The phenomenon of planetary motion, as observed and recorded quantitatively (measured angles) by Tycho Brahe and Kepler, was used by the latter to *induce* his three laws. Eq.(6.60) represents Kepler's first law:

$$r^{-1} = p^{-1}\,[1 + e\cos(\phi+\omega)] \qquad (6.60)$$

where r and ϕ are plane polar coordinataes; p is the *latus rectum* of the elliptic orbit; e is the orbit eccentricity, and ω is the perihelion longitude in its orbit. The sun coincides with one foci of the ellipse. Eq.(6.61) represents Kepler's second law:

$$\tfrac{1}{2}\, r^2\, d\phi/dt = \tfrac{1}{2}\, h = \text{constant} \qquad (6.61)$$

where h is the orbital angular momentum of the planet per unit mass. The variable t is the coordinate time. Eq.(6.62) represents Kepler's third law:

$$a^3/T^2 = K = \text{constant} \qquad (6.62)$$

where T is the planet's period of revolution around the sun, and a is the semi-major axis of the elliptical orbit.

Eqs. (6.60) through (6.62) constitute a *clear* representation of the obscure, mysterious or disorganized set of measured distances and angles. These equations represent *empirical formulas*, i.e., a mathematical synthesis of quantitative astronomical observations. This is an ordered, neat and short representation of observed astronomical facts, and most importantly a representation which can be handled *logically*. Kinematically, the radial and transversal components of the acceleration in plane polar coordinates are given by:

$$a_r = d^2r/dt^2 - r(d\phi/dt)^2 \qquad (6.63)$$

$$a_\phi = r^{-1}[d(r^2 d\phi/dt)/dt] \qquad (6.64)$$

Now we are ready to determine the acceleration of a planet using the logic and techniques of mathematics. We will behave as mathematicians. Differentiating eq. (6.61), with respect to time t, we get:

$$[d(r^2 d\phi/dt)/dt] = 0 \qquad (6.65)$$

Comparing eqs. (6.64) and (6.65), we conclude that from Kepler's second law, we deduced that the planet "shows" no transversal acceleration. We can say that it is not acted on by a transversal force; but this way of expressing ourselves will suggest that we know about, or believe in the existence of an agent which can act on a planet in some particular way. If this happens, we call that agent the *cause of.* We, then, keep on sinking into a bottomless pit, because now we must explain the "mechanism" through which this strange agent acts on the planet. Obviously, in saying "the planet *shows* no transversal acceleration," we have not escaped questioners, because "to show" implies "to see," and we do not see transversal acceleration. We do not see acceleration; we do not observe in nature abstract definitions such as $d^2\mathbf{r}/dt^2$. The only thing we observe in the night is a little spot of light which changes its position after many nights of observations. We continue to watch our little spot with respect to another background of other little groups of bright spots. The elliptical trajectory of a planet is not an empirical fact. We do not observe ellipses on a clear night. The ellipse is in our minds, and all we can hope is that in "reality" the path of a planet is elliptical. Let us now determine the radial acceleration of a planet. Differentiating eq. (6.60) twice with respect to t, and using eq. (6.61) we obtain:

$$a_r = -(h^2/p)/r^2 \tag{6.66}$$

The *minus sign* in eq.(6.66) is *interpreted* as indicating that $\mathbf{a}_r = -\mathbf{u}_r(h^2/p)/r^2$ points in the opposite direction of the unit vector $\mathbf{u}_r$ which points away from the origin of our polar coordinate system. From the planet's position, the vector $\mathbf{a}_r$ points to the focus of the ellipse from where we say the sun is.

Let us now use Newton's theory of classical mechanics to determine the force F_p on the planet. This force, according to Newton's second Axiom of Motion, is $\mathbf{F}_p = m_i\,\mathbf{a}$, where by convention we have agreed to qualify the mass m as m_i, i.e. as *inertial mass*. Thus, according to Newton's "dynamical methodology," from "the phenomenon of motion" of a planet (Kepler's laws) we have obtained $\mathbf{a}_r = -\mathbf{u}_r(h^2/p)/r^2$ and from Newton's second Axiom of Motion we can write

$$\mathbf{F}_p = -\mathbf{u}_r(m_i h^2/p)/r^2 \tag{6.67}$$

In eq. (6.67) m_i is *logically* the inertial mass of the planet. Arbitrarily, of course, by means of another *convention,* we can agree to change *inertial mass* for *gravitational mass*. In other words, we agree on the substitution m_i for m_g, where m_g is the so-called *gravitational mass.* Logically, this substitution is unacceptable unless it is proven with the use of $m_i = f(m_g) = m_g$ which in turn has to be proven previously. But how can we prove that $m_i = m_g$? Certainly we cannot use the arguments of pendulum experiments as we saw in section 1. Nor can we use Newton's gravitational law because we are now trying to derive it. We are left, then, with ontological arguments about the being of matter and its attributes, such as its *mass.*

Thus, it seems that if we insist on distinguishing two kinds of *"masses"* which differ in their essences, we are practicing an act of arbitrariness in a field which is *beyond* physics. Once we are back into physics with two *names*, inertial and gravitational mass, we should know we are using two *different names* to refer to *one* and the *same thing*. Hence, the principle of Identity is very useful when investigating the careless and unreasonable work done by generations of physicists.

6.3.5 FORCE "ON" A PLANET

The semantics of our primitive written language forces us to accept undesirable metaphysical responsibilities. Eq. (6.67) gives us the force "on" a planet which moves in an elliptical trajectory. Now let us consider Newton's third Axiom of Motion. Newton himself, in introducing the names "action" and "reaction" in the statement of his third Axiom of Motion, opened the door to a vast field of ignorance, speculation, discussion, controversy and confusion. According to this Axiom, we must have an *even* number of forces in the universe. This fact allows us to have one half of the bodies in the universe as *passive* bodies which *suffer* the actions of the other half which are called the *active* bodies. Now according to Newton's third Axiom of Motion, the *passive* bodies *mysteriously* become *active* bodies and the active bodies, in turn, have now *mysteriously* become *passive* bodies. Mendel Sach is right [82] in saying:

> "On the other hand, if man has thus far been unable to formulate conceptually simple, though subtle, natural laws in equally simple language, this is not God's fault! It appears to me that this inability is rather due to the relatively primitive stage of man's intelligence."

Now, let us return to mathematico-physics. Integrating eq. (6.61) for a whole period T of the planet, we get: (the semi-major axis of the elliptic trajectory is "a")

$$h^2=4\pi^2 a(1-e^2)(a^3/T^2) \qquad (6.68)$$

We know that:

$p=a(1-e^2)$ and $^{A3}/T^2=K$

Thus eq. (6.68) becomes:

$$h^2/p = 4\pi^2K = 4\pi^2a^3/T^2 = N_o = \text{constant} \tag{6.69}$$

Introducing the last equation in eq. (6.67), the *magnitude* of the force $\mathbf{F}_p$ on the planet becomes:

$$F_p = m_ia_p == m_i(N_o/r^2), \text{ from which } a_p = N_o/r^2 \tag{6.70}$$

We must insist that the presence of inertial mass m_i of the planet in eq. (6.70) is a *logical* consequence of using Newton's second Axiom of Motion. In order to finally determine the mathematical structure of F_p, we must use Newton's third Axiom of Motion or Axiom of Action and Reaction. Let us make a pedagogical parenthesis at this point, and paraphrase a statement made by H. Hertz in his book *Mechanics*:

> The great majority of physics teachers cover the subject of Newton's Laws of Motion as fast as they can, in order to avoid embarrassing questions from their students.

This is particularly true with the Axiom of Action and Reaction. In order to apply this axiom we must first identify the bodies which participate in the interaction. After this identification, we must tell our students that *the reaction is always applied on the body which caused the action*. This elementary explanatory note will protect the teacher from embarrassing questions, and the students will be considerably benefitted. Hence, if we consider F_p the action exerted by the sun on the planet, then the reaction F_s is applied on the sun. According to the axiom of action and reaction, the magnitude of these forces are equal. Now, using Newton's second axiom of motion in which only the *inertial mass* appears, we can write:

$$F_p = m_ia_p$$
$$F_s = M_ia_s$$
$$F_s = F_p$$

$M_i a_s = m_i a_p$
$a_s / m_i = a_p / M_i$

where a_s is the sun's acceleration and M_i is the *inertial mass* of the sun. Introducing in the last equation the mathematical expression of $a_{p,}$ given by eq. (6.70), we get:

$$a_s / m_i = (N_o / r^2) / M_i$$

Being that r is the relative distance between the planet and the sun or between the sun and the planet, we might try the "educated guess" that $a_s = (N_o^* / r^2)$. Introducing this expression for a_s in the previous equation, and simplifying by r^2, we get:

$$N_o^* / m_i = N_o / M_i = G = \text{constant} \qquad (6.71)$$

From the last equation we get:

$$G = [4\pi^2 a^3 / T^2] / M_{ii} \qquad (6.72)$$

The last equation defines a gravitodynamic constant G. Eq. (6.76) shown below, introduces a correction to eq.(6.72). From eq. (6.71) we deduce that:

$N_o^* = G m_i$
$N_o = G M_i$

Introducing the last two equations in the planet's and sun's accelerations, respectively, we get:

$a_p = G M_i / r^2$
$a_s = G m_i / r^2$

Finally, we can write:

$$F_p = m_i a_p = m_i (G M_i / r^2) = G M_i m_i / r^2 \qquad (6.73a)$$

$$F_s = M_i a_s = M_i (G m_i / r^2) = G M_i m_i / r^2 \qquad (6.73b)$$

A point we must strongly emphasize, over and over again, is that in Newton's planetary gravitational force-law, we only have *inertial masses* as shown in eqs. (6.73). Another point we must strongly emphasize is the following. Eq. (6.67) is another *form* of Newton's second Axiom of Motion:

$$\mathbf{F}_p = m_i\mathbf{a} = m_i\,[-(h^2/p)\mathbf{r}/r^3] \qquad (6.74a)$$

$$\mathbf{F}_p = m_i\mathbf{a} = m_i\,[-(4\pi^2a^3/T^2)\mathbf{r}/r^3] \qquad (6.74b)$$

The last equation clearly shows that the acceleration **a** is still a *kinematical* concept consistent with its kinematical definition $d^2\mathbf{r}/dt^2 = d\mathbf{v}/dt$, i.e., equal to the *total time derivative* of the velocity of the planet. Some authors claim that gravitodynamics, established by analogy with some electrodynamics, like Weber's for instance, allows them to derive Newton's second Axiom of motion: $\mathbf{F} = m_i\mathbf{a}$, from a particular application of their gravitodynamics. These people do not realize that every force-term of Weber's electrodynamics must be reduced to "mass times acceleration." It is vitally necessary that we come back later to this subject in section 6.4.

Now we face a delicate ontological problem. Can we demonstrate, deduce, conclude or derive that the very nature or essence of this kinematical acceleration is actually a gravitational acceleration? If we think that the demonstration, deduction, conclusion or derivation, can be done with the logic of mathematics, then the answer is in the negative; because mathematics, being a formal science, is forever condemned to remain silent in matters of ontology. Then, what do we mean by gravitational acceleration? First of all, we should recall that in Kinematics, we study the motion of bodies with respect to reference systems, making no attempt to identify any other body which might be interacting with the body that is moving. Kepler's laws constitute a typical set of geometro-kinematic laws, which say nothing about the existence of force or forces acting on the planet. When we combine the law of inertia with Kepler's first law, we must conclude that a force must be acting on the planet, because the planet does not move uniformly in a straight line but in an elliptical orbit. Thus, Newton's law of inertia allows us to make an ontological statement about the *existence* of a force acting on the planet. Newton's second axiom, after we have determined the mathematical structure of the acceleration of the planet, allows us to determine the magnitude of the force acting on the planet. Finally, Newton's axiom of action and reaction allows us to make another ontological statement about the *material cause* which produces a dynamical effect on the planet. Thus, Newton's third axiom allows us to *identify* the material body which *causes* the action on the planet. This identification is *very important* because it permits us to know where the reaction force of the planet is acting and on what. Whether positivist people like it or not, we see that Newton's Dynamics uses more than one ontological principle. At this point, we must go back to Newton's *Principia*. In the Scholium of Proposition V, Theorem V, Book III, he wrote:

> "The force which retains the celestial bodies in their orbits has been hitherto called centripetal force; but it being now made plain that it can be no other than a gravitating force, we shall hereafter call it gravity."

In this Scholium of Newton, we find the essence of Einstein's *Principle of Equivalence*. The reader must be acquainted with the enormous methodological difference between a Scholium and a Principle. Had Einstein studied the *Principia,* he never would have proposed his Principle of Equivalence, as we will discuss in 6.4. Thus, the gravitational force is the *weight* (*gravitas*) of the planet with respect to the sun. We must recall that Newton wrote the *Principia* in Latin. Newton sophisticated the pedestrian concept of weight when he transferred it to the heavens and applied it to celestial bodies. Hence, the weight of any body is not an intrinsic property of a body, but a relative attribute of one body relative to another. *The concept of weight refers to the interaction of two material bodies*. We can never overemphasize, when addressing our students, that forces are *caused* by material bodies which *act on* material bodies. This weight-force (gravita-tional force), acting on the planet, *induced* or produced *(caused)* by the sun, divided by the mass of the planet provides the gravitational acceleration. Later on, when the concept of *field* is introduced into physics, this gravitational acceleration will be called *intensity* of the gravitostatic field of the sun. When teachers lecture, and authors of textbooks write, that the "weight of a body is the gravitational pull of the earth," what they are saying is a tautology: "The weight of a body is the weight of the body exerted by the earth."

There is an objection to our derivation of Newton's planetary force-law. When we invoked Newton's third Axiom of Motion, we concluded a reaction-force was acting on the sun. Thus, when we have forces acting on the planet and on the sun, neither of these bodies can remain at rest. This conclusion contradicts our assumption of the sun being at rest on one focus of the elliptical path of the planet. To avoid this contradiction, we would refer the motion of the planet to the sun-planet's center of mass, and describe the planet's motion in terms of the distance sun-planet in order to use Kepler's laws. This procedure introduces one change in our previous mathematico-physical derivation which consists of replacing the mass of the planet m_i by $m_i(1+m_i/M_i)$. Eq. (6.73) then becomes

$$F_s = M_i\, a_s = \mu_i(N_o/r^2) = F_p$$

with $\mu_i = m_i(1+m_i/M_i)^{-1}$, and $N_o = 4\pi^2a^3/T^2$. Now we can write:

$$F_p = m_i(1+m_i/M_i)^{-1}(4\pi^2a^3/T^2)/r^2, \qquad \text{or}$$

$$F_p = \{4\pi^2a^3M_im_i/[T^2(M_i+m_i)]\}/r^2 \tag{6.75}$$

Let us define:

$$G^* = [4\pi^2a^3/T^2]/[M_i(1+m_i/M_i)] = G/(1+m_i/M_i) \tag{6.76}$$

Hence, eq. (6.74) becomes:

$$F_p = G^*M_i\,m_i/r^2 \tag{6.77}$$

To this last derivation, we raise another objection. Why do we use the inertial masses and not the gravitational masses in the definition of the center of mass sun-planet? There are two reasons: First, we had no need to introduce a "strange" concept into our discussion. Second, because of the following rational use of Newton's theory of classical mechanics. The application of Newton's second Axiom of Motion to the sun and planet is given as

$$m_i\, d^2\mathbf{r}_p/dt^2 = \mathbf{F}_p$$

$$M_i\, d^2\mathbf{r}_s/dt^2 = \mathbf{F}_s$$

where $\mathbf{r}_p$ and $\mathbf{r}_s$ are the position vectors of planet and sun, respectively, from the origin of an inertial system of reference. Adding the last two equations and using the assumption of constant mass in classical mechanics, we get:

$$d^2(m_i\mathbf{r}_p+M_i\mathbf{r}_s)/dt^2 = \mathbf{F}_p+\mathbf{F}_s$$

or

$$(m_i+M_i)d^2[(m_i\mathbf{r}_p+M_i\mathbf{r}_s/(m_i+M_i)]/dt^2 = \mathbf{F}_p+\mathbf{F}_s$$

or $(m_i+M_i)d^2\mathbf{r}_c/dt^2 = \mathbf{F}_p+\mathbf{F}_s$

where, by definition, the position vector of the center of mass is given by:

$$\mathbf{r}_c = (m_i\mathbf{r}_p+M_i\mathbf{r}_s)/(m_i+M_i)$$

In this definition, we only have *inertial masses.*

6.3.6 GENERALIZATION OF NEWTON'S PLANETARY FORCE-LAW

To generalize requires a great deal of faith. To generalize requires the handling of arguments which go far beyond *experimental philosophy.* Finally, to generalize requires a great deal of audacity, imagination, flexibility of mind and most of all a great capacity to frame hypotheses of all kinds: "metaphysical," "physical," "of occult qualities" or "mechanical." If the final product of this adventure of the mind permits experimental verification of consequences derived from it, then we must possibly admit the irrational or transrational origin of the theoretical structure.

Eq. (6.73) represents Newton's planetary force-law. To go from eq. (6.73) to eq. (6.78) shown below, is to make a tremendous leap.

$$F = Gm_1m_2/r^2 \tag{6.78}$$

where G is Cavendish's universal gravitational constant; m_1 and m_2 are the respective masses of any two bodies of negligible geometrical dimensions compared to r which is the distance between the inter-acting bodies. Is there any *logical path to prove* that eq. (6.78) is a logical consequence deduced from eq. (6.73)? We know such a logical path *does not* exist. Hence, eq. (6.78) is not rational, i.e., we cannot "explain" it, we cannot produce a chain of logical syllogisms with which to obtain eq. (6.78) as a rational conclusion. M. Born [83, p. 61], referring to Newton says,

> "What a prodigious imagination it required to conceive the motion of the planets about the sun or of the moon about the earth as a process of falling that takes place according to the same laws under the action of the same force as the falling of a stone released by my hand."

But eq. (6.78) is not *demonstrated* from the phenomena of nature. A. Koyré [84, p. 273] tells us about an objection Roger Cotes raised about Newton's theory of attraction:

> "Cotes found that Newtonian attraction implied the attribution of 'attractive forces' to bodies and that Newton, tacitly, made that 'hypothesis,' or 'supposition'... Newton's reaction to Cotes's 'difficulties' is rather interesting. He first enlightens Cotes about the meaning of the word 'hypothesis,' then he tells him that universal attraction is not a 'hypothesis' but a truth established by induction, and that the mutual and mutually equal attraction of bodies is a case of the third fundamental law or axiom of motion, that of the equality of action and reaction as is explained already in the *Principia.*"

Newton's answer to Roger Cotes was, indeed, very clever. Newton's *tacit* metaphysical hypothesis is not contained in his attractive gravitational force but precisely in his third Axiom of Motion as we mentioned earlier. On the other hand, Newton's defense of his "Hypotheses non fingo" was always based on *induction.* After almost three centuries, we have a better understanding of the meaning of induction. In a free translation of the author, Louis de Broglie in his *Sur les sentiers de la science* [85, p. 354], says:

> "Upon breaking, by means of irrational leaps, the rigid circle in which deductive reasoning encloses us, induction founded on imagination and intuition is unique in permitting the giant steps in the conquest of thought where the origin of all true progress in science is founded. Having this capacity is why human thought seems to be definitely superior to all machines, which can calculate or classify better, but can neither imagine or perceive."

The generalization of Newton's planetary force-law is better than an irrational act. It is an act which is transrational and through which Newton arrived at the universal gravitational law represented by eq. (6.28).

6.3.7 ACCELERATIONAL FIELDS AND GRAVITATIONAL FIELDS

We have seen that in determining the theoretical expression of the weight of bodies with respect to planet earth, or the weight (*gravitas*) of planets with respect to the sun, the only mass that appears in the final mathematical expressions must logically be the inertial mass. Therefore, eq. (6.28) should be written as follows:

$$F = Gm_{i1}\, m_{i2}/r^2 \qquad (6.79)$$

Let us write $m_{i1} = M_i$ for the earth's mass, and $m_{i2} = m_i$ for the falling body. Eq. (6.79) becomes:

$$F = GM_i\, m_i/R^2 \qquad (6.80)$$

where R is the radius of the earth. Now, if we use Newton's second axiom of motion, $F = m_i a$, from the previous equation, we get:

$$m_i a = m_i(GM_i/R^2) \qquad (6.81a)$$

If we simplify this last equation by m_i, we see no mystery in concluding that the "falling acceleration" *a* is independent of any attribute, accident or property of the falling body, including the inertial mass m_i. Then - why did Newton run experiments with pendulums with bobs of different chemical compositions, but using the same weights? - Because, after we simplify eq. (6.81a) by m_i, we get

$$a = GM_i/R^2 \qquad (6.81b)$$

Newton knew, very assuredly, that the previous equation was a *theoretical* or metaphysical conclusion without any experimental verification. Thus, to verify the experimental validity of the previous equation, Newton conducted experiments with pendulums. Before the advent of Newton's universal gravitational law, one would have agreed to consider the equality of the accelerations of two falling bodies of equal weights and different chemical compositions as a "profound physical problem." But, after Newton's universal gravitational law was introduced in classical mechanics, that "profound physical problem" should have been solved long ago in the minds of physicists. The last equation is the Newtonian seed of the Principle of Equivalence of Einstein. The acceleration "a" in equation (6.81b) should be called *intensity of an accelerational field* while the right hand side of the same equation is called *intensity of a gravitational field.* We will come back to reconsider eq. (6.81b) in the following section, when we discuss Einstein's Equivalence principle.

6.3.8 EÖTVÖS' EXPERIMENT AND THE EQUIVALENCE PRINCIPLE

If there is a true or fundamental difference in nature between m_i and m_g, then Eötvös' experiment is an extremely delicate, intelligent and admirable approach to detect this difference. Eötvös had in mind to submit, simultaneously, two bodies of the *same weight*, but different chemical compositions, to two different force fields. One field is the gravitational force field of planet earth. The other force field is inertial, represented by the horizontal component of the local centrifugal force due to the earth's rotation. He suspended the bodies from a torsional balance. Being that the weights of the bodies are equal, then the gravitational mass of the bodies are equal. However, Eötvös must have thought there was no guarantee that the inertial mass of the two bodies were equal. If these inertial masses are different, thought Eötvös, he would detect a rotation of the torsional balance. But Eötvös did not detect any rotation. He determined, if there was a difference in the inertial masses of the two bodies, this difference had to be less than 0.000 000 005 times the inertial mass of one of the bodies. In other words, $m_g/m_i = 1 \pm 5 \cdot 10^{-9}$

Nevertheless, if Eötvös would have done our analysis of section 6.3 on the identity of inertial mass and gravitational mass, he would have never conducted his experiment. On the other hand, if Eötvös had done our analysis of section 6.3 immediately after he conducted his experiment, he would have realized he was testing the accuracy of the equality of the two weights of the bodies hanging from the torsional balance. Philosophers would have criticized Eötvös for trying to verify, experimentally, the ontological principle of identity: *every thing is identical to itself.* There are many cases, in different sciences, in which the *same* thing has two different names. Depressingly still, each of the *concepts* associated to the names are hypostatized differently by the human mind, i.e., each of them are assigned different essences. After this illusory or noumenal "reification," everyone organizes experiments to search for an entity, m_g, which has no real existence. This is what happened with the concept of *mass*.

In 1957, Bondi proposed to distinguish between passive gravitational mass and active gravitational mass. Ten years later, in 1967, Sciama proposed to distinguish between passive inertial mass and active inertial mass. In the early years of the 1960s, Dicke in U.S.A. and Braginsky in Russia, initiated new experiments to increase the accuracy of

Eötvös' experiment. A new element was introduced in the design of an Eötvös-type experiment. Einstein's mass-energy equivalence, $E = m_i c^2$, was considered at this time The mass in the last relation corresponds, of course, to inertial mass. Let us analyze this type of experiment but with our conclusions of section 6.3 in this chapter. In this section, we did not establish the equality in magnitude of *inertial mass* and *gravitational mass*. We established *the identity* of the inertial mass and the gravitational mass. We established that these two names m_i and m_g correspond to *one and the same entity*. We established that the *mass* of any body seems to be a unique characteristic or attribute of the body, no matter how many names we use to designate it. We established that the nature, the essence, the being of these two names: m_i and m_g, is one and the same, because any nature, any essence, any being is identical to itself.

Dicke [86], introduced the novelty of submitting material bodies of the *same weight* to the accelerational field of the sun which is provided by the centrifugal force of the orbiting terrestrial laboratory around the sun. For simplicity, let us assume the presence of two bodies in Dicke's experiment. If the weight of these two bodies is the same, then the magnitude of their gravitational masses is equal:

$$m_{g1} = m_{g2}$$

According to section 3, we know the identity of the gravitational mass with the inertial mass. Therefore, from the previous equation we must conclude that:

$$m_{i1} = m_{i2}$$

If body 1 is made out of gold, and body 2 is made out of aluminum, then the binding nuclear energy of these two bodies is different, i.e.:

$$m_{i1N} = E_{1N} / c^2 \quad \mathrm{K} \quad E_{1N} / c^2 = m_{i2N}$$

Now, when Nature acts and shows a natural effect, she does it by integrating everything. She does it independently of the partial knowledge of terrestrials. This means that to preserve the previous equality of inertial masses, we must write the following equations:

$$m_{i1} = \Sigma m'_{i1k} + m_{i1N} = \Sigma m'_{i2k} + m_{i2N} = m_{i2}$$

where the summations are different, but each of them include all the other sources which contribute to the *total inertial mass* of a body. Thus, according to our identification of gravitational mass with inertial mass, the equality in the weights of two bodies of different chemical compositions assures immediately the equality of their inertial masses. However, not Dicke nor Braginsky knew about our identification of these seemingly and "fundamentally" different masses pertaining to the same material body.

Principle of equivalence. Not knowing the *identity* of inertial mass with gravitational mass, it was rational to increase the accuracy in the experimental determination of the ratio m_g/m_i. For Einstein, it was absolutely necessary to have this ratio identical to unity. But in physics, we cannot attain this kind of accuracy, unless, the numerical equality is reduced to an identity of the real nature of m_g and m_i. In ontology, we do not quantify "the real nature" (essence) of entities, as Einstein [87, p. 57] clearly tried to convince his
readers. Einstein, in his book, *The Meaning of Relativity,* writes our equation (6.81a), given by:

$$m_i a = m_i GM/R^2$$

in a form similar to the following one:

(Inert mass)·(Acceleration) = (Gravitational mass)·(Intensity of the gravitational field)

For us, it is very easy to simplify eq. (6.81a) by the inertial mass. But for Einstein, it was not so easy. He had to be very sure of the equality of the magnitude of gravitational mass with the magnitude of inertial mass to a great degree of accuracy. This equality of m_g with m_i is a necessary and sufficient condition to conclude that the *Kinematic Acceleration is equal to the Intensity of the Gravitational Field.* In italics, we have presented the essential statement of Einstein's equivalence principle. In other words, this is exactly what Newton wrote in his Scholium of Proposition V, Theorem V, Book III. Let us repeat this Scholium again:

> "The force which retains the celestial bodies in their orbits has been hitherto called centripetal force; but it being now made plain that it can be no other than a gravitating force, we shall hereafter call it gravity."

Obviously, the statement of the Scholium refers to absolute space or to an inertial reference system. If instead of a planet, we consider a laboratory in orbit around the sun with a noninertial reference system attached to the laboratory, then the previous Scholium is not valid. Inside the laboratory the weight of a body, with respect to the sun, is compensated exactly by the *centrifugal force* inside the revolving laboratory. This centrifugal force is called *fictitious force* because it is not *caused* by any other material body. For this reason, Newton's axiom of action and reaction is completely invalid in any accelerated reference system. Now, let us rephrase Einstein's equivalence principle:

> It is impossible to establish, in an hermetically closed laboratory, any physical difference, when the laboratory K is at rest on the surface of a planet, or when the same laboratory K' is being accelerated with respect to a stationary reference system fixed at infinity (absolute space).

Thus, a gravitational field is *almost* totally equivalent to an "accelerational field" inside an accelerated laboratory. This is an easy conclusion to draw from our Newtonian eq. (6.81a). Every physics textbook, in the world, establishes that a laboratory on planet earth does not constitute an inertial or stationary reference system. The reason is that the earth is rotating. If we could stop the earth's rotation, so they say, the earth's laboratory would constitute an inertial reference system. This is not true at all, even if the earth is not rotating. According to the equivalence principle, a terrestrial laboratory, on a nonrotating earth, is *equivalent* to the same laboratory being accelerated at 9.8 m/s^2, with respect to the stellar reference system. After this last assertion, any student who wonders about the bio-chemical processes in a human body, traveling at a speed close to the speed of light, only has to check with his family doctor about differences in metabolism. All terrestrial creatures are in an equivalent cosmic space voyage, and all of them are traveling simultaneously with different speeds starting from the day they were born. Is this another paradox? Presently, there are many subdivisions of Einstein's equivalence principle, depending on the type of physical phenomena being considered.

What is the purport of this principle of equivalence? If we use Faraday's language of "lines of force," we have to conclude the following geometrical statements. The lines of force in a gravitational field, inside a laboratory sitting on a spherical plane, *are convergent.* It makes no difference how small the laboratory is. The lines of force in an accelerational field, inside an accelerated laboratory, *are always parallel to each other*. The central concept of this principle is that it establishes a very good equivalence but never an identity between a gravitational field and an accelerational field. This is the reason, we keep saying, *these two fields are almost equivalent.*

6.4 Mach's Definition of Mass and Operational Definition of Inertial Reference System.

Mach's definition of mass is a clever way to use Newton's axiom of action and reaction in the interaction of two material bodies. Undoubtedly, Mach's definition of mass is related to *Mach's Principle.* As a matter of historical fact, it was Einstein [88], who in 1918, gave the name *Mach's Principle to the field equations of General Relativity Theory.* On that occasion, Einstein wrote in a footnote his definition of Mach's principle:

> "I have chosen the name 'Mach's principle' because this principle is a generalization of Mach's requirement that inertia can be traced back to an interaction of bodies."

According to Newton's second axiom of motion, force is mass times acceleration when the mass is constant. Thus, if body S interacts with body A, according to the axiom of action and reaction, the *magnitude* of the forces are equal:

$$F_{SA} = m_{SA}\, a_{SA} = m_{AS}\, a_{AS} = F_{AS} \qquad (6.82)$$

from which we get:

$$a_{SA}/a_{AS} = m_{AS}/m_{SA} \qquad (6.83)$$

If we chose m_{SA} as the *standard unit mass,* the last equation can be written as follows:

$$m_{AS} = a_{SA}/a_{AS} \qquad (6.84)$$

Hence, the mass of body A is numerically equal to the inverse ratio of the accelerations when body A interacts with the standard body of unit mass. This last statement, expressed mathematically by eq. (6.84), is in essence Mach's definition of mass or "Mach's requirement that inertia can be traced back to an interaction of bodies," as Einstein has said. Mach, in his *Science of Mechanics* [89:303], proposed to replace Newton's *principles* of dynamics by a particular set of "gedanken" experimental propositions and definitions. Mach seems to ignore the difference between the *general* character of a principle and the *particular* character of a set of empirical facts. This obsession of Mach to replace Newton's *axioms* (principles) of mechanics does not contribute to any simplification of classical mechanics. Ironically, Mach's experimental propositions lead to an *operational definition* of a Newtonian inertial reference system. Mach's set of experimental propositions and definitions, to replace Newton's axioms of dynamics, is the following:

"*a. Experimental Proposition.* Bodies set opposite each other induce in each other, under certain circumstances to be specified by experimental physics, contrary *accelerations* in the direction of their line of junction. (The principle of inertia is included in this.)

b. Definition. The mass-ratio of any two bodies is the negative inverse ratio of the mutually induced accelerations of those bodies.

c. *Experimental Proposition.* The mass-ratios of bodies are independent from the character of the physical states (of the bodies) that condition the mutual accelerations produced, be those states electrical, magnetic, or what not; and they remain, moreover, the same, whether they are mediately or immediately arrived at.

d. Experimental Proposition. The accelerations which any number of bodies A, B, C . . . induce in a body K, are independent of each other. (The principle of the parallelogram of forces follows immediately from this.) [This proposition contradicts Mach's proposition P1 in section 6.5]

e. Definition. Moving force is the product of the mass-value of a body into the acceleration induced in that body. The theorems *a* to *e* were given in my note "Über die Definition der Masse" in Carl's *Repertorium der Experimental Physik,* IV, 1868; reprinted in *Erhaltung der Arbeit*, 1872, 2nd ed., Leipzig, 1909. *Cf.* Also Poincaré, *La Science et l'Hypothese*, Paris, pp. 110 *et seq.*"

We must cautiously study Mach's criticism of Newton's *Principia*, because he did not understand that Newton *verified experimentally a theoretical law* which establishes that the mass of a body is proportional to its weight. This piece of knowledge is not a definition or an *assumption*, as Mach would want us to accept. This confusion of Mach is very clearly established in reference [90, p. 265]. In section 6.5, we will return again to this important point of weight and mass. Let us now formulate some comments on the previous quotation of Mach.

On Experimental Proposition a. Mach says, in this proposition, that two interacting bodies *induce* acceleration on each other. Mach avoids the word *cause*. He could have said that two interacting bodies *cause* acceleration on each other. If a body is at rest in an *inertial reference system,* and from one moment to the next begins an accelerated motion, then there is nothing wrong in asking what is the body that *causes* the acceleration of the body which was at rest. But Mach never would have used the ontological principle of cause and effect because he was a fanatic positivist. Mach would do anything possible to eliminate, avoid or hide *metaphysical obscurities* in his discourses. According to Mach's positivistic intolerance, atoms could not exist. Mach's influence concealed the word *cause* from physics from the end of the 19th century until the beginning of the 21st century. In essence, Mach is expressing, in this experimental proposition "a", Newton's principle of *action and reaction* in term s of accelerations instead of *forces*.

This author told his students that people, including physics teachers, of course, believe that forces are like locomotives. We unhook the locomotive from one train to hook it up to another. For this reason, this author made the following pedagogical explanatory notes to understand Newton's axiom of action and reaction.

Explanatory note # 1. Newton's axiom of action and reaction is *only valid* with respect to *absolute space*, or with respect to an inertial laboratory which moves uniformly and rectilinearly with respect to absolute space. In other words, Newton's axiom of action and reaction *is not valid* in rotating laboratories or in linearly accelerated laboratories. This last statement was established by Newton in 1687. In the 20th century, Einstein accused Newton's mechanics of violating the principle of action and reaction inside accelerated laboratories. Is this a valid accusation? Einstein knew better as we will soon see.

Explanatory note # 2. Real forces come into existence when at least two material bodies (neutral or electrically charged) inter-act with each other.

Explanatory note # 3. Action and reaction, of the same nature, never act on the same body. Action and reaction acts on different and separated bodies.

Explanatory note # 4. The force of reaction always acts on the body which *causes* the action.

Explanatory note # 5. *Practically* always, action and reaction are collinear, i.e., are in the line which joins the center of mass of the interacting bodies. Lorentz's electrodynamic force is a clear example of non-Newtonian action and reaction.

On Definition b. Mach's Definition b is given by eq. (6.83) which is a logical conclusion in the context of Newton's dynamics. Mach does not contribute with anything new in this definition "b."

On Experimental Proposition c. This is an obscure proposition. In this proposition, Mach is definitively wrong. He was acquainted with the work of Gerber at the end of the 19th century [91, p. 235]. In 1898, Gerber published an explanation of the anomalous motion of planet Mercury by using a gravitodynamics in which the acceleration of planet Mercury not only depended on its relative distance to the sun but on the kinematic state of the planet itself. Mach, also, was acquainted with the German electrodynamics. Thus, we must reject his proposition that "The mass-ratios of bodies are independent of the character of the physical states (of the bodies)." This proposition of Mach's is easily rejected on the grounds of GRT as we saw in section 2 of this chapter. Grassmann's gravitodynamic force is a clear example of Mach's mistake. Poincaré [92, p. 92], in his book *Science and Hypothesis,* wrote about his own generalization of the principle of inertia:

> "The acceleration of a body only depends on its position and that of neighboring bodies, and on their velocities."

Mach and Poincaré are trying *a priori* to establish the mathematical structure of Newton's force **F** in Newton's second axiom of motion, d(m**v**)/dt=**F**, for electrodynamic and gravitodynamic interactions, ignoring completely Newton's dynamical methodology established in the preface of his *Principia.* Poincaré points out that the principle of inertia can be deduced from his general principle quoted above. It is an elementary mathematical work to deduce, from Newton's second axiom of motion, the *erroneously* called "principle of inertia." It is a simple corollary.

On Experimental Proposition d. This proposition is false also. Because the acceleration a_{KA} is not independent from the accelerations a_{KB}, a_{KC} . . ., as Mach himself establishes in proposition P1 in section 6.5 of this chapter.

On Definition e. This definition of Mach's is a *disguised plagiarism* of a particular case of Newton's second axiom of motion when the mass of the body is constant.

6.4.1 MACHIAN OPERATIONAL DEFINITION OF INERTIAL REFERENCE SYSTEMS

Let us again recall what Einstein said about Mach's requirement:

> "I have chosen the name 'Mach's Principle' because this principle is a generalization of Mach's requirement that inertia can be traced back to an interaction of bodies."

Inertia, of course, is a synonym of mass in the writings of Mach and Einstein. On the other hand, if material bodies would not interact with each other, human beings would never have had the concepts of inertia, mass and force. Thus, it is superfluous to say that the mass of a body can be traced back to an interaction with another body. This is exactly what eq. (6.82) is saying in the language of mathematics. But eq. (6.82) is a consequence of physical observations and the use of Newton's second and third axioms of motion. When Newton established that the weight of a body is proportional to its mass, he was saying that the *gravitas* (weight) of the body is *caused* by planet earth when the body interacts with it. This is exactly what Newton said in the Scholium of Proposition V, Theorem V, Book III. Thus, there is nothing new added to Newton's *Principia* either by Mach or by Einstein when Einstein said *"that inertia can be traced back to an interaction of bodies."*

We have to assume that the accelerations in Mach's "Experimental proposition c" are measured with respect to an *inertial reference system* S^*, though Mach did not mention it. If the accelerations are measured inside a laboratory that is moving with a certain acceleration $\mathbf{a}^*$ with respect to S^*, then (6.84) becomes:

$$m_{AS} = (a_{SA} - a^*)/(a_{AS} + a^*) \tag{6.85}$$

If we exchange the positions of the interacting bodies S and A, then eq. (6.84) becomes:

$$m'_{AS} = (a_{SA} + a^*)/(a_{AS} - a^*) \tag{6.86}$$

The quotients m_{AS} and m'_{AS} will be different. Only when a^* is zero will the mentioned quotients be equal and, consequently, the laboratory will be experimentally determined as an *inertial reference system.* This sequence of laboratory *operations* well warrant the name of *Machian Operational Definition of Inertial Reference Systems.* This operational definition will stop the vicious cycle, invented by Einstein [93, p.58], and repeated by relativists *ad nauseam,* to discredit Newton. Einstein writes:

> "The weakness of the principle of inertia lies in this, that it involves an argument in a circle: a mass moves without acceleration if it is sufficiently far from other bodies; we know that it is sufficiently far from other bodies only by the fact that it moves without acceleration."

This statement of Einstein is very astute. However, when Einstein says: a mass moves without acceleration if it is sufficiently far from other bodies, this statement is not necessarily true. If the mass of the body is that of a spacecraft, *sufficiently far from other bodies*, the spacecraft can be in an accelerated motion caused by a self-propulsion system. Newton's principle of inertia, to begin with, is not a *principle*, but a necessary (logical) consequence of Newton's second true *principle* of motion. Secondly, and most importantly, the law of inertia does *not* mention *gravitational forces*. For these reasons, the previous quotation is unacceptable. The weakness of the previous quotation is based on the lack of knowledge about the law of inertia with respect to *absolute space*, or equivalently, with respect to the distant "fixed stars." In section 6.5.2 we will prove the equivalence of absolute rotation with relative rotation to the fixed stars.

What Einstein could have manufactured was a homemade *accelerometer*, consisting of a cork attached to a string inside a bottle filled with water. Then, he could have entered a hermetically sealed spacecraft carrying his accelerometer. If Einstein had detected an acceleration inside this hermetically closed spacecraft, this acceleration would have to be *absolute,* because it is determined in respect to nothing material outside the spacecraft. If, on the other hand, Einstein had not detected any acceleration with his homemade accelerometer, his hermetically closed spacecraft would have to be an inertial reference system.

6.5 Mach's principle according to Einstein and others.

Ernst Mach in his book, *The Science of Mechanics*, wrote many propositions about the concept of *inertia* (mass) and, particularly, about *inertial forces.* The most outstanding propositions made by Mach [94, pp. 282-284] are the following two:

P1. "The motion of body K can only be estimated by reference to other bodies A, B, C, . . . It might be, indeed, that these isolated bodies A, B, C, . . . play merely a collateral role in the determination of the motion of the body K, . . When we reflect that we cannot abolish the isolated bodies A, B, C, . . . that is, cannot determine by experiment whether the part they play is fundamental or collateral, that hitherto they have been the sole and only competent means of the orientation of motions and of the description of mechanical facts, it will be found expedient provisionally to regard all motions as determined by these bodies."

P2. "The principles of mechanics can, indeed, be so conceived, that even for relative rotations centrifugal forces arise."

Concerning proposition P1, we may replace the expression "isolated bodies A, B, C, . . ." by *universal matter*. Thus, paraphrasing Mach we can say - *When we reflect that we cannot abolish the universal matter* - we see in Mach's mind the embryonic seed of the Principle of Inseparability presented in chapter 1. Mach reaffirms this cosmic inseparability when he writes - . . . *it will be found expedient provisionally to regard all motions as determined by these bodies* - or universal matter. In other words, Mach's concept reduces to the following proposition P3:

P3. The motion of bodies is determined by the total distribution of matter in the universe.

Mach never expressed the last proposition in the language of mathematics. However, this proposition P3, qualitatively coincides completely with Einstein's mathematical statement of Mach's conception. Let us rephrase proposition P3 as follows:

P4. The motion of bodies is ruled by the metric tensor, which is determined by the universal matter-energy tensor of Einstein's GRT.

P4 is our version of Mach's mathematical principle created by Einstein and given by proposition P6 below. About proposition P2, Mach should have said:

P5. The principles of mechanics can, indeed, be conceived in such a way that even for relative rotations, **new gravitational forces arise**, *analogous* to inertial forces like Coriolis, centrifugal and Euler forces.

These quasi-inertial forces should be called *Einstein's gravitational forces.* In respect to propositions P2 and P5, as we saw in section 2 of this chapter, when Einstein's field equations are solved with a first degree of approximation, we get a *kind* of Coriolis force, but not exactly a Coriolis force. In the work of Lense and Thirring [46], we also find a *kind* of centrifugal force, but not exactly a centrifugal force. Besides, we also get an *axial force,* which does not exist in classical rotating reference systems. In 1918, Einstein [95] called his field equations of GRT *Mach's Principle.* On that occasion, Einstein wrote the following proposition P6, which was about his Mach's mathematical principle:

P6. "The G-field is 'completely' determined by the masses of the bodies (in the universe). Since mass and energy are the same according to the results of the special theory of relativity, and the energy is described formally by the symmetric energy tensor $T_{\mu\nu}$, hence this says that the G-field is caused and determined by the energy tensor of matter."

After this statement of Mach's principle, Einstein added a footnote, trying to justify or explain the reason he had to name his field equations as "Mach's principle." Einstein writes:

> "I have chosen the name «Mach's principle» because this principle is a generalization of Mach's requirement that inertia can be traced back to an interaction of bodies."

Mach's requirement, according to Einstein, is Mach's definition of mass, given by eq. (6.84). The generalization of Mach's requirement was made by Mach, as it was clearly established above by proposition P3. The name Einstein gave to his field equations is very appropriate and pays a great tribute to Mach. Let us now compare Einstein's original definition of Mach's Principle, with other definitions of the same, collected by Mario Speiser in 1965.

6.5.1 DIFFERENT DEFINITIONS OF MACH'S PRINCIPLE

Speiser [96], in his Master Thesis *Mach's Principle*, collected 47 statements of this principle written by different physicists. This principle has been surrounded by a cryptic fog which blinds the minds of whomever writes about it. Speiser found that out of these 47 statements of Mach's Principle, only five coincided with Einstein's original statement of this principle. This means that 89% of the other statements are confused versions of Einstein's original definition of Mach's Principle. Speiser, in the introduction to his thesis, writes:

> "Throughout its long history, this principle has been ignored, scorned, buried, resuscitated, neglected, ridiculed, respected, admired, and even acclaimed."

Let us compare the five statements of Mach's Principle which coincide with Einstein's original one.

Wheeler [97]: "Formulation 2: the geometry of space-time and therefore the inertial properties of every infinitesimal test particle, are determined by the distribution of energy and energy flow throughout all space."

Pirani [98]: "The gravitational field (metric tensor) is 'determined' by the material content of space-time (energy-momentum tensor)."

Tolman [99]: "The general hypothesis that the metrical field is determined by the distribution of matter and energy may be called Mach's principle."

Pauli [100]: "We put forward this postulate: The G-field is to be determined in a unique and generally covariant manner, solely by the values of the energy tensor ($T_{\mu\nu}$)."

Whittaker [101: "the field represented by the ten potentials $g_{\mu\nu}$ is solely determined by the masses of bodies. 'Mass' here is understood as in the theory of relativity, i.e., as equivalent to energy."

Among the other 42 statements of Mach's Principle, collected by Speiser, he concludes: ". . . that the most common version of the principle appears to be:"

> "The inertial mass of a body is determined by the total distribution of matter and energy."

In the opinion of this author, Mach's proposition P3 has more physical meaning than the previous "most common statements" of Mach's principle. These were obtained by Speiser out of 42 different statements of the same principle. In proposition P3, Mach writes about the motion of a body. Mach does not write about the inertial mass of the body. On the other hand, Einstein's field equations do not contain explicitly the inertial mass of the body which is in motion, following a geodesic path in tetra-dimensional space-time.

There are other different statements of Mach's principle. Weisskopf [102], for example, has his own personal interpretation of Mach's principle. In this respect he writes:

> "Mach's principle, as usually expressed, amounts to a denial of the detectability of absolute rotation in space."

Usually Mach's principle is not expressed as Weisskopf claims. Anyway, in the next section, we will rationally prove that *absolute rotation* is equal to stellar relative rotation, ending, in this way, a long controversy started by Berkeley against Newton.

6.5.2 OPERATIONAL DEFINITION OF ABSOLUTE ANGULAR VELOCITY

The content of this section is a summary of an unpublished essay of 31 pages written in Spanish in 1981. At this time, the author was working as a researcher in Coro, Venezuela, in the *Area de Tecnologia*, at *La Universidad Francisco de Miranda*. In this section, we will see the most absurd *metaphysical* arguments presented by Mach against Newton's concept of absolute rotation. Mach wrote his *Science of Mechanics* with the incredible purpose of eliminating *metaphysical obscurities* from the science of mechanics.

Newton, in his *Principia*, presents two examples of rotating material systems. One system embodies a rotating bucket with water. The other system consists of two material globes, tied by a chord and rotating in an empty universe. By *absolute determination,* we understand the measurement or determination of the numeric magnitude of a physical quantity when *absolutely* no reference is made to anything external to the system. A *relative determination* is made when we make reference to external material objects of the system. From now on, we will be writing about absolute rotation and relative rotation. In this section, for the sake of brevity, we will take as synonyms the word *rotation and* the concept *angular velocity*. We will refer to absolute rotation as *dynamic rotation*, and we will refer to relative rotation as *kinematic rotation.*

Newton's concept of *absolute space* is a pure concept, i.e., it is an abstract creation or conception of Newton's mind. Absolute space has a noumenal (mental) existence; it has a metaphysical existence. Absolute space has no real existence, because Newton makes no reference to the external world of things (reality) when he defines absolute space. In his *Principia,* Newton presents two thesis: T1, and T2.

T1. Absolute rotation can be determined dynamically. The case of the two globes is a good example.

T2. Centrifugal forces do not appear in relative rotations. The case of the rotating bucket with water is another good example.

George Berkeley, with respect to T1, was totally unable to *imagine* a rotation when nothing material exists except the two globes. Berkeley was right, but he was referring to kinematic rotation. He proposed to use the "fixed" stars (stellar space) to represent absolute rotation by means of stellar rotation. This stellar rotation would be relative or kinematic rotation. Berkeley understood nothing about Newton's dynamic determination of absolute rotation. On the other hand, Mach understood, exceedingly well, Newton's dynamical approach to absolute rotation, and strongly criticized Newton's conceptions on this subject. In the following quotation of Mach [103, p.337], we see his advice of not confusing "ability to imagine" or "ability to conceive" absolute motion with sensorial "knowledge" of motion. We also see his well known positivist philosophy in action:

> "Conceivability and knowledge of absolute motion are not to be confounded. Only the latter is wanting here (*loc. Cit.*, pages 120, 164) . . . . Now it is precisely with knowledge that the natural enquirer is concerned. A thing that is beyond the ken of knowledge, a thing that cannot be exhibited to the senses, has no meaning in natural science."

Mach is right in the first sentence. Not every entity which exists in our minds, having been conceived by them, must exist in the external world of things or reality. On the other hand, to remain exclusively with sensorial empirical data is analogous to collecting a pile of hard facts, "bricks," but never building a theoretical house. In this way, any positivist like Mach kens a lot, but does not understand anything.

Going back to absolute and stellar spaces, Mach [104, p.280], in one instance, says that Newton did not accept the representation of absolute space by stellar space, but in another occasion Mach, in the same page of his book, says that Newton accepted such representation. If Newton had known the result of Foucault's experiment, conducted in Paris in 1851, perhaps he would have accepted such sensible representation of absolute space by stellar space. Foucault determined the absolute rotation of the earth with his famous pendulum. This absolute rotation was numerically equal to the relative stellar rotation of planet earth. This equality would have convinced Newton that stellar space would have been an excellent representation, but not an identity of absolute rotation with stellar kinematic rotation. Today, we know that by taking Foucault's pendulum to the bottom of a mine or cave, we can determine the absolute rotation of planet earth without considering or observing any external star in the sky. Thus, the equality in magnitude of the absolute rotation of the earth, with its stellar rotation, is an admirable fact which requires a rational explanation.

About thesis T1, Mach presented a solid ontological argument against Newton's assumption of a dark universe without any material stars. Mach claimed that the universe is given one time to human understanding, and in this unique mode of the universe, there are stars whether Newton likes it or not. Mach's argument is irrefutable. Nevertheless, Newton did not need to assume an empty universe to calculate the absolute angular velocity of the two rotating globes. In the near future, astronauts in space can measure the tension F in the chord, the mass m of the globes and the length L of the chord to calculate ω from $F = m\,\omega^2(L/2)$. The astronauts do not need to look at the stars to know the value of ω. This dynamic method proposed by Newton was never understood by Berkeley and other philosophers like Leibniz.

About thesis T2, Mach presented two preposterous criticisms against Newton. Both of them refer to the example of the rotating bucket with water. Newton's followers agree that centrifugal forces appear in the water only when the bucket rotates with respect to absolute space or with respect to the fixed stars, after Foucault's experiment. Mach says that if we stop the bucket and *rotate the heavens of fixed stars,* then Newton's advocates must prove the absence of centrifugal forces in the water. In Mach's [105, p. 279] own words we read:

> "Try to fix Newton's bucket and rotate the heaven of fixed stars and then prove the absence of centrifugal forces."

Newton's advocates should have replied to Mach - *You (Mach) try to rotate the heavens of fixed stars and prove the presence of centrifugal forces in the water.* - "To rotate the heavens of fixed stars" is a *metaphysical statement.* It is an unreasonable or fatuous statement. This is not an intelligent but a preposterous statement coming from an anti-metaphysicist like Mach. But Mach came around to say that - *. . . the Ptolemaic and Copernician view is 'our' interpretation, but both are equally actual.* - Thus, according to Mach's conceptions, we can draw the following absurd conclusions. The reference system of the fixed stars must be *simultaneously* rotating with different angular velocities and with different axes of rotations. This should be so, because we know that some planets are oblate and their axes of rotation are not parallel among themselves. Another absurd consequence of assuming the equivalence of the Ptolomaic reference system with the Copernican one, is that Newton never would have established his gravitational law. Kepler never would have discovered his observational astronomical laws in a maze of Ptolemaic cycles and epicycles. We wonder how Mach and his followers might set in rotation the material universe. Even in the noetic realm, this conception is pure metaphysical insanity. It is interesting to observe that the same Machian conceptions are defended by Einstein's advocates of GRT and by Weberian gravitodynamists, simply because they have run out of time to meditate properly. They have misinterpreted the mathematics of Einstein's theory. Consider this other insanity which is deduced logically from an unreasonable conjecture, the assumption that we could rotate the fixed stars with respect to *absolute space*. If we stop the earth, with respect to *absolute space*, then we have the same relative rotation between the rotating starry universe and the nonrotating earth. We need to force the universe to make one complete revolution (with respect to earth) in 24 hours. What is the meaning of this? The meaning of this supposition is that the matter of the starry universe will have a tangential velocity much greater than the speed of light with respect to absolute space. As a matter of simple calculations, the whole material universe should have a tangential velocity, whose magnitude is approximately a *trillion times the speed of light*. Of course, this simple result is obtained with classical concepts. A different result is obtained using GRT.

Operational Definition of Absolute Angular Velocity. For centuries people have been saying that any relative rotation, kinematically determined with respect to the "fixed stars," is equal, identical or is an excellent representation of absolute rotation dynamically determined. These concepts were reasoned after the numerical equality of the absolute angular velocity of the earth which was determined with Foucault's pendulum, with the relative stellar angular velocity of the earth. This equality is factual, is experimental but it has no rational or theoretical explanation. In what follows, we will prove the following proposition or thesis:

Thesis: The absolute angular velocity of any body is the limit of the relative velocity of the same body when the distance to the object, with respect to which the relative rotation is measured, tends to infinity.

The proof of this thesis is based on classical concepts, and will constitute the theoretical explanation of the empirical equality of the absolute angular velocity of the earth, with the relative angular velocity of the same with respect to the so-called fixed stars.

Let us assume we release a beacon from one of the globes of the rotating system of two material globes in Newton's example. Let us also assume two systems of reference. One absolute $S^*(X^*,Y^*,Z^*)$, and another $S(X,Y,Z)$ attached to the rotating globes. Finally, let us assume the origin of S^* and S coincides with the center of the chord, which holds the two globes that rotate in the plane X^*Y^*. Now, the beacon is released when the chord coincides with axis X^*.

The beacon begins to move in a rectilinear trajectory parallel to the Y^* axis, with a constant velocity $\mathbf{v}^*$ equal in magnitude to the tangential velocity of the globes. If the length of the chord is 2R, then after a time t, the parametric equations of the beacon with respect to absolute space S*, is given by:

$$x^* = R$$
$$y^* = v^*t$$

After the same interval of time, t, the X axis has rotated an angle $\phi = \omega^*t$, with respect to the X* axis, where $\omega^*$is the absolute angular velocity of the globes. The position vector of the beacon with respect to S forms an angle θ with the X axis, given by:

$$\tan\theta = y/x$$

Taking the time derivative of the last equation, we get:

$$\omega = d\theta/dt = (xdy/dt - ydx/dt) / (x^2 + y^2) \qquad (6.87)$$

Now, the parametric equations of the beacon's trajectory, relative to the rotating reference system S, is given by:

$$x = x^*\cos(\omega^*t) + y^*\sin(\omega^*t)$$

$$y = -x^*\sin(\omega^*t) + y^*\cos(\omega^*t)$$

Introducing the last two equations in eq. (6.87), we obtain:

$$\omega = [Rv^* - \omega^*(R^2 + v^{*2}t^2)]/(R^2 + v^{*2}t^2) \qquad (6.88)$$

The distance r from the center of the chord to the beacon at time t satisfies the following equation:

$$r^2 = R^2 + v^{*2}t^2$$

Introducing the last equation in eq. (6.88), we get:

$$\omega = (Rv^* - \omega^* r^2)/r^2$$

This last equation represents the relative angular velocity ω, of the beacon, with respect to the rotating globes. Now, the relative angular velocity ω_R of the globes, with respect to the beacon, satisfies the following equation: $\omega_R = -\omega$. Introducing this last relationship, in the previous equation, we get:

$$\omega_R = \omega^* - K/r^2 \quad (6.89)$$

where K = Rv* is constant. Eq. (6.89) shows that ω_R approaches w*, asymptotically, as the distance r increases. Therefore, ω_R tends to ω* when r tends to infinity. Another way to express this last concept is to write eq. (6.89) in the following way:

$$\omega^* = \lim_{r \to \infty} (\omega_R + K/r^2) = \omega_R \qquad \text{Q.E.D.} \quad (6.90)$$

This last equation proves our thesis and converts the thesis into the *operational definition of absolute angular velocity*. Eq. (6.90) also shows that Berkeley's intuition was correct, and Mach's insistence on representing absolute space by something that is exhibited to the senses, like the fixed stars, was also correct. Distant stars in our own galaxy, Andromeda galaxy or the extra galactic nebulas, are extremely distant and excellent objects to measure the absolute angular velocity of planet earth. Eq. (6.90) also explains why we have identified the *Stationary Reference System fixed at infinity*, of GRT, with Newton's *absolute space*. Finally, we see that classical mechanics can perfectly well explain, theoretically, the empirical equality of the stellar rotation of planet earth with the absolute rotation of the same planet.

6.6 Newtonian Relativistic Gravitodynamics and the Cosmic Collective Potential Energy.

In chapter 4 section 4, we proved that Einstein was mistaken in accusing Newton's dynamical methodology. Einstein wrote: "*. . . the laws of force cannot be obtained by logical and formal considerations, so that their choice is a priori to a large extent arbitrary.*" The way Newton deduced the gravitational force was by no means arbitrary. This was just another false accusation by Einstein. The proof of this false testimony is clearly established in this chapter, Sections 3.5 and 3.6. Such false testimony reveals complete lack of knowledge of Newton's dynamical methodology. The great majority of high school, undergraduate and graduate students believe that, perhaps, Newton had a mystical experience when the "apple fell on his head." At that moment, students may think an angel presented the equation of the force of gravitation to Newton. Newton's dynamical methodology begins with empirical observations (Kepler), but not with mystical experiences. In any case, we accepted Einstein's challenge to obtain the gravitational force by logical and formal considerations. The result was extraordinary. We proved that Newton's theory of dynamics is extremely powerful. The new Newtonian relativistic gravitodynamics, in reference to an inertial reference system, is given by:

$$\mathbf{F}^* = -CMm\, r^{-3}\{\mathbf{r}+c^{-2}\,[\mathbf{r}\{\alpha \mathbf{v}^{*2}+\beta r^{-2}(\mathbf{r}\cdot\mathbf{v}^*)^2+\gamma(\mathbf{r}\cdot\mathbf{a}^*)\}+\delta\mathbf{v}^*(\mathbf{r}\cdot\mathbf{v}^*)+\varepsilon r^2\mathbf{a}^*]\} \quad (6.91)$$

where

$\mathbf{v}^* = \mathbf{v} - \mathbf{v}'$, is the Galilean relative velocity of the interactive particles.

$\mathbf{v}$ is the velocity of the test particle.

$\mathbf{v}'$ is the velocity of the particle which creates the gravitational field.

$\mathbf{r} = \mathbf{R} - \mathbf{R}'$ is the relative vector position of the test particle with respect to the source particle.

$C = G = 1/(4\pi\varepsilon_o^*)$

ε_o^* can be called the gravitostatic permittivity of the cosmic energy field. We can also define a gravitodynamic permeability μ_o^* of the cosmic energy field, given by:

$$\mu_o^* = 1/(\varepsilon_o^* c^2) = 4\pi G/c^2 \quad (6.91^*)$$

Obviously, this Newtonian relativistic gravitodynamics has to be submitted to experimental verification, in the course of which we will have the opportunity to empirically adjust the numeric values of the Greek parameters. In addition to the secular variations of the perihelia of planets, there are other secular variations of orbital parameters.

Since the 19th century, physicists and astronomers have been looking for new gravitational forces beyond Newton's only gravitostatic term. Jefimenko's book, *Causality, Electromagnetic Induction, and Gravitation* [106], is an unusual book. It could also be a textbook for the future generations of students who truly want to understand the *physics* of *"the rather involved mathematical method"* of GRT. Einstein himself recognized this mathematical involved methodology of GRT. From scientific and pedagogical points of view, Jefimenko's book is an admirable physical introduction to Einstein's linearized GRT, and the so-called Asterisk's gravitational theory which we presented in Section 2.1 of this chapter. Jefimenko takes the reader from electromagnetism and electrodynamics to gravitodynamics. Heaviside [107, p. 96], in 1893, was not the first scientist to suggest that a second force field may be involved in gravitation beyond Newton's gravitostatic force. According to Assis [108, p 203], it seems that in 1870, Holzmuller was the first in using a Weber's gravitodynamics. Then, in 1872 Tisserand [109] used

Weber's gravitodynamics. Later, in 1890, M. Lévy [110, vol 2, p. 527] used a linear combination to represent a new gravitokinetic potential energy U, given by $U = (1 - \alpha)W + \alpha R$, where W refers to Weber, and R to Riemann. However, it was Newton, in 1687, who added a second gravitational term, inversely proportional to the cube of the distance, as we saw in section 1 of this chapter. Thus, Heaviside had forerunners when he suggested another gravitational force-term beyond Newton's. What seems certain, as Jefimenko declares, is that Heaviside was the first in introducing into physics the concept of *Gravitodynamic Induction* **B***, or as Jefimenko calls it *Cogravitational Field* **K**.

Given different numerical values to the Greek coefficients in eq. (6.91), we can obtain Lorentz's gravitodynamics, or Ritz's gravitodynamics, or Weber's gravitodynamics. For this reason, we will again refer the reader to the uncommon book written by Assis [111], titled *Weber's Electrodynamics*. Assis's book is also a textbook. It offers scientific novelties, and pedagogically is highly recommended to undergraduate students as a clear introduction to linearized Einstein's geometrodynamics (GRT) via Weber's gravitodynamics as developed by Assis. His book, from a purely scientific point of view, is a gold mine of bibliographic references for those interested in serious research on any kind of Newtonian Relativistc Gravitodynamics, especially in relation to the "anomalous motion of planet Mercury." When referring to Mach's principle and the concepts of gravitational and inertial mass, we are sure this prolific young author will review them in future editions of his book, which is otherwise excellent.

Cosmic Collective Potential Energy. The cosmic collective potential energy K* is completely analogous to the quantum collective potential energy which was established in chapter 2 of this book. Thus, K*, in analogy with (2.18), is given by:

$$K^* = - \{\partial S/\partial t + [1/(2m)] (\nabla S)^2 + U\} \qquad (6.92)$$

where S is the energy-momentum potential of the moving material body of mass m; U is the gravitostatic potential energy, given by $U = - GMm/r$. Eq. (6.92) can be rewritten as follows:

$$\partial S/\partial t + [1/(2m)] (\nabla S)^2 + U + K^* = 0 \qquad (6.93)$$

We call the previous equation *The Gravitational Bohm-Hamilton-Jacobi Equation*. The cosmic collective potential energy K* is the synergistic action of the entire material universe on two material bodies, interacting with potential energy U. K* can also be expressed, like Bohm's quantum collective potential energy, by the following equation:

$$K^* = -[\hbar^{*2}/(2m)](\nabla^2 R)/R \qquad (6.94)$$

where * is a cosmic constant to be determined experimentally. If gravitation is an electrodynamic phenomenon, as we will show later, then we should expect $h^* = h$. If we restrict ourselves only to our solar system, then we can use eq. (6.93) to study the *natural* (not anomalous) motion of planet Mercury around the sun. In this particular case K* is given by:

$$K^* = -\Sigma GMm/|\mathbf{r}_1 - \mathbf{r}_k|;\ k = 2, 3, 4, \ldots, 9 \qquad (6.95)$$

Introducing eq. (6.95) in eq. (6.93), we can determine the total perihelic rotation of planet Mercury. Of course, we can include in eq. (6.95) all the stars of the Milky Way which are contained in a spherical volume of 10 LY (Light Years) of radius from the sun. Later on, we may compare both results concerning the total perihelic rotation of planet Mercury. If we find a slight astrometric difference with the first results, then we will begin to understand one of the many concepts assigned to many misinterpretations of "Mach's Principle." The cosmic collective potential K* is the holistic action of the material universe on each one of the individual material bodies of the entire universe. But, is this cosmic action of K* a significant one? In other words, do the other galaxies exert any action on the members of the solar system? Let us estimate the gravitostatic potential ϕ^* of the nearest galaxy on our Milky Way.

The nearest galaxy is Andromeda which is around two millions light-years from the Milky Way. This enormous distance is equal to 19,000,000,000,000,000,000 kilometers. This galaxy is so far away that most likely it may not influence our sun or planets in our planetary system. However, Andromeda's gravitostatic potential ϕ^* has the same magnitude as the solar gravitostatic potential ϕ_o at a distance of 189 million kilometers from the sun. This distance from the sun is almost half way between the orbits of planet earth and planet Mars. As we proceed away from Mars, the solar gravitostatic potential keeps on decreasing, while the gravitostatic potential of Andromeda remains practically constant over the entire region of our planetary system. This means the gradient of ϕ^* is zero, and therefore, there is no gravitostatic force acting on any of the planets of our solar system. This conclusion, of course, only refers to the gravitostatic potential. It does not refer to the gravitokinetic potential, which causes gravitodynamic forces proportional to acceleration (gravitational radiation). In other words, our solar system is immersed in a fluctuating flow of a gravitodynamic density of energy. This source is not localized in any particular place in the universe, because it is omnipresent in the entire universe.

Here we are in the presence of *quantum gravitodynamics.* It would be interesting to calculate the probability of a cosmic tunneling effect on planet earth. If the geometrical configuration of all the galaxies of the universe is a very special one, then the gradient of K* could be greater than the solar attraction of the earth and could be in the opposite direction. This improbable event would take planet earth away from the solar system. The distant material galactic universe contributes to the increase of the gravitational energy density in the neighborhood of our solar system, but this distant cosmic matter of the universe does not act with any force on any planet of our solar system. This is so because the potential energy of the entire universe, including the extra-galactic nebulas, is constant in the neighborhood of the solar system, and hence, there is no gradient of cosmic gravitational potential energy.

If we introduce eq. (6.94) in eq. (6.93) we get *Bohm-Schrödinger's Equation* of quantum gravitodynamics, in total analogy with eq. (2.42):

$$- (\hbar^{*2}/2m)\ \nabla^2\psi + U\psi = \quad {}^*i\ \partial\psi/\partial t \qquad (6.95b)$$

Let us finish this section with a metaphysical (unrealistic, absurd) cosmic experimental proposal. Let us diffract one galaxy by making it collide with another galaxy. Analyzing the "cosmic diffraction pattern" of the scattered stars, we may determine the numeric value of $\hbar^*$. To be sarcastic, we must say that this experimental proposal would be much easier to conduct, than to set the entire universal galactic system into rotation around planet earth at rest, without rotation. This latter metaphysical (unrealistic, absurd) experimental proposal of Mach, celebrated by Einstein, his advocates and some new so-called "dissident" physicists, is to prove the existence of an authentic centrifugal force on planet earth when the galaxies rotate around it. We must not forget that the rotation of the distant galactic matter, as well as the non-rotating earth, is with respect to a *stationary* reference system *fixed at infinity*. This reference system fixed at infinity is identical to Newton's *absolute space*, as we proved before.

6.6.1 PHYSICAL MISINTERPRETATIONS OF EINSTEIN'S GEOMETRO-DYNAMICS.

On Mach's Principle. There are, indeed, too many different versions of Einstein's original mathematical Mach's principle. Assis, in his recent book of 1999, titled *Relational Mechanics* [112, p. 121], has another list of eleven statements of Mach's principle. Two of the authors mentioned by Assis are on Speiser's list. Thus Speiser's list is reduced to 45 statements of Mach's principle. If we add Assis's eleven statements to Speiser's list, we end up with a total of 56 statements of Mach's principle. Only seven statements out of 56 coincide with Einstein's original mathematical statement of Mach's principle. In other words, only 12.5 % of the quoted renowned physicists understand rigorously that *Mach's Principle is nothing but Einstein's field equations!* How then, can we explain that 87.5 % of our *well-known* physicists do not understand that Einstein, himself, called his field equations of GRT by the name *Mach's Principle?* Perhaps, the only rational explanation of this confusion is the existence of another Mach's principle. According to Assis [113, p. 119], M. Schlick, in 1915, was the first to mention some conceptions of Mach with the terms *Mach's principle* and *Mach's postulate*. Schlick's statement of his Mach's principle is that *"the cause of inertia must be assumed to be an interaction of masses."* Thus, we have at least three Mach's principles.

1. Mach's metaphysical principle, created by Mach himself, is given by proposition P3 of section 6.5.
2. Mach's mathematical principle, created by Einstein, is given by Einstein's field equations.
3. Mach's ontological principle, created by Schlick, establishes that the inertia (mass) of any body is *caused* by the interaction of the body with all other material bodies.

Schlick's statement should be included in the 87.5% of physicists who claim the inertia (mass) of a body is determined or caused by the total distribution of matter in the universe. From now on, we will not distinguish these three Mach's principles from each other. Whenever we mention *Mach's principle,* we will be referring to Einstein's field equations. Mach's concepts constitute a variety of statements, not too clearly expressed, and what is most unfortunate, they are misinterpreted by a great majority of physicists. We have seen that Einstein's Mach's principle is confounded with Schlick's Mach's principle.

The operational procedure to determine the motion of a particle in a gravitational field according to GRT, is the following:

1. We determine the energy-momentum tensor $T_{\mu\nu}$.

2. We solve Einstein's field equations in order to determine the metric tensor $g_{\mu\nu}$. In other words, we use *Einstein's Mach's principle* to determine the components of the gravitational potential given by the components of the metric tensor $g_{\mu\nu}$.
3. We determine the components of the affine connection $\Gamma^{\mu}{}_{\alpha\beta}$.
4. Finally, we introduce the affine connection in *Einstein' equivalence principle*, given by the following equation of motion:

$$d^2x_\mu/ds^2 = [- \Gamma^{\mu}{}_{\alpha\beta} (dx_\alpha/ds)(dx_\beta/ds)] \tag{6.96}$$

Multiplying the last equation by the inertial mass of the body, moving in a gravitational field, we get:

$$F_\mu = m_i\, d^2x_\mu/ds^2 = m_i [- \Gamma^{\mu}{}_{\alpha\beta} (dx_\alpha/ds)(dx_\beta/ds)] \tag{6.97}$$

This last equation is Newton's second axiom of motion, generalized by Einstein to tetra-dimensional space-time. In going back from eq. (6.97) to eq. (6.96), it is obvious that we have simplified eq. (6.97) by the inertial mass of the body in motion. In eq. (6.96) there is no trace of the inertial mass of the body in motion. In the equivalence principle of Einstein, given by eq. (6.96), we have nothing about the chemical (atomic), or nuclear composition of the inertial mass of the body in motion. Einstein's equivalence principle leaves no trace of the inertial mass of the body in motion; i.e., GRT is not supposed to say anything about the ontology of the inertial mass of the body in motion. Let us not mix Schlick's misinterpretation of Mach's concepts with Einstein's GRT. If we insist on saying the mass of a body is caused by its interaction with the entire material universe, then we must conclude with Mario Bunge, that the mass of the universe is zero, because the material universe does not interact with anything material outside the universe. Mach's definition of mass, as a ratio of acceleration, is a laboratory method to determine the mass of a body with respect to an *inertial reference system*. Mach never said what Schlick said he did: the inertia (mass) of a body *is caused* by its interaction with all material bodies. For Mach's positivist philosophy it was an extremely iconoclastic act to mention the word *cause*. For this reason, he would never say the inertia of a body *is caused* by the rest of the material universe. Mach's proposition P3, in section 6.5, refers to the motion of a body. It does not refer to the inertia (mass) of the body. Even Einstein believed that, according to Mach, the inertia of a body has to increase in the vicinity of other material bodies. Einstein [114, p. 102] writes an equation similar to the following one:

$$(d/dl)[(1+\sigma^*)\mathbf{v}] = \nabla\sigma^* + \partial\mathbf{A}/\partial l - \mathbf{v}\text{x}(\nabla\text{x}\mathbf{A}) \tag{6.98}$$

where σ^* is the gravitostatic potential; $l = x_4$ in 4-dimensional space-time; **A** is the geometrodynamic vector potential, or as we call it in this book, gravitodynamic vector potential. The last equation is the equation of motion of a body of inertial mass "m" in a linearized gravitational field of GRT. The right hand side of Eq. (6.98) is Lorentz's gravitodynamic force. We should pay serious attention to the last equation. The mass m of the moving test body never appears. Eq. (6.98) is the solution of Einstein's linearized field equations, or maybe we should say, is the solution of the Linearized Mach's principle. On the other hand, the left hand side of eq. (6.98) represents the kinematical acceleration (accelerational field) *of something*. The right hand of the same equation represents the different gravitational accelerations (gravitational field). Therefore, eq. (6.98) represents the mathematical expression of the *Einstein Equivalence Principle*. Thus:

> The solution of Mach's principle (Einstein's field equations) is Einstein's equivalence principle (equality of an accelerational field to a gravitational field).

However, eq. (6.98) is not a dynamical equation until we multiply it by the mass m of the testing body:

$$(d/dl)[m\mathbf{v} + m\sigma^*\mathbf{v}] = m\nabla\sigma^* + m\,\partial\mathbf{A}/\partial l - m\,\mathbf{v}\mathrm{x}(\nabla\mathrm{x}\mathbf{A}) \qquad (6.99)$$

Einstein makes the following interpretation of the square bracket on the left hand side of the last equation:

$$[m\mathbf{v} + m\sigma^*\mathbf{v}] = [(m + \sigma^* m)\mathbf{v}]$$

It is clear that the mass m increases in the quantity σ^*m when the body of mass m is moving with velocity **v**. What happens when the velocity **v** is zero? In this case, nothing can be said based on eq. (6.99). This eq. (6.99), as is written above, offers this other interpretation. The gravitational field *transfers* linear momentum, $\sigma^*m\mathbf{v}$, to a moving body.

Still another interpretation of eq. (6.99) is possible. Let us rewrite eq. (6.99) in the following form:

$$(d/dl)[m\mathbf{v}] = m\nabla\sigma^* + m\,\partial\mathbf{A}/\partial l - m\,\mathbf{v}\mathrm{x}(\nabla\mathrm{x}\mathbf{A}) - (d/dl)[m\sigma^*\mathbf{v}] \qquad (6.100)$$

Obviously, the last equation represents Newton's axiom of motion and a few gravitational terms belonging to a linearized solution of Einstein's field equations. The last term of eq. (6.100) indicates the existence of a gravitational force proportional to the acceleration of the test body. Our Newtonian Relativistic Gravitodynamics, given by eq. (6.91), contains such a term proportional to the acceleration of the test body in a gravitational field. Eq. (6.98) is the result of a mathematical manipulation and interpretation which is very similar to the work we did with Marinov's hybrid electrodynamics given by eq. (4.36).

In summary, when we read literature written by many scientists and a few philosophers, saying that Einstein's GRT does not satisfy Mach's principle, we must ask - which Mach's principle? - If they say it is Einstein's Mach's principle, then the answer they give is equivalent to this absurd statement: *Einstein's field equations do not satisfy Einstein's field equations!* Or this other one: *Mach's principle of Einstein does not satisfy Mach's principle of Einstein!*

On Inertial Forces. An inertial force is not caused by any material body. The fact of multiplying the mass of a body by its acceleration, with respect to a laboratory, does not assure we have a force **F**=m**a** caused by another material body. This will be the case only if the laboratory constitutes an *inertial reference system*. In this case, the laboratory is at rest or moves with a constant vector velocity with respect to absolute space. This is equivalent to a stationary reference system fixed at infinity. Only in this case, Newton's axioms of dynamics are valid. Only in this case, there is a reaction to the action force **F**=m**a** which is acting on the material body which caused the action force **F**=m**a**.

If the laboratory is moving rectilinearly with acceleration **a***, with respect to absolute space (stationary reference system fixed at infinity), then the laboratory constitutes a *noninertial reference system*. In this type of laboratory, *Newton's axiom of action and reaction is not valid*. We may call the force **F**=m**a** an action force where "a" is the kinematic acceleration of the body of mass m with respect to the walls of the laboratory. But in this case, the action force **F**=m**a** is not caused by any material body. By the way, the magnitude of "a" is equal to the magnitude of a*. In this case, the force F=ma is called *inertial force*. To any inertial force there corresponds no reaction force in the context of Newton's theory of classical dynamics. Now, if the laboratory is rotating with an angular velocity ω*, with respect to absolute space (stationary reference system fixed at infinity), then we have other types of inertial forces which are not caused by material bodies. The names of these forces are Coriolis' inertial force, centrifugal inertial force and Euler's inertial force. To each of these three forces, there are no corresponding reaction forces.

Mach, in his metaphysical speculations, thought that: *"The principles of mechanics can, indeed, be so conceived, that even for relative rotations centrifugal forces arise."* We should generalize this speculation of Mach by saying - *The principles of mechanics can, indeed, be so conceived, that even for relative rotations inertial forces arise*.- Mach's statements have remained in his mind and relativistic books as a disgraceful metaphysical obscurity until today. Is it true that centrifugal forces on planet earth are the *relative inter-actions* between planet earth and the distant extra-galactic material nebulas? This question introduces, in any cosmological discourse, an extremely dark metaphysical obscurity. We can be absolutely certain that it is impossible, for any arrogant little terrestrial being, to stop the rotation of the earth with respect to absolute space (stationary system fixed at infinity). It is, also, impossible to physically set into rotation the whole entire material universe to show *pragmatically* the existence of centrifugal forces on planet earth. This is the most ludicrous proposition which, of course, we can never accept in this new millennium.. Even Einstein [115, p. 102] claims, that inside a rotating hollow sphere, a Coriolis force and a centrifugal force are deduced from GRT. This conclusion, Einstein said, has been shown by Thirring. But Thirring [116], in the conclusion of his paper, was very clear as we have mentioned before. He declared that these two Einsteinian gravitational forces are *analogous* to the Coriolis force and the centrifugal force. These two Einsteinian gravitational forces *are not identical* to the inertial forces of Coriolis and centrifugal in classical mechanics.

Thirring, in the introduction of his work *On the Effect of Distant Bodies in Rotation According to Einstein's Gravitational Theory*, asked the following question:

> "Is the new theory [GRT] free of the deficiencies of Newtonian theory, such that the rotation of distant bodies, according to its equations, produce gravitational fields equivalent to a centrifugal field?"

Equivalent (equal value or magnitude) does not mean *identical*. Thirring, in his work of 1918, tried to explain the presence of the geometrodynamic axial force proportional to the square of the angular velocity, like the centrifugal force. In this respect he wrote:

> ". . . the approximation of the celestial sphere by a hollow sphere infinitesimally thin, is incorrect. Even if we improve our approximation (having a space mass distribution), we never would obtain a field that would be equivalent to a true centrifugal field."

Let us repeat Thirring's conclusion, quoted already in the comments on Axiom 3, in section 4.3:

> "Through a concrete example it is shown that in the gravitational field (of Einstein) produced by distant masses in rotation, appear forces that are *analogous* to the centrifugal and Coriolis forces." [Italics added]

The final conclusion about the inertial forces and the Einsteinian geometrodynamic forces is that Einstein created a gravitational field theory from which it is perfectly possible to deduce gravitodynamic terms. These terms may go way beyond the only classical gravitostatic term in Newton's gravitation theory. Einstein's gravitodynamic theory is simply the solution of Einstein's field equations. The other final conclusion is that these different gravitodynamic forces of Einstein's GRT *are not identical* to the inertial forces of Coriolis, centrifugal and Euler. Similar geometrodynamic forces of Einstein are reproduced by our Newtonian Relativistic Gravitodynamics given by eq. (6.91), which is a consequence of Axiom 5, in section 4.2. Let us bring to this section the statement of Axiom 5:

> The mathematical structure of the forces of interaction between two particles in motion is directly proportional to the mathematical structure of the inertial accelerations: Coriolis, centrifugal and Euler; and inversely proportional to the relative separation of the particles.

Thus, from the very beginning of our Newtonian Relativistic Gravitodynamics, the new gravitodynamic forces *are not equal* to the inertial forces of Coriolis, centrifugal and Euler, but *directly proportional* to the inertial accelerations of Coriolis, centrifugal and Euler, and *inversely proportional* to the relative distance of the interactive particles. Now, the gravitodynamic terms of our Newtonian Relativistic Gravitodynamics are equal to the corresponding geometrodynamic terms of Einstein's GRT except for a numerical factor in the constant C of eq. (6.91). Consequently, Thirring's conclusion of his work of 1918 on GRT is completely corroborated by our Axiom 5 of the Newtonian Relativistic Gravitodynamic.

On Derivations of Newton's Axioms of Motions. After Newton established the foundations of classical dynamics, other physicists like Mach, and mathematicians like Lagrange, Hamilton or Jacobi, plus philosophers like Kant have produced derivations of all or some of Newton's axioms or principles of dynamics. Recently, some physicists claim the derivation of Newton's axiom of motion from gravitodynamics, which were obtained by analogy, with electrodynamics of the 19th century. We call these derivations *naive derivations* because, in one way or another, these authors of naive derivations of **F** = m**a**, do not realize that their starting point of the so-called "naive derivation" is precisely **F** = m**a**. In other words, their conclusions are derived from their conclusions. In what follows, let us make some brief comments on these naive derivations of Newton's axiom of motion. Let us recall that an axiom of any theory is a proposition which cannot be derived logically, or deduced from any other previous proposition. An axiom of one theory could be a theorem (logical conclusion) of another theory with another set of principles. One more explanatory note. *The principle of inertia*, in Newton's theory, *is not a principle* but a logical consequence (physical theoretical law verified experimentally) of Newton's axiom of motion **F** = m**a**.

On Kant's derivation of the law of inertia. Kant's derivation [117, chap.7], of Newton's law of inertia, is only a metaphysical deduction based on the principle that "every change [effect] must have a cause." After some scholastic discourse, Kant reaches the declaration of the inertia law: "*Every change in matter has an external cause.*" Of course, we can modify this metaphysical statement in many ways. For example, we can say that "Every change in the motion of a material body must have a cause." Let us improve on this last statement: "Every change in the quantity of motion (linear momentum) has a cause." Obviously, we can keep modifying Kant's original statement until we arrive at **F** = m**a** =d(m**v**)/dt, from which the law of inertia is deduced. In this scholastic process, we need a synonym of *change.* This synonym is *time* according to the ancient philosopher Heraclitus. Kant never intended to do this last work.

On Mach's derivation of Newton's axioms of motion. We have already seen how assertive Mach was in proposing "gedanken" experiments using Newton's axiom of motion (F=ma), and Newton's axiom of action and reaction in order to define the mass m of a body from m/m' = a'/a, choosing m'=1. Perhaps, this derivation of Newton's axiom by Mach is not naive, but subtle and very shrewd. The practical consequence of Mach's scientific masquerade is a Machian operational definition of a Newtonian inertial reference system.

On Analytical Dynamics. Newton's axiom of motion is a second order differential equation. We can integrate this differential equation, with respect to time, obtaining the vector law of the change of momentum equal to the total impulse. As it is easier to mathematically manipulate scalar quantities, we can integrate Newton's axiom of motion with respect to space displacement, obtaining the law of variation of kinetic energy, equal to the total work function. This work function can be replaced by the variation of potential energy. Analytical dynamics was born using this scalar approach, and after the Calculus of Variations was created by mathematicians. Ever since this author was an undergraduate, he has always had problems understanding this analytical method. When Lagrange equations are deduced, in order to apply them to practical physical problems, we must first determine the Lagrange function L or Lagrangian defined by L = T - U. T is the kinetic energy, of the material system, expressed as a function of generalized coordinates, as well as, the potential energy function U.

Now, the vicious circle begins with U. To determine U, we must first know the force *F.* Knowing the mathematical structure of force *F,* we determine the function U as the space-displacement integral of *F.* Now, we insert U in the Lagrangian. Now we can solve Lagrange's equations to determine the motion of a body of mass m. But the vector equation of motion is given by m**a** = **F**. Therefore, the analytical method of Lagrange begins by using **F** = m**a.** Then we conclude that m**a** = **F**. Is there any other way to determine U so as to avoid this absurdity of Lagrange's equations? Yes, but it is not too elegant.

This other method is called, by this author, "Mexican Piñata Math." The candies in the hanging bag are mathematical functions which might correctly represent U, but this would be only by chance. In other words, electrokinetic or gravitokinetic potential energy functions U are invented by mathematical physicists. Wesley [118, p. 214] is very accurate when criticizing Lagrange's method. Let us repeat his quotation from section 3.1:

> "The Lagrangian requires a knowledge of the energy integral (the kinetic energy and the potential energy) to start with, so one has to essentially solve the problem before one can even state the problem by the Lagrangian method."

This is totally equivalent to saying, we need **F** = m**a** to use Lagrange's method in order to deduce or derive **F** = m**a**. Wesley drastically proposes to ignore Lagrange's method. From a logical point of view, Wesley is absolutely correct. But from a guessing point of view, we can continue playing the " Mexican Piñata Math." Many professors and authors presume that *Analytical Dynamics* is a branch of Mathematics. Lanczos [119, p. 7] is very explicit by saying that:

> "Analytical mechanics is a completely mathematical science."

Such an assertion clearly shows the confusion of the substance of the subject matter with the language used to express it. No one would accept as true the following assertion: "The statement of the principle of 'least action' is a completely literary work," simply because it has been expressed in words. Lanczos, [120, p. xxviii] himself, interprets Mach's positivist comments on Lagrange's method by saying:

> "According to this philosophy the variational principles of mechanics are not more than alternative mathematical formulations of the fundamental laws of Newton, without any primary importance."

Both statements of Lanczos are very badly expressed. The first one should say that - Analytical mechanics is the application of the Calculus of Variations to Newton's axioms of dynamics. - The second quotation of Lanczos expresses Mach's conception of the variational principle to mechanics. The variational principle of Calculus of Variation is, indeed, a branch that completely belongs to mathematical science. On the other hand, the fact that Newton's axiom of motion is an ordinary differential equation of the second order, does not make Newton's axiom of motion a branch of ordinary differential equations, belongimg to mathematical science.

On naive derivations of Newton's axiom of motion. It seems that Mach's conceptions bewitched the minds of subjective relativists (Einstein's advocates) and objective anti-relativists (Einstein's opponents). Twentieth century authors of books and papers about gravitation seem to have taken for granted the metaphysical obscurities of Mach. Everyone wants to set the entire material universe into rotation with respect to absolute space (stationary reference system fixed at infinity). Thus, everyone who wrote about gravitation in the last century, compares his conclusions with the undeserved "great authority" of Mach. There are some people who write in a very assertive manner about *Mach's Mechanics.* Mach's mechanics never existed. Finally, there are others who claim to have *derived* Newton's axiom of motion from some gravitational theories. Let us address these claims now.

In 1991, J.P. Wesley [121, p. 187], after converting, by analogy, Weber's electrodynamics into Weber's gravitodynamics, analyzed *his personal Mach's principle.* He concluded that from Weber's gravitodynamics he derived Newton's second law ($\mathbf{F} = m\mathbf{a}$). This claim is correct if we start from Weber's electrodynamics, but what is the genesis of Weber's electrodynamics?

In 1993, A.K.T. Assis [122, p. 218], and again in 1999, Assis [123, p. 182], claims he has derived Newton's second law of motion from Weber's gravitodynamics. He obtained this gravitodynamics by analogy from Weber's electrodynamics. This claim is similar to Wesley's, and it is, also, correct if we start from Weber's electrodynamics. But we again repeat, what is the genesis of Weber's electrodynamics?

Both authors, Wesley and Assis, forgot the valuable lesson taught by Wesley himself, in his criticism of Lagrange's method of Analytical Dynamics which we quoted previously. These claims of deriving Newton's axiom of motion are simple illusions. The reason for these misconceptions is the following. Weber's electrodynamics was *derived* from a potential energy function which was *invented.* Then we have to use Lagrange's method to get Weber's electrodynamics. At this point, we realize that Lagrange's method is another mathematical way to express Newton's axiom of motion. Thus, Lagrange's scalar equations of motion represent Newton's vector axiom of motion but expressed in a rather abstract way by Lagrange. It is no wonder that at the end of the Lagrangian process we *recuperate* Newton's axiom of motion. We hope, in the future, these types of claims will disappear from scientific literature.

Even Einstein's GRT has the imprint of Newton's mark in the equations of motion which were calculated by Einstein and others, such as Thirring. The connection with Newton is through Gauss's law, and the solid angle inversely proportional to the square of a distance, as it is in Newton's gravitational law, based on Kepler's laws. These steps lead to Poisson's equation which inherited the universal gravitational constant G. After the generalization of Poisson's equation into Einstein's field equations, everything is fine from a metaphysical point of view. At one point, Einstein's GRT is forced to descend into reality from the *Olympus of Trancendental Symbolism of the Aprioristic Realm*. This mundane bridge is created by Einstein when he identifies his constant κ (kappa) with Kepler's astronomical observations of the real world, but through Newton's work. At that moment, the metaphysical GRT becomes a physical theory to be verified later. The conversion of κ is the following one:

$$\kappa = 8\pi G/c^2 \tag{6.101}$$

Einstein's constant κ doubles our constant $\mu_0^* = 1/(\varepsilon_0^* c^2) = 4\pi G/c^2$, given by eq. (6.91a). When the equations of motion of a body of mass m are immersed in a gravitational field, and are obtained from Einstein's field equations and multiplied by m, then at that moment, we recuperate Newton's axiom of motion. In a particular case of a weak gravitational field in "vacuum," we get the components of Newton's equations of motion in terms of Einstein's new gravitational terms: This recuperation of Newton's axiom of motion is very subtle. It is not astute nor is it naive.

6.7 Starlight deflection by the solar energy field.

Now we will apply the Primordial Energy Field Theory, developed in chapter 5, to analyze the starlight deflection caused by the gravitostatic energy field of the sun. In this first analysis, we will not consider the gravitokinetic solar energy, nor the solar electromagnetic field which surrounds the sun. Later on this assumption may have significant consequences. However, from a methodological point of view, any problem to be solved, for the first time, should be reduced to its simplest statement, without, of course, changing the essence of the problem.

Starlight Deflection by the Sun. In 1922, Prof. C.L. Poor [124] in his book, *Gravitation versus Relativity,* describes the rejection of 28 photographic plates out of 35 which were taken in 1919 on the occasion of a solar eclipse. The rejection was based on great discrepancies with Einstein's theoretical prediction. The results were presented in London, in November of the same year, acclaiming the astronomical work done by British scientists led by A. Eddington. He *verified* the theoretical prediction of the German scientist, Albert Einstein. None of the seven "best" photographic plates, presented on that memorable meeting according to Prof. Poor, had a discrepancy less than 20% in respect to Einstein's prediction.

In 1924, Capt. T.J.J. See [125], professor of mathematics in the U.S. navy and government astronomer at Mare Island, claimed that Einstein's theoretical calculations of the solar deflection of starlight were in error. Professor See's work deserves some serious review, particularly after the corrections made in 1974 by professor Merat of the University of Paris. An interesting quotation in Capt. See's work when he mentioned the calculation of starlight deflection by Von Soldner of Munich in 1801. Soldner found a deflection of 0.84" for a ray grazing the solar surface. Soldner did not multiply by two in order to get the total deflection equal to 1.68". According to Cpt. See, Einstein made the same mistake in 1911.

In 1974, Prof. P. Merat [126] from Paris analyzed most of the observations of the starlight deflection, near the solar limb, which had been published up to that year. He writes in his paper: "*Although the actual number of stars is somewhat inferior, our data comprises a total of 297 stars' deflections resulting from nine groups of observations during six total solar eclipses.*" Merat's results indicate that there is an observational excess of starlight deflection of 10-15% in respect to Einstein's GRT prediction for $R_o \circ r \circ 5 R_o$, where R_o is the solar radius. For $5R_o \circ r \circ 13R_o$, there is a close agreement between Einstein's prediction and the observational one. Our conclusion of Merat's investigation is that GRT should improve the analysis of the interaction between electromagnetic waves and geometrodynamic fields for the case of rays passing close to the solar limb. Table 6-V shows Merat's modifications to the astronomical determination of starlight deflection by the sun.

In Table 6-VI, we display Einstein's prediction of starlight deflection with Merat's deflection δ(Merat) $\pm \Delta\delta$, where $\Delta\delta$ is the root mean square deviation of all the astronomical observations. We see in this Table that in the first three rows, Einstein's predictions fall outside the astronomical interval of accuracy. The astronomical measurements are in excess of Einstein's theoretical magnitudes. In the rest of the rows of Table 6-VI, we see an acceptable agreement between Einstein's predictions and the astronomical data.

From Table 6-VI, we see that Einstein only has four predicted values inside Merat's intervals of accuracy. Any theoretical amendment, to the inverse law of Einstein's deflection of starlight by the solar gravitational field, should explain the excess of deflection in the proximity of the solar limb. The same requirement is demanded for any other theory.

The distance r is expressed in units of the radius of the sun R_o. Einstein's starlight deflection by the gravitational field of the sun, according to GRT, were calculated by the following equation:

$$\delta = 4GM_o/(mR_o) \tag{6.102}$$

where M_o is the sun mass, and $mR_o = r$. Merat concludes that Einstein's deflection may be different from the inverse distance law given by eq. (6.102) for $r < 5R_o$. We observe in row 4 of Table 6-V that in Merat's column the numeric value 0.40" seems to be out of place. A simple linear interpolation between the values 0.58" and 0.41", provides the value 0.495". This could be a statistical anomaly.

Starlight Deflection by the Sun Energy Field as a Refraction Phenomenon.
The sun, like any other celestial body, is immersed in a universal primordial field of energy. To this field is superimposed the galactic field of energy caused by the galactic matter of the universe. The stellar field of energy is superimposed onto the previous one produced

Table 6-V. Comparison of Einstein's predictionof starlight deflection with Merat's astronomical corrections

Row	(R_o units)	Einstein(")	Merat(")
1	2.09	0.84	1.02 ± 0.11
2	3.12	0.56	0.67 ± 0.08
3	4.02	0.44	0.58 ± 0.04
4	5.10	0.34	0.40 ± 0.07
5	6.06	0.29	0.41 ± 0.04
6	7.11	0.25	0.31 ± 0.04
7	7.84	0.22	0.24 ± 0.04
8	9.51	0.18	0.20 ± 0.06
9	11.60	0.15	0.16 ± 0.03

Table 6-VI. Comparison of Einstein's prediction of δ^* with Merat's law: $\delta \pm \Delta\delta$. The RMSD is $\Delta\delta$

Row	r	Merat δ-$\Delta\delta$. (")	Einstein δ^*	Merat δ+$\Delta\delta$. (")	Merat's Interval (")
1	2.09	0.91	0.84	1.13	OUT
2	3.12	0.59	0.56	0.75	OUT
3	4.02	0.54	0.44	0.62	OUT
4	5.10	0.33	0.34	0.47	IN
5	6.06	0.37	0.29	0.45	OUT
6	7.11	0.27	0.25	0.35	OUT
7	7.84	0.20	0.22	0.28	IN
8	9.51	0.14	0.18	0.26	IN
9	11.60	0.13	0.15	0.19	IN

by the stellar matter of our own galaxy. Finally, the energy field of our sun is superimposed onto all the previous fields of energy. Thus, the total energy density ρ in the proximity of our sun is given by:

$$\rho = \rho^* + \rho_S \tag{6.103}$$

where

$$\rho_S = GM_o^2/(8\pi r^4) \tag{6.104}$$

ρ^* is the universal or cosmic energy density. Eq. (6.104) expresses the gravitostatic energy density of our sun. The purpose of this sub-section is to determine the numeric value of the cosmic eñergy density ρ^*, using the starlight deflection by the energetic envelope of the sun. To perform this task, we have two methods to accomplish it. One is to use the eikonal equation of chapter 5, or Snell's law of refraction. To use the eikonal equation, is analogous to killing a mosquito with a gun. We will use the refraction method because of the insignificant angle of deflection of the starlight, and because we will not evade the physical phenomenon involved in the solution of this problem.

If we consider the total cosmic energy field, it is very easy to understand the light deflection by the sun. Every celestial body is surrounded by an invisible envelope of gravitostatic energy caused by the matter of the body and given by eq. (104). To proceed with a colorful description, let us assign a yellow color to the sun's gravitostatic energy. Let us picture the background cosmic energy with a bluish color. Now we can see, in our imagination, that the sun is surrounded with a green atmosphere of energy. The green color fades away into a bluish color as we recede from the sun. The GRT starlight deflection is very small; at two solar radii from the solar limb, its value is 0.58 (arc-seconds), according to Einstein's eq. (6.102). This insignificant magnitude means that the starlight refraction must occur where the solar green energy envelope is almost bluish. If the light from the distant stars passes close to the solar limb, the starlight refraction should increase, because the energy density increases as we get closer to the sun. Thus, the Primordial Energy Theory, developed in chapter 5, seems to explain the starlight deflection by elemental geometrical optics. Now we have to translate this allegoric picture into the language of mathematics. We must do this in order to quantify this "metaphysical obscurity," to calculate the numeric value of the cosmic energy density ρ^*.

Let us assume an impact parameter $p = NR_o$, with $N \geq 1$. To determine the radius "r" of the solar energy sphere at which the ray of the starlight begins to refract, we need to know the *sagitta* "s" of the great circle which contains the center of the sun, the point where the refraction begins to take place, and the point where the astronomical observatory is located on planet earth. In mathematical terms we have:

$$r = p + s.$$

To estimate a reasonable value for s we assume:

$$s = f R_o$$

where f is a numeric factor. Now, the previous equation becomes:

$$r = NR_o + f R_o = (N + f)R_o \qquad (6.105)$$

Assigning different numeric values to N and f we will determine the cosmic energy density ρ^*. Let us now proceed with the analysis of the refraction phenomenon. Let α and β be the angles of incidence and refraction, respectively. Then:

$$\sin\alpha = n\sin\beta \qquad (6.106)$$

where $n = n(r)$ is the index of refraction, as we demonstrated in chapter 5. If there is no refraction of the starlight ray, the ray would be a geometrical secant which would go across the great circle we mentioned before. However, the ray is slightly refracted, and runs closely to the chord determined by the secant. The index of refraction n, for all practical purposes, remains constant inside a layer of energy. The starlight ray will be refracted again when leaving the energy sphere of radius r. At that moment, the ray suffers another refraction which bends the ray again. The first refraction deviates the ray from the secant direction an angle ε:

$$\varepsilon = \alpha - \beta$$

The total final deflection from the secant direction is given by:

$$\delta = 2\varepsilon = 2(\alpha - \beta) \qquad (6.107)$$

From trigonometric considerations in the great circle mentioned before, we get:

$$\sin\alpha = 1/(1 + f/N) \qquad (6.108)$$

From equation (6.106), we get:

$$\sin\beta = (1/n)\sin\alpha \qquad (6.109)$$

In principle, the last three equations solve the problem, theoretically, if we know the index of refraction n. The rest is purely applied mathematics. From chapter 5, we get the index of refraction n, given by:

$$n = n(r) = \pi\{1 + GM_o^2/(8\pi\rho^*r^4)\} \qquad (6.110)$$

After some substitutions and the definition of K, the last equation becomes:

$$n = n(r) = \pi\{1 + K/[(N + f)^4\rho^*]\} \qquad (6.111)$$

where $K = GM_o^2/[8\pi R_o^4] = 4.48525 \times 10^{13}$ (ISU) (6.112)

Now, using the method of trial and error, and the previous equations with N = 4.02, δ(Merat) = 0.58" from Table 6-V, we tried different values for f to find the best theoretical fit to Merat's observational law.

The best value is f = 5.296. With these numeric parameters, we get the cosmic energy density ρ* from Snell's and Merat's law:

$$\rho^* = 1.094291 \times 10^{15}\ (J/m^3) \text{ or } (N/m^2) \qquad (6.113)$$

With the previous equations and numeric parameters, we also calculated the starlight deviations with our theory of the primordial energy field. In Table 6-VII, we show these results in comparison with Merat's astronomical observational law of starlight deflection by the solar energy field. Our results are better than Einstein's prediction. Out of the nine observational numeric values of Merat's δ, we fall in his empirical accuracy interval with seven deflections. In row 5 of Table 6-VII, we have 5% discrepancy with the lower limit of the accuracy interval. In row 9 of the same Table, we have a discrepancy of 15% from the low limit of the accuracy interval determined by Merat.

Table 6-VII. Comparison of this author's predictions of δ^* with Merat's law: $\delta \pm \Delta\delta$. The RMSD is $\Delta\delta$

Row	r (R_o units)	Merat $\delta - \Delta\delta$. (")	Author δ^* (")	Merat $\delta + \Delta\delta$. (")	Merat's Interval
1	2.09	0.91	0.91	1.13	IN
2	3.12	0.59	0.73	0.75	IN
3	4.02	0.54	0.58	0.62	IN
4	5.10	0.33	0.44	0.47	IN
5	6.06	0.37	0.35	0.45	OUT
6	7.11	0.27	0.27	0.35	IN
7	7.84	0.20	0.23	0.28	IN
8	9.51	0.14	0.16	0.26	IN
9	11.60	0.13	0.11	0.19	OUT

From Table 6-VII, we see this author has seven predicted values (77%) inside Merat's intervals of accuracy, while Einstein has only four of his predicted values (44%) inside Merat's intervals of accuracy . For the value $\delta^*=0.35$, we have a small discrepancy of 5% in respect to Merat's value. For $\delta^*=0.11$, the discrepancy is 15%. The identification of the luminiferous ether, with a primordial field of energy, has provided us with a theory with which we are able to explain the empirical corrections, introduced byMerat, to Einstein's theoretical law of starlight deflection by the gravitational field of the sun. An *empirical work* could be an experimental work or an observational work. Both are based on εμπειρια (Gr. *empeiria* = experience). Experimental works are controlled by human beings or instruments designed by them. Observational works are never controlled by human beings. Kepler's laws are typical observational laws. The conclusions, obtained by Merat, after his analysis of 297 astronomical observations of starlight deflections, well deserves the name of *Merat's empirical law of starlight deflection by the sun*. Thus, any future theory must compare its theoretical predictions, of the solar deflection of starlight deflection, with Merat's empirical law. In obtaining the deflections shown in Table 6-VII, we also determined the numeric value of the cosmic energy density ρ^* equal to 1.094291×10^{15} (J/m^3).

6.8 Cosmological red shift and big bang theory.

The Big Bang Never Happened is the title of an extraordinary book written by Eric J. Lerner [127], and published in 1991. We believe there is too much audacity in the minds of metamathematicians who bring onto paper their metaphysical and mathematical obscurities. We also believe there is too much irreverence in the writings of metamathematicians who dare to tell us how the universe was created, if it ever *was* created. Once again we feel the rational necessity to paraphrase George Berkeley who, in the 18th century, said: "W*e will stop talking nonsense (absurdities) in Natural Philosophy, the day we will have the mental capacity of distinguishing distinctively and clearly, the difference between mathematical hypotheses and the essence of things.*" The Big Bang (BB) theory is the child of one *interpretation*. The interpretation was the cosmic or cosmological red shift of the spectral lines of distant stars. Astrophysicists accepted the *interpretation* or theoretical explanation that the cosmological red shift was caused by receding distant stars. These receding stars, they say (they speculate), causes a Doppler effect which is detected in terrestrial astronomical observatories. Thus, a whole planet of "thinkers" are supposed to stop thinking that the cosmological red shift could have another cause! The receding stars were taken for granted, and it was very easy to infer that *the universe is expanding*. Then it was elementary to conclude that some time, in the past, the universe was concentrated into a point. Then one day, when time did *not* exist, a huge BB took place with no apparent cause. In this way, the Big Bang (BB) theory was born, but only in the minds of cosmologists. Since then the BB only has had a noetic existence We have only one question to ask the advocates of the BB theory. Was the proto-universe a black hole? This question is extremely metaphysical, because any black hole has only a noetic existence; i.e., the black holes only exist in the minds of their advocates.

If we consult *The Cambridge Dictionary of Philosophy* to learn what *Cosmology* is, we become dismayed when reading - "*cosmology* - See METAPHYSICS." Then we go to METAPHYSICS and read that cosmology is "the science of the nature, structure and origin of the universe as a whole." Even in the very definition of cosmology, we find the assumption that the universe had an origin. In this book, we will take as synonyms the meaning of the words *cosmology* and *astrophysics*. When this author was an undergraduate student, he defined *astrophysics as a respectable bridge between science fiction and science*. The justification of this definition is extremely simple. No one can run experiments with stellar objects. Astrophysics is not an experimental science. Astrophysics is an empirical science based on pure observations. But observations are registered perceptions which must be *interpreted* in the context of a particular theory, or primitive pre-conceptions. In astrophysics or cosmology, or sociology, we cannot repeat the same natural phenomena, at will, in any laboratory. Hence, is there another interpretation of the cosmological red shift? The answer is in the affirmative. It has been since 1873. In Volume II of J.C. Maxwell's [128]: *A Treatise on Electricity and Magnetism*, page 446, equation (2) (Dover Publication, NY, 1954), we find the following equation written in the terminology of 19th century:

$$d^2F/dz^2 = \mu K d^2F/dt^2 + 4\pi\mu C df/dt \qquad (6.114)$$

which in our modern notation and nomenclature of Roberto A. Monti [129], and units of the ISU, look like this:

$$\nabla^2 F = \mu_o \varepsilon_o\, \partial^2 F/\partial t^2 + \sigma_o \mu_o \partial F/\partial t$$

or

$$\nabla^2 F - \mu_o \varepsilon_o\, \partial^2 F/\partial t^2 = \sigma_o \mu_o \partial F/\partial t \qquad (6.115)$$

The last equation is a nonhomogenous D'Alambert equation. F may represent one of the components of the electromagnetic field of a propagating wave with a decaying amplitude. The constant σ_o is the electrical conductivity of the cosmic energetic medium (ether) which pervades the entire cosmic vacuum. The general solution of the last equation is:

$$F = F_o \exp[- R^*\sigma_o\, r/2] \exp[(2\pi i/\lambda)(r - c^*t)] \qquad (6.116)$$

where $R^* = (\mu_o/\varepsilon_o)^{1/2}$ is the wave resistence of the energetic vacuum (ether), and c* is the light velocity in intergalactic space. Perhaps many professors of physics were told by obnoxious colleagues how stupid it was to use the International System of Units, in which we have the electric properties of vacuum, such as μ_o and ε_o. They never mentioned the other property σ_o of vacuum. If one truly and strongly believes that an electromagnetic wave can propagate in the "nothingness," of the vacuum of interstellar space, then these obnoxious professors will soon make their first step to wisdom. For these reasons, these unreasonable people refuse to use the old Rational MKS System of Units, because they are completely ignorant of the being of vacuum except that vacuum has no reality, does not exist in the external world , has *no being.* Let us emphasize once more, the lack of knowledge of the existence and nature of the cosmic energetic ether engendered the BB theory. The cosmological red shift can be *interpreted* in an elegant and rigorous way.

The first time this author learned about this other interpretation was in a paper by A. Ya. Kipper titled *Nature of the Cosmological Red Shift*, translated into English from Astrofizika, Vol. 10, No. 2, pp. 283-293, April-June, 1974 (Plenum Publishing Company, NY, 1975). In the physics community, more jokes were made about the new concept "photon fatigue," introduced by Kipper, than intelligent comments about the interaction of an electromagnetic wave with the energetic ether. The interested reader, in this new interpretation, must study a scholarly paper written by Professor Roberto A. Monti who has already been mentioned. He published his work in *Physics Essays*, Vol. 9, No. 2, p. 238, June (1996). The title of this paper is *Theory of Relativity: A Critical Analysis*. The mathematical clarity of the exposition, of the new interpretation of the red shift, is absolutely excellent. Using eq. (6.117) and the photo-electric law, we determine the energy associated to an electromagnetic wave:

$$\xi = F_o^{\,2} \exp[- R^*\sigma_o\, r] = K\nu = Kc^*/\lambda \qquad (6.117)$$

where

$$K = h(n + ½) \qquad (6.118)$$

The last equation represents the energy associated with the oscillation of an elementary electromagnetic wave; n is an integer such that $n \geq 0$. The case n=0 introduces the concept of the energetic ether from the point of view of quantum mechanics. In this case of n=0, the wave does not carry any energy susceptible to measurements, and, therefore, it is undetectable in a direct way in any experiment. According to the last two equations, when n=0, the wave energy exists and is called *a wave of the zero-point state,* and the corresponding energy is called *the zero-point energy*. This quantum analysis leads to the conclusion that the entire universe is pervaded with electromagnetic waves of the zero-point state and also of excited waves when n K0. Here we are in the presence of another path with which to arrive at the existence of the cosmic energetic ether. This is called the *zero-point energy of vacuum* from a quantum mechanical point of view of an electromagnetic wave.

Using eq. (6.117) for the energy associated to a propagating electromagnetic wave from the source at r=0, to a distance r from the source, we have for r=0:

$$\xi^* = F_o^{\,2} = K\nu^* = Kc^*/\lambda^*$$

The ratio of the energies associated to the same electromagnetic wave, after traveling a distance r, is given by:

$$\xi/\xi^* = \exp[- R^*\sigma_o\, r] = \nu/\nu^* = \lambda^*/\lambda \qquad (6.119)$$

Now, the red shift is defined as follows:

$$z = \Delta\lambda/\lambda^* = (\lambda - \lambda^*)/\lambda^* = \lambda/\lambda^* - 1 \qquad (6.120)$$

Introducing eq. (6.119) in the last equation, we get:

$$z = \exp[\, R^*\sigma_o\, r] - 1$$

or $$r = [1/(R^*\, \sigma_o)]\, \ln(z + 1) \qquad (6.121)$$

The last equation establishes a logarithmic law relating the distance r with the red shift. Monti, after comparing Hubble relativistic linear law and the logarithmic law, deduced from Maxwell's electromagnetic wave equation and Planck-Einstein's hypothesis, he concluded that:

> "... in any case, the logarithmic law fits experimental data much better than the linear law[12,34-37]; moreover, it has no problems with the age of the universe."

Thus, in terms of Kipper's and Monti's papers about another probable explanation of the cosmological red shift, the existence of the cosmic energetic ether finds another solid pragmatic verification. From eq. (6.117), we have:

$$\nu = \nu^* \exp[- R^*\sigma_o\, r]$$

Multiplying the last equation by K we get:

$$\xi = \xi^* \exp[- R^*\sigma_o\, r] \qquad (6.122)$$

From this last equation, we can deduce the *loss of energy* $\Delta\xi$ of an electromagnetic wave after it has propagated a distance r, interacting with the ubiquitous energy field of the universe (energetic ether or zero-point energy of vacuum):

$$\Delta\xi = \xi^* - \xi = \xi^*[1 - \exp(-R^*\sigma_o\, r)] \qquad (6.123)$$

The last equation shows, with meridian clarity, that the cosmological red shift can be explained, perfectly well, by means of the loss of energy of an electromagnetic wave which interacts with the cosmic energy field over astronomical distances. Eq. (6.122) is formally identical to Beer's absorption law of light by transparent solvents. We will call eq. (6.122) Beer's cosmological law. In this case, the absorbing transparent media is the zero-point energy of vacuum (ether: *identical* to cosmic energy primordial field).

Thus, the works of Maxwell (1873), Kipper (1974), Lerner (1992), and Monti (1996) constitute the foundations of a modern cosmology based on the ancient conception of an ether, of a modern ether, of an energetic ether with physical attributes. Hence, we may say, not with arrogance nor with fanaticism, that the cosmological red shift is probably and mainly caused by the interaction of a propagating electromagnetic wave with interstellar and intergalactic energy fields. If this is the case, then we have to enjoy Lerner's book *The Big Bang Never Happened.* Adding the plasma cosmology, initiated by Hannes Alfvén, to these other conceptions, we come to an agreement with Lerner when he said that plasma cosmology leads our understanding to a new conception of the universe:

> "... without a Big Bang, without any beginning at all, a universe that has always existed, is always evolving, and will always evolve, with no limits of any sort."

If the universe is Spinoza's universe or Einstein's universe, as we will see in the next and last chapter of this book, then the material visible universe and the energetic invisible universe, two aspects of the Being of all entities, cannot have a beginning nor an end. It must be eternal, outside the flow of time, outside the human mind. Now, let us consider the following question: Is dark matter rather bright?

Olbers' paradox states that we should have no nights on planet earth. This statement is a theoretical conclusion deduced by Heinrich W. Olbers [129*], in 1823. This conclusion is disproved every night when we look at the stars. The sky at night should be bright instead of dark, according to Olbers. The assumptions used by Olbers were two. The universe is infinite, and the number of stars in this universe is infinite. Because of the uncountable number of stars in the universe, Olbers concluded the nights should look like our daylight. Over the years scientists have provided many ephemeral *explanations* of this absurd theoretical conclusion. We can very well explain Olbers' paradox, using eq.(6.121) and the following parameter:

$$\sigma_o = 2.85 \times 10^{-29} \quad [1/(\text{ohm} \times \text{m})]$$

$$R^* \sigma_o = 1.095445 \times 10^{-26} \ (1/m)$$

Using eq. (6.121) and the value $z = 4.9$, we get $r = 17.12 \times 10^9$ (LY). This distance corresponds to the radius of the visible universe, according to Beer's cosmological law. Monti [129] reports the recession velocity v, according to the theory of the expanding universe, in terms of the red shift parameter z:

$$v = [(z+1)^2 - 1]/[(z+1)^2 + 1] \ c \qquad (6.123^*)$$

where c is the speed of light. Obviously, the last equation proscribes any distant galaxy or cluster of galaxies to move faster than the speed of light c. Let us determine the value of z for $v=0.99c$. The result is $z= 16.29$. Now, let us go back to Beer's cosmological law.

According to the Maxwell-Kipper-Monti (MKM) theory of light energy absorption in cosmic space, we should expect that after a critical distance r_c, the light emitted by stars in galaxies, beyond the critical distance, will never reach our solar system. Even in a finite universe with a radius $R>r_c$, we should find very *bright "dark" matter.* Thus, we do not have bright nights simply because the light of bright super distant galaxies is absorbed by the luminiferous ether, or primordial cosmic energy field, or zero-point energy of space (*of space* is better than saying *of vacuum*). Olbers' conclusion has been wrong not because the number of stars is finite in an infinite universe, but because, even in a finite universe with a finite number of stars, the cosmic transparent medium absorbs the energy of traveling electromagnetic waves according to Beer's cosmological law, based on Maxwell-Kipper-Monti's theory. Hence, here we bring a new source of dark matter that is very bright in itself.

6.9 Is gravitation an electrodynamic phenomenon?

Why do heavy bodies fall to the ground? Because the natural place for heavy bodies is on the ground. This was the answer given by Aristotle twenty-three centuries ago. What is gravity? Newton, in one of many letters sent to Bentley in 1692-93, wrote:

> "Gravity must be caused by an agent acting constantly according to certain laws; but whether this agent be material or immaterial, I have left to the consideration of my readers."[130, p.634]

Faraday speculated in 1850 that any type of force could be a different manifestation of one unique fundamental force. From this class of deliberation he came forth with this statement:

> "[It] has made me often think upon the possibility of establishing, by experiment, a connexion between gravity and electricity "[131].

Faraday's experimental results were negative, but not his mood, because he wrote: " *They do not shake my strong feelings of the existence of a relation between gravity and electricity.*" We will see in the following pages that Faraday's intuition, in this respect, was right.

At the end of the19[th] century, there were a series of attempts to reduce, *formally,* gravitation to gravitodynamics in analogy to electrodynamics. It appears that in 1938, O'Rahilly [132, vol. 2, p. 544] was the first one to write:

> "It is easy to see the *possibility*, on Ritz's theory, of explaining gravitational attraction as residual statistical forces between groups of moving charges; no such possibility is available from the Lorentz-Liénard theory. These forces must be due to terms of a high order and the forces will be small relative to the first-order forces familiar to us in electromagnetics."

Because of this conception of O'Rahilly, the author of this book, in 1985, initiated some "gedanken" experiments with the help of a computer. The diameter of a hydrogen atom is 1.056 Å (1 Ångstrom = 10^{-10} m). He discovered that two neutral hydrogen atoms, separated by 100 Å, were repelled in some cases, attracted or showed no interaction at all in other cases. The dynamical resultant interaction depends on the relative orientation of the orbital angular momentum of the revolving electrons in the hydrogen atoms. Coulomb's force was the only one used in these "gedanken" computerized models. After the two hydrogen atoms were separated by a larger distance, the same three results were observed, to the point that it was necessary to extend the computer precision to 17 significant decimal digits. Beyond a certain separation, it was necessary to extend the computer precision up to 50 significant decimal figures. In 1985, some universities of the area had extended the precision up to 30 significant decimal figures. Today, we can have any number of decimal significant figures in our desktop computers. Thus, this author dropped this mathematical model in 1985. At that time, he never saw that a hydrogen atom, in Bohr's model, could be represented by a rotating electric dipole moment which is equivalent to two orthogonal oscillating electric dipole moments. Only when this author read the extraordinary paper of André Assis [133], published in 1992, did he realize the meaning of *residual statistical force* produced by an ensemble of variable dipole moments in time. Before we comment on Assis' work, let us articulate what gravitation is in terms of GRT, and also mention the origin of Van der Waal's forces. Gravitation is a phenomenon of *a curvature of tetra-dimensional spacetime due to the presence of matter-energy* in a specific locality of the expanding tetra-dimensional universe.

To explain the deviations of real gases, from the *ideal gas law,* Van der Waal proposed the existence of intermolecular forces between neutral atoms and molecules. These forces arise in the interaction of (1) dipole-dipole, (2) dipole-induced dipole and ion-induced dipole, and (3) induced dipole-induced dipole (dispersion). There is a variety of intermolecular potentials. Perhaps, the Lennard-Jones intermolecular potential is the most popular. Any theoretical approach to reproduce the radial distribution function of real gases or liquids, determined by X-Ray or neutron diffraction, depends heavily on the mathematical structure of the intermolecular potential. The derivation of experimental thermodynamic properties of the system is the crucial test of the molecular theory and the intermolecular potential used, see J.C. Curé [134] . Any non-polar atom or molecule constitutes an ensemble of variable (in time) dipole moments. This concept of interacting electric dipoles is O'Rahilly's and Assis' basic starting points from which to demonstrate, as Assis did, that gravitation is an attractive electrodynamic statistical residual force. To accomplish this incredible ontological identification between gravitodynamic forces and electrodynamic forces, is a gigantic leap for mankind and for the future of Stellar Astronautics. This intellectual feat of Assis will be remembered as the greatest accomplishment of the human mind in post diluvian times. Now, let us briefly comment on this significant paper of Assis [135] which he titled *Deriving Gravitation from Electromagnetism* (1992).

Assis' paper begins with a historical introduction about the conception that the gravitational phenomenon, somehow, must be connected to electrical phenomenon. A heuristic motivation is presented when he concluded that Grassmann's magnetic force is proportional to v^2/c^2 when two conductors, carrying the same electric currents, interact with each other. Assis concludes that the electrostatic force is a *zeroth-order* (v^0/c^0) *force*, while Grassmann's magnetic force is a *second-order* (v^2/c^2) *force*. He then ponders that maybe the gravitational force is a *fourth-order* v^4/c^4, or *a sixth-order* v^6/c^6 *force*. If this is the case, Assis decided to generalize Weber's Electrokinetic potential energy U:

$$U = [qq'/(4\pi\varepsilon_o r)]\,[1 - \alpha(v/c)^2 - \beta(v/c)^4 - \gamma(v/c)^6 - \ldots] \qquad (6.124)$$

Obviously, with $\alpha = ½$ and the rest of the Greek coefficients equal to zero, the last equation is reduced to Weber's Electrokinetic potential energy. Then, Assis writes:

> "The general idea is to calculate, using [7], the force between two neutral dipoles. Each dipole is supposed to consist of a positive charge at the center and a negative charge oscillating harmonically around the positive charge, as usual."

Eq.(7), in Assis' paper, is the force derived from Assis' equation represented by our eq. (6.124). The mathematical expression of Weber's generalized force can be obtained from eq. (6.124) by using Lagrange's analytical methodology. After this preparation, Assis launched his mind into calculating the force between two dipoles. When the reader arrives at eq. (28) of his paper, there is the realization that Assis may be a *Tibetan monk* because of his patience to first deduce, and later analyze this equation which on the right hand side has a total of 71 terms. Assis considered a different space orientation of two interacting electric dipoles in which he found, in some cases, a total force equal to zero, and in other cases a resultant force different from zero. It is not our intention to reproduce Assis' results of the *time average* interaction between two ensembles of electric dipoles. Our intention is to invite the reader to study this uncommon work of Assis. The gravitostatic force is incredibly small when it is compared with the electrostatic force in the interaction of two electrons separated by the same distance. This ratio is given by:

$$F_G/F_E = 4\pi\varepsilon_o G/(e/m_e)^2 = 2.4 \times 10^{-43} \qquad (6.125)$$

In spite of the insignificant magnitude of the gravitostatic force between two electrons, its essence is electrodynamic. Assis' inference is:

> "In conclusion we may say that in this model of generalized Weber electrodynamics we obtain: electrostatic as a zeroth-order effect , magnetism and Faraday's induction as a second-order effect, gravitation as a fourth-order effect, and inertia and precession of the perihelion as a sixth-order electromagnetic effect."

After his conclusion, Assis acknowledges the works of Dragone [136], (1990), and Jaakkola [137], (1991), on the same subject of gravitation identified with electrodynamic phenomena. It is the opinion of this author that the four "essentially" different forces have been reduced to one. In this section, we have shown that gravitodynamics has been identified with electrodynamics. In chapter 4, we identified the weak nuclear force with a Newtonian relativistic electrodynamic force. In this same chapter 4, we suggested that the strong nuclear force is another Newtonian relativistic electrodynamic force between neutrons and ionized neutrons. If we reproduced all these results by independent researchers, then this would be an excellent indication that we have started walking into the 21st century on the right foot.

6.10 Einstein-Hamilton-Jacobi's equation and Bohm-Hamilton-Jacobi's equation.

In this section, we will restrict ourselves to an outline of two extraordinary mathematical works done in functional space. The first work was done in 1962 by A. Peres [138], titled *On Cauchy's Problem in General Relativity - II.* In the summary of his work Peres writes:

> "The explicit construction of initial data for General Relativity then reduces to the Lichnerowicz scalar equation, and can be achieved with arbitrary accuracy. This method can be considered as a first step towards a Hamilton-Jacobi formalism."

At the end of his work, Peres writes:

> ". . . the Hamilton-Jacobi equation reduces to
>
> 35. $gP+(g_{mn}g_{rs} - 1/2\, g_{mr}g_{ns})(\delta S/\delta g_{mr})(\delta S/\delta g_{ns}) = 0,$
>
> holding for any $g_{mn}(x)$. The solution of (35) seems to lie far beyond our present mathematical techniques, and will probably require integration in functional space([14])."

In eq. (35), $g = \det[g_{mn}]$, P is the curvature invariant, and S is "Hamilton's principal function," or as we call it "the energy-momentum potential." In 1969, Ulrich H. Gerlach [139] published a paper titled *Derivation of the Ten Einstein field Equations from the*

Semi-classical Approximation to Quantum Geometrodynamics. It seems that it was Gerlach who gave, to eq. (35) of Peres, the name Einstein-Hamilton-Jacobi's (EHJ) equation written in superspace. Gerlach added to his study what he calls the principle of *constructive interference* which is our Madelung-Bohm's hypothesis given by eq. (2.2): ψ=Rexp{(iS/(h/2π) In doing so, Gerlach is ready to develop *Quantum Geometrodynamics* in hyperspace, or "the superspace of three geometries" in which the scalar *functional* S is defined at each "point" of three-geometry $S({}^{(3)}G)$. In section X, Gerlach shows the basic two "covariant-Hamilton equations of the 3+1 formulation of general relativity" from which it allows him to derive the ten Einstein field equations of vacuum:

$$G_{\mu\nu} = 0, \text{ with } \mu, \nu = 0, 1, 2, 3$$

In principle, it seems possible to deduce Einstein's field equations from Newton's second axiom of motion via the collective quantum gravitodynamic Bohm-Hamilton-Jacobi equation. The interested reader should consult chapter 11 of the book *The Undivided Universe: An ontological Interpretation of Quantum Theory*, written by Bohm and Hiley [140, p. 240-247], in 1993. Another important book, from a mathematical point of view, is *Gravitation* [141, p. 424], written by Misner, Thorne and Wheeler.

6.11 Nonlinear Electrodynamic Field Theory as a Relativistic Time Bomb.

In 1975, this author developed a Parametrized Newtonian Relativistic Electrodynamics (PNRED) and Gravitodynamics (PNRGD) which were presented in chapter 4 of this book. By 1982, this author in an oneiric experience perceived the concept of using the Millikan Apparatus to verify experimentally a conclusion deduced from PNRED. This conclusion was about the probable existence of a new electrodynamic force proportional to v'^2/c^2, where v' is the speed of a source charge which causes the electrodynamic field. The author realized that a paper announcing this theoretical conclusion, based on a different theory than Einstein-Maxwell-Lorentz's theory, would not have a chance for publication in journals of the "establishment." For this reason, he based the paper on an analogy pertaining to the probable existence of a new electrodynamic force-field and GRT-Gravitation. In 1982, the paper was published in a journal of the establishment under the title *A Modified Version of the Millikan Oil Drop Experiment to Test the Probable Existence of a New Electrodynamic Field* [142]. The results of such an experiment were reported, for the first time, in chapter 4 of this book

In 1986, the chairman of the Department of Physics of a prestigious university in California, after reading the abstract of the published paper, asked this author if he realized the *consequences of this paper*. This author answered affirmatively, and told the chairman that the paper also had another name, the *Time Bomb.* This author told him not to worry because the bomb would not explode until the middle of the 21st century. (At that time, the author had no way of knowing that the Internet was just around the corner). Concerning the analogy between geometrodynamics (GRT) and electrodynamics, in 1982, this author wrote:

> "If the new electrodynamic field really exists, then GRT will extend the knowledge we have about electromagnetic phenomena and at the same time GRT will have the opportunity to verify experimentally at laboratory level many of its theoretical conclusions via new electrodynamic laws obtained through this analogy."

If this new electrodynamic field-force is corroborated by other independent researchers, performing different experiments, then Einstein's General Relativity Theory will prove that Einstein's Special Relativity Theory is wrong because it is incomplete. This is the Relativistic Time Bomb! In chapters 3 and 4, we mentioned the theoretical Newtonian relativistic electrodynamic force-field, given by eq. (4.23a):

$$F_z = 2\pi(\alpha-\gamma)kq\,[\,I^2/(|\rho_e|A)]\,Rz/(R^2+z^2)^{3/2}$$

Introducing in the previous equation $k = 1/(4\pi\varepsilon_0 c^2)$ and the approximation $R/z \ll 1$, we get for the Newtonian electrodynamic force- field:

$$F_z = (\alpha-\gamma)[q/(2\varepsilon_0 c^2)](z/R^2)[\,I^2/(|\rho_e|A)] \qquad (6.126)$$

From eq. (8) of Curé's paper of 1982, we can get the z-component of the *Einsteinian Electrodynamic field* on a point on the axis of a coil of radius R, carrying a steady current I. The force $F_z = qE_z$, is given by:

$$F_z^* = (-3/2)[q/(2\varepsilon_0 c^2)](z/R^2)[\,I^2/(|\rho_e|A)] \qquad (6.127)$$

q is the charge on the coil axis, $|\rho_e|$ is the absolute value of the electron density in the conductor of cross section A. If we do not pay attention to the coefficient $(\alpha - \gamma)$ and $(-3/2)$, of the last two equations, the *identity* of the physical parameters enclosed by the square brackets in both equations is remarkable. This identity *cannot be a coincidence*. Eq. (6.126) comes from classical Newtonian relativistic electrodynamics, while eq. (6.127) comes from Einsteinian relativistic electrodynamics, obtained by analogy from GRT. This fantastic identity is not due to synchronicity, but to a deeper ontological unification.

In the case of Weber's electrodynamics, as well as for Liénard-Schwarzschild's electrodynamics, the term (α-γ) is equal to zero. For Spencer-Gauss' electrodynamics, we have from eq. (3.34), $\alpha = ½$, and $\gamma = -½$. Thus, the force given by eq. (6.126), i.e., Spencer-Gauss' force is an attractive force on static electrons. The coefficient (α-γ) is greater than zero for Ritz's force field, indicating that the Newtonian electrodynamic force is attractive for electrons on the axis of the coil. In chapter 3, we saw a vast collection of electrokinetics and electrodynamics. Some of these forces provide attractive or repulsive Newtonian electrodynamic force-fields on electrons. Einstein's force F_z^*, given by eq. (6.127), is a repulsive force when acting on electrons on the axis of the coil. Ritz's electrodynamics, given by eq. (3.26), and eq. (3.25) with $\alpha = A' = (3 - m)/4$, and $\gamma = -½$, is the most flexible of all Newtonian relativistic electrodynamics. Giving the proper value to parameter m, we can accommodate (α -γ) to be equal to -3/2 for m = 11, in Einsteinian electrodynamic force, given by eq. (6.127). The experimental result reported in chapter 4, shows that the new electrodynamic force, acting on electrons on the axis of the coil is attractive to the center of the coil. This experimental result is consistent with the Newtonian relativistic electrodynamic force given by eq. (6.26).

We must emphasize that we do not need a GRT-electrodynamic analogy to develop a totally new Newtonian Relativistic Electrodynamics. Nonetheless, we need GRT analogy to save time and effort when developing an advanced nonlinear Maxwell's field theory. Einstein's field equations, in gravitation, can be converted by analogy in Einstein's field equations in electromagnetics. If we do so, to our astonishment we will see that Einstein's field equations always had contained a *Unified Field Theory*. As we know, Einstein dedicated his life to the search for a unified field theory, from 1919 until his death in1955. His failure to see, in his own field equations, the unification of all the forces is a consequence of his consistent belief that Maxwell's field equations, and Maxwell's electrodynamics (Lorentz's force) were the *correct, complete and eternal equations* of the electromagnetic field. After all, this was the belief of Einstein which was the foundation for his beloved SRT. But now it is time for relativists to wake up from this tetra-dimensional illusion of Minkowski. Relativists, at the entrance of the 21st century are facing an unpleasant dilemma. (1) They will never accept that GRT has already demonstrated, *by analogy*, that SRT is wrong because it is *incomplete*. Because of this, they must leave to dissidents the task of developing the new Newtonian Relativistic Electrodynamics and

Gravitodynamics. This development will be theoretical and experimental. If this is the alternative the relativists choose, then their work will become obsolete in a short time. (2) They may accept that GRT has already demonstrated *by analogy* that SRT is wrong because it is *incomplete*. In this case they will change the noetic-mathematical background of GRT for an ontologico-experimental background, developing the Einsteinian Nonlinear field theory of electromagnetism. If relativists decide for the alternative (2), they will unite eclectic efforts with the rest of the community of natural philosophers. In 1982, this author [143], referring to eq. (8) of his paper, wrote about these new Einsteinian electrodynamic forces:

> "This new field has a radial centrifugal component and another axial component."

Ten years later, in 1992, Assis [144] published a paper titled *Centrifugal Electrical Force*. Assis derived this force from Weber's electrodynamics. Assis considers a moving electric charge inside a charged spherical shell in rotation. This problem was completely solved by Thirring [145] in 1918, in gravitation using GRT. In 1986, this author, working for Neo-Dynamics Corporation, in Miami, along with Dr. E. Greaves and Dr. V. Varela, discovered a very interesting theoretical result in a computer model of a circular coil of radius R, carrying a steady current. According to Ritz's electrodynamics, the radial force on the plane of the coil for $r > R$, alternates from a centripetal force to a centrifugal force, and vice versa, as the charge q recedes from the coil. This dynamical behavior corresponds to potential energy wells in the radial direction. Greaves developed a very clever computer program with which to study Ritz's electrodynamic field in three dimensions around the coil. We also discovered that inside a (first?) cone with its axis collinear with the coil axis, and the vertex of the cone coinciding with the center of the coil, Ritz's force was attractive for electrons. Outside the cone, the Ritz force was repulsive on electrons. We also noticed the existence of an axial component of the Ritz force that is not contained in Weber's force, but is contained in the Einsteinian-analog electrodynamics. In 1986, Varela and this author proved that Ritz's field is irrotational. This is the first time these results are being reported. This author hopes that young physicists will continue this work of Greaves and Varela in the experimental and theoretical fields, respectively.

In the last decade, the most prolific of young physicists, André Assis from Campinas University, Brasil, has written about many theoretical aspects and proposed experiments related to Weber's electrodynamics and Assis-Weber's gravitodynamics. In 1993, Assis [146] published a paper with the title *Changing the Inertial Mass of a Charged Particle*. In this paper, Assis proposes an experiment entirely analogous, as we said before, to Thirring's [147] linearized solution (1918) of Einstein's field equations. Thirring's solution was expressed finally by eq. (22) of his paper, in which we have a kind of Coriolis' force and a kind of Centrifugal force. By analogy, Coriolis' force corresponds to Grassmann's force in electrodynamics. This kind of centrifugal electrodynamic force is contained in all of the Parametrized Newtonian Relativistic Electrodynamics. Weber's force does not contain the axial component proportional to ω^2. In no instance do these comments on Thirring's work diminish the importance of Assis's work. On the contrary, Assis' work shows that the incredible minute gravitational effects can be multiplied by a factor of 10^{43}, and bring GRT to terrestrial laboratories in the form of an Advanced Nonlinear **Electromagnetic** theory and an Advanced Nonlinear **Electrodynamic** theory.

The latest amazing experimental result is that it seems possible to *interpret it* as the change of the inertial mass of electrons by an electrostatic potential, as predicted by Assis [148] in 1993. In 1999, V.F. Mikhailov [149], published a paper titled *The Action of an Electrostatic Potential on the Electron Mass*. Assis' theoretical prediction for parameters used by Mikhailov is $m_w/m_o = 2.0 \times 10^{-3}$. Weber's mass m_w, according to Assis, is given by:

$$m_w = qU/(3c^2) \qquad (6.128)$$

Mikhailov's experimental result is $m_w/m_o = (3.0 \pm 0.3)10^{-3}$, and the effective mass of the electron is $m = m_w - m_o$. In the comments on Mikhailov's results, Olivier Costa de Beauregard addressed Mikhailov in the following terms: "*Your experiment is superb, very ingenious, very elegant. Its results is of the top rank scientific importance. Congratulations.*" Indeed, the results are a magnificent confirmation of a magnificent prediction by Assis. Other commentators are more critical in the analysis of the experimental results. Could it be possible that there is another interpretation of the theoretical prediction and the experimental results? It is possible that the electrodynamic fields, in the experiment, transfer linear momentum to the electrons, as we showed in chapter 4, leaving the inertial mass of the electrons unaltered. However, Assis and Mikhailov deserve our highest intellectual congratulations for having closed the 20th century in such a magnificent way. In their minds and in their hands, physics and natural philosophy have a wonderful future.

Conclusions

The conclusions of this long chapter are listed below in relation to their main sections. Before the conclusions of each section we have listed the corresponding subsections.

CONCLUSIONS OF SECTION 6.1

6.1 Newton's Explanation of the Anomalous Motion of Planet Mercury.
6.1.1 Newtonian Dynamical Methodology.
6.1.2 Proposition XXIV, Theorem XIV.
6.1.3 Gravitodynamics and Geometrodynamics.

Einstein was right when he declared that *"Classical Mechanics is only a general scheme: it becomes a theory only by explicit indication of the force laws (d) as was done so very successfully by Newton for celestial mechanics."* All we need in order to create a better Newtonian theory of gravitation is to improve the astronomical measurements of the planet's motions, as Leverrier did more than a century ago. In this respect, Kepler's second empirical law should be submitted to a serious analysis of random, as well as systematic errors, using recent or new astronomical measurements. If we, eventually, verify the theoretical conclusion that the orbital angular momenta of the planets are not constant, then gravitational theories will provide new force terms, among which, an exponential short-range term will appear. Similar consequences can be obtained in electrodynamics.

The Newtonian solution presented here is slightly better, in the root-mean-square deviation sense, than the relativistic solution. Because of this fact, the so-called "non-Newtonian" gravitational term is no longer justifiable, nor the accusation that Newtonian dynamics is powerless to account for the excess perihelic rotation of the planets. On the contrary, all the elements needed to solve this problem are contained in the *Principia*, published in 1687, more than three centuries ago. Einstein was guilty because of his false accusation against Newtonian mechanics in relation to the so-called anomalous motion of planet Mercury.

CONCLUSIONS OF SECTION 6.2

6.2 Angular Momentum of the Sun.
6.2.1 Lense-Thirring Results of 1918.
6.2.2 Intrinsic Angular Momentum of the Sun (IAM).
6.2.3 Excess of Perijovian Rotation of Jupiter's Satellites.

Our main objective when determining the numeric value of the IAM per unit mass of the sun was fully accomplished. Our solution avoids expensive space missions to accomplish the same results. This calculated solar IAM is much larger than the IAM of the sun calculated with optical astrometric observations. This result proves that the thesis of the late Prof. Dicke has been proven in this work: *the sun has an inner core rotating much more rapidly than its surface*. We also proved that Einstein's original solution of the excess perihelic rotation of planet mercury was absurd, because the sun was treated as a point-like particle. The same idea of a point-like sun allowed anti-relativists and some relativists to admit that GRT is wrong if the sun is oblate. We proved that this is a false accusation against GRT. We fully demonstrated that GRT is able to account for the excess perihelic rotations of the planets, even in the case of a real oblate sun. We also proved that a Newtonian gravitodynamics, classified in this work as *asterisk gravitational theory,* provides the same results of GRT in the case of excess of perihelic rotation of the planets with an oblate sun. This is a very simple approach to gravitation which can be taught at an undergraduate level. In this elementary gravitational theory, we showed that our easily understandable gravitodynamic field **B*** replaces the GRT interpretation of the same. It substitutes as an induced *dragging rotation* of the local reference system with respect to another reference system *fixed at infinity* (absolute space) and caused by the rotating distant matter. No one should be deceived by the impressive name of "*reference system fixed at infinity*." This reference system is *identical* to what Newton called *Absolute Space*, and what is considered today as an abhorrent concept in relativistc circles.

We applied the same theory of Lense-Thirring (Asterisk theory) to the extremely oblate planet Jupiter. We found very intriguing results. However, more interesting results will be found when young Einsteinian astro-geometrodynamists or Newtonian astro-gravitodynamists extend Lense-Thirring results to higher degrees of approximation. They will find axial forces parallel to the axis of rotation of the central body forcing the fastest satellites to almost orbit in the equatorial plane of Jupiter. This is the case of all the Galilean satellites. They will also find the centrifugal-type gravitodynamic force which explains why the fastest satellites have circular orbits.

CONCLUSIONS OF SECTION 6.3

6.3 Inertial Mass, Gravitational Mass, and the Equivalence Principle.
6.3.1 The Weight of a Body Is Proportional to its Inertial Mass.
6.3.2 Acceleration of Falling Bodies of Different Chemical Substances.
6.3.3 Newton's Dynamical Methodology.
6.3.4 Acceleration of a Planet.
6.3.5 Force "on" a Planet.

6.3.6 Generalization of Newton's Planetary Force-Law.
6.3.7 Accelerational Fields and Gravitational Fields.
6.3.8 Eötvös' Experiment and the Equivalence Principle.

We have shown that a logical empirical derivation of Newton's planetary force-law necessarily brings us to having *inertial* masses in the final mathematical expression of Newton's gravitational force. This formal fact shows, in turn, the arbitrary character of forcing the introduction of gravitational mass, by *definition,* into Newton's planetary force-law.

We have also shown that Newton's universal gravitational law is not deducible from Newton's Axioms of Motion and Newton's planetary force-law. That is, unless we resort to philosophical arguments based on the hypothesis about the essential identity of the weight of a body, and the gravitational attraction exerted by the earth, in *analogy,* with the planet-sun phenomenon.

We have shown that the equality in the acceleration of falling bodies in vacuum, even having different chemical compositions, is a necessary (logical) consequence of classical mechanics. We have concluded that $W = m_i g$ is an experimentally verified *theoretical conclusion* and not an arbitrary definition.

We have shown that the planet's kinematical acceleration a_k is by definition a function of time. This planetary acceleration can be written as being inversely proportional to the square of the distance r between the sun and planet. Therefore, from a logical point of view, r must be a function of time. Through a metaphysical hypothesis, contained implicitly in Newton's third Axiom of Motion, we have shown that the sun is the *cause* of the force acting on the planet. In doing so, we accept the validity of the Principle of Causality. We call this force "gravitational force" and we say it is a "real force." By "real force" it is, or at least it should be understood, as a real force *caused* by a *material* object. In $F = ma$, we assign the same nature, or *essence* of the force, to the kinematical acceleration a_k and we call it "gravitational acceleration." Later on, we will forget we have assigned or attached that essence to the unessential but operational definition of acceleration.

We saw that this *metamorphosis* of the planet's essential kinematic acceleration, into the induced sun's essential gravitational acceleration, is a consequence of "studying the motion of bodies in order to discover the forces acting on them." We also saw that in order to avoid this metamorphosis, we have to frame two hypotheses: (1) Gravitational acceleration has a separate and independent existence from kinematic acceleration. (2) The magnitudes of gravitational and kinematic accelerations are equal. Only by accepting these hypotheses can we apply Newton's procedure so as to investigate the forces acting on bodies in motion. But we have seen that when we proceeded in this way, we have not avoided a mathematico-physical problem created when we combined this Newtonian procedure with the concepts of gravitostatic force-field theory. It is this problem which impels the author to suggest that the science of Kinematics is probably nothing more than another branch of mathematics, and more precisely, differential geometry. What we should conclude is the gravitational force exerted by the sun on a planet plays the role of centripetal force, as Newton proposed.

Finally, we should emphasize that Newton's *inductive* universal gravitational law is a generalization of Newton's *dynamic* particular gravitational planetary-law. Therefore, when we apply Newton's dynamic gravitational law, we must satisfy *similar conditions* such as those which existed when the generalization was performed. Cavendish experiments *are not dynamic but static experiments.* In this sense, we should distinguish between two gravitational constants: astronomical or dynamic gravitational constant G_A, and Cavendish or static gravitational constant G_c.

As $G_A M_o(1+m_p/M_o) = 4\pi^2 a^3/T^2$, the only way to determine the solar mass M_o, once the fraction m_o/M_o has been neglected, is to identify the *dynamic* gravitational constant G_A with the *static* or *Cavendish* gravitational constant G_c. Is this identification or equality, $G_A = G_c$, a proper one? If the question is not absurd, then its answer will have an interesting implication on the value of the solar mass, for example, and the claimed variation of the gravitational constant.

After this historical, semantic, etymological and ontological analysis, we concluded that two different names and natures (essences) have been assigned to *one and the same entity*. This entity is the *mass* of bodies. Now, if we insist on calling the *mass* of a body by two different names, associating them with two fundamentally (essentially) different masses, like m_i and m_g, then we must qualify this viewpoint as an unfortunate and unscientific attitude. We should never forget that Newton called the centripetal force, acting on a planet, the *gravitas* of the planet with respect to the sun.

Once the *identity* of the gravitational mass with the inertial mass is ontologically and logically established in the context of Newton's dynamics, any experiment of the Eötvös type is reduced to the paradox of trying to verify experimentally the ontological Principle of Identity. Finally, the identity of m_g with m_i demonstrates that Einstein's Principle of Equivalence belongs to Newton's theory of dynamics.

We find Einstein guilty because his accusation against Newton's dynamics offers no explanation for the equality of m_g with m_i. Classical mechanics can do more than demonstrate an equality; it can demonstrate an identity of m_g with m_i.

CONCLUSIONS OF SECTION 6.4

6.4 Mach's Definition of Mass
6.4.1 Machian Operational Definition of Inertial Reference Systems

Mach's definition of mass, as the quotient of acceleration, is a clever or, perhaps we should say, an astute way to give the impression that he discovered a new foundation of mechanics, from which Newton's axioms are deduced as simple corollaries. The best application we found of Mach's definition of mass is *an operational definition* of "inertial reference system." Mach's definition of mass is the source of metaphysical obscurities surrounding the concept of Mach's principle, as we will see in the next section. People who do not understand the difference between the being of an entity and the quantification or magnitude of one of the properties of the entity, confuse the *matter of a body* with the *mass of a body*. This is the reason why Mach did not accept Newton's definition of mass as the quantity of matter of a body.

The operational definition of mass in an inertial reference system, offered by Mach, is an excellent tool to define quantitatively the presence of an inertial reference system. This operational definition of an inertial or stationary reference system fails to fall in the contradiction, created by Einstein, where he discredited Newton's classical mechanics.

CONCLUSIONS OF SECTION 6.5.

6.5 Mach's Principle According to Einstein.
6.5.1 Different Definitions of Mach's Principle.
6.5.2 Operational Definition of Absolute Angular Velocity.

There are some physicists who wonder if Einstein's field equations of GRT contain Mach's principle. This ludicrous speculating is equivalent to asking - Is Mach's principle contained in Mach's principle? - It was Einstein himself who named his field equations of GRT *Mach's principle*. Mario Speiser collected 47 definitions of Mach's principle, out of which only 10% agreed with Einstein's original definition of Mach's principle. Thus, 90% of the physicists offer their own definition of Mach's bewildering concepts, calling their personal interpretation "Mach's principle." According to Speiser, the most common version in the 90% is: "The inertial mass of a body is determined by the total distribution of matter and energy." However, Mach says: "*The motion of body K can only be estimated by reference to other bodies A, B, C. . . .*" Mach decided that the rest of the bodies may not play a collateral role in the motion of body K, but a determinant role. Thus, we can rephrase the last statement of Mach's as follows: The motion of bodies is determined by the total distribution of matter in the universe.

It is, indeed, so simple to see that Mach refers to *the motion of a body* as being determined by the universal matter. Mach **does not** refer to the *inertial mass* as being determined by the universal matter. We should mention here that Mach hides the verb "to cause" behind the verb "to determine." We must reject all these pseudo-Mach's principles, and reassure ourselves there is only one Mach's principle, represented by Einstein's field equations of GRT. Finally, we proposed, in this section, an operational definition of absolute angular velocity as the limit of the galactic relative angular velocity when the distance to the galaxies tend to infinity.

CONCLUSIONS OF SECTION 6.6.

6.5 Newtonian Relativistic Gravitodynamics
and Quantum Gravitodynamics.
6.6.1 Physical Misinterpretations of Einstein's Geometro-dynamics.

In chapter 4, we deduced theoretically, i.e., formally, noumenally, logically, a Parametrized Newtonian Relativistic Gravitodynamics, disproving Einstein's assertion, that Newton's dynamical methodology is incapable of obtaining theoretically the mathematical structure of gravitational forces. In this section, we introduced the *cosmic collective potential energy* in gravitation by means of a nonlinear partial differential equation which we proposed to call *Bohm-Hamilton-Jacobi's equation.* The most important conclusion of the cosmic collective potential energy is when it includes gravitokinetic potentials, in addition to gravitostatic potentials. In this case, we have variable gravitodynamic forces proportional to the acceleration of the entire matter in motion in the universe (gravitational radiation). Thus, our solar system is immersed in a fluctuating flow of a gravitodynamic density of energy. The source of fluctuating energy is not localized in any particular place in the universe, because it is omnipresent in the entire cosmos. We finish this subject with an equation we called *Bohm-Schrödinger's Equation* of "quantum" gravitodynamics.

There are two serious misinterpretations in Einstein's geometro- dynamics. One is about Mach's Principle (MP) and the other one pertains to inertial forces. About MP, we distinguish at least three versions. The first being Einstein's mathematical MP which is the set of field equations of GRT. Another is Mach's metaphysical MP which establishes that the motion of a body is determined (caused) by the universal matter. Finally, we have Schlick's ontological MP which establishes that the inertial mass of a body is caused by its interaction with the rest of the bodies of the universe. Schlick's MP is based on the mistake of identifying the being of an entity with the quantification of one of its attributes. Mach's MP is very close to Einstein's MP, because Mach writes about the motion of a body in a gravitational field. Mach does not refer to the mass of any body being *determined* by the rest of the universal matter. Indeed, we can rephrase Mach's MP by saying that "the motion of bodies in gravitational fields is ruled by the metric tensor, which is determined by the universal matter-energy tensor of Einstein's GRT." The motion of material bodies in a gravitational field is **independent** of the mass of the moving body. Schlick's MP should be eradicated from any gravitational theory.

Einstein's GRT does not explain the inertial forces as metaphysically speculated by Mach. After the work of Thirring in 1918, it was very clear that GRT provided geometrodynamic forces which are similar, analogous to the inertial forces in a rotating laboratory, but **are not identical** to them. Our Parametrized Newtonian Relativistic Gravitodynamics provides what we call pseudo-inertial forces or new gravitodynamic forces. Our gravitodynamic theory also provides axial forces, parallel to the axis of rotation and pointing to the equatorial plane. These axial forces are contained in GRT, but not in Weber's gravitodynamics, nor in a classical rotating laboratory.

We finally criticized some authors for claiming, like Mach once did, that they can deduce Newton's axioms of motion. These false claims, not in the case of Mach, is based on the application of Lagrange's analytical dynamics to invente gravitokinetic potential energy functions. These authors did not realize that Lagrange's mathematical theory, as well as Hamilton-Jacobi's mathematical theory, is a generalization of Newton's second axiom of motion.

CONCLUSIONS OF SECTION 6.7.

Starlight Deflection by the Solar Energy Field

In this section, we applied the Primordial Energy Field Theory, elaborated in chapter 5, to the problem of starlight deflection by the gravitational field of the sun. First, we presented the analysis made by Merat of 297 astronomical determinations of solar starlight deflections. We called this work *Merat's empirical law*. Einstein's theoretical prediction, of the solar starlight deflections, shows great discrepancies with Merat's empirical law, particularly for $r<5R_0$, where R_0 is the optical radius of the sun. We then used Snell's law of refraction of a starlight ray going through the solar energy field. We decided against using the Eikonal Equation to solve this problem. The classical method of geometrical optics, which we used, gave us a solution considerably better than Einstein's solution of the same problem. The agreement of our elementary solution with Merat's empirical law of solar starlight deflection is excellent when compared with Einstein's prediction. We also determined the numeric value of the cosmic energy density equal to 1.094 x 10^{15} (J/m^3). This numeric result indicates we must review the mathematical expression of the Fresnel-Fizeau dragging coefficient, in relation to our solution of Michelson-Morley's interference problem.

CONCLUSIONS OF SECTION 6.8.

The Cosmological Red Shift and the Big Bang Theory.

If the cosmological red shift is interpreted as a Doppler effect, then the extra-galactic light sources must be receding from terrestrial astronomical observatories. This in turn means that sometime in the past all the stars in the universe were concentrated in an extremely small volume with an extraordinarily high density of matter and energy. Then, a Big Bang (BB) occurred, and the rest is cosmological history. In this section, we presented another interpretation of the cosmological red shift based on the Cosmic Primordial Energy Field which we developed in chapter 5. We concluded that intergalactic vacuum is not a synonym of *nothingness* or void. We concluded that space vacuum has physical properties. In 1873, J.C. Maxwell introduced the concept of conductivity of vacuum. In 1974, A.Ya. Kipper developed a new explanation called "photon fatigue" to explain the cosmological red shift. In 1996, R. Monti, unaware of Kipper's work, developed a similar explanation of the cosmological red shift. Presently, we should call these explanations Monti-Kipper's cosmological explanation of electromagnetic absorption by intergalactic vacuum. Let us propose the name *cosmological Beer's law* of absorption of light by the transparent zero-point-energy (energetic ether) of intergalactic, interstellar and interplanetary space.

Obviously, Beer's cosmological law provides another theoretical interpretation of the universe. In this new interpretation, the Big Bang never happened, as E.J. Lerner, in 1991, said in the title of his book *The Big Bang Never Happened.* Let us finish the conclusions of this section, again quoting Lerner, by referring to the new cosmological plasma theory in which we have a universe: "*without a Big Bang, without any beginning at all, a universe that has always existed, is always evolving, and will always evolve, with no limits of any sort.*"

CONCLUSIONS OF SECTION 6.9.

Is Gravitation an Electrodynamic Phenomenon?

The conception of a unified field theory seems to belong to Michael Faraday when he said, in 1850, that perhaps there is a connection between gravity and electricity. In 1938, O'Rahilly thought that gravitational attraction may be a residual statistical force between groups of moving electric charges. In 1985, this author discovered, in computer models, that two interacting neutral hydrogen atoms, separated by more than 100 Å, show attractive forces, repulsive forces and no force at all. These forces, or the lack of forces, depended on the relative orientation of the orbital angular momentum of the electrons. Only when this author read the paper of A. Assis, *Deriving Gravitation from Electromagnetism*, published in 1992, did it come to his mind that a hydrogen atom can be treated as a rotating electric dipole.

The paper of Assis acknowledges two forerunners on the same subject: The works of L.R. Dragone in 1990, and T.Jaakkola in 1991, will be landmarks in the history of physics. We believe that our Newtonian Relativistic Electrodynamics contains the weak and strong nuclear forces. If our miniature hydrogen atom is Eddington's neutron, then the strong nuclear forces should be the resultant interaction of neutrons and ionized neutrons. Thus, the use of Hamilton-Jacobi-Bohm's equation, along with the nuclear quantum potential energy, will provide a feasible nuclear theory as a simple application of the Newtonian Relativistic Electrodynamics to the atomic nucleus.

CONCLUSIONS OF SECTION 6.10.

Einstein-Hamilton-Jacobi's Equation and Bohm-Hamilton-Jacobi's Equation

In 1969, it seems that U.H. Gerlach gave the name Einstein-Hamilton-Jacobi equation to a Hamilton-Jacobi formalism created by A. Peres in 1962. Gerlach added a principle he called *constructive interference*, which corresponds to our Madelung-Bohm's functional transformation. Gerlach was able to derive Einstein's ten field equations of vacuum, $G_{\mu\nu} = 0$. From a mathematical point of view, there is nothing wrong in making Einstein's energy-momentum tensor equal to zero, $T_{\mu\nu} = 0$. From a physical point of view, to make $T_{\mu\nu} = 0$ is to completely destroy the unbroken wholeness of the entire universe represented by the cosmic collective potential energy. From an ontological point of view, to make $T_{\mu\nu} = 0$ is equivalent to denying the existence of the entire universe, including all physicists. We left another problem for the young generation. This problem was to try to deduce Einstein's field equations from Newton's second axiom of motion via the quantum gravitodynamic Bohm-Hamilton-Jacobi's equation.

CONCLUSIONS OF SECTION 6.11.

Nonlinear Electrodynamic's Field Theory as a Relativistic Time Bomb

The most important conclusion of this section is to convert Einstein's nonlinear field equations of GRT into a nonlinear electromagnetic theory, from which we may deduce our Newtonian Relativistic Electrodynamics. Curé's paper of 1982 showed the probable existence of an electrodynamic force proportional to v'^2/c^2. This force was obtained by analogy from a geometrodynamic force of Einstein's GRT. The same electrodynamic force can be deduced from our Newtonian Relativistic Electrodynamics. In chapter 4, we presented the experimental evidence of this new electrodynamic force.

In 1993, Assis predicted the change of the inertial mass of electrons moving inside a charged spherical shell. His prediction was based on Weber's electrodynamics which provided a pseudo-centrifugal electrodynamic force. The same result can be predicted with Einstein's geometrodynamics, once it is converted, by analogy, into an electrodynamics. The action of a rotating massive spherical shell on a speck of matter moving inside the shell was accomplished by Thirring, in 1918, using GRT. Thirring found analogous forces to Coriolis force and to centrifugal force. In 1999, Mikhailov reported the experimental evidence of Assis's prediction. We began to accumulate theoretical and experimental evidence to thoroughly investigate the conversion of Einstein's GRT into a wonderful nonlinear electromagnetic theory. With these new Einsteinian nonlinear field equations, of an electromagnetic phenomena, we will discover new electrodynamic force fields, and also, we will amplify 10^{43} times the geometrodynamic (gravitational) effects of GRT by performing electrodynamic experiments at laboratory levels.

Naturally, some changes must be done to accomplish this unification. However, in this way we will finally verify Thoth's lesson: "*As above so below, for the fulfilment of unity.*

"In chapter 6, Einstein was found guilty on two counts:

1. "[Relativity theory] has already explained a result of observation in astronomy against which Classical Mechanics is powerless."

2. “The equality of these two masses, so differently defined, is a fact which is confirmed by experiments of very high accuracy (experiments of Eötvös). Classical mechanics offers no explanation for this equality.”

There is a minor accusation against classical mechanics which was made by Einstein when he discussed the “principle” of inertia. Einstein phrased this “principle” in a very clever way in order to show that it contains a vicious circle:

> “The weakness of the principle of inertia lies in this, that it involves an argument in a circle: a mass moves without acceleration if it is sufficiently far from other bodies; we know that it is sufficiently far from other bodies only by the fact that it moves without acceleration.”

We rejected this statement of the law of inertia, invented by Einstein, by creating an operational determination of inertial reference systems. The “principle” of inertia is not a principle in the context of Newton’s theory of dynamics, but a deducible law.The most important conclusion in this chapter is to investigate whether GRT contains the formality of nonlinear electromagnetic and electrodynamic theories. This author is indirectly convinced that Einstein, indeed, created *The Nonlinear Unified Field Theory* in 1916 . Evidently, we are going to be forced to change the ontological and geometrical background, or change the foundations of GRT. One consequence of this unification is that GRT will prove that SRT is wrong because it is very incomplete.

References

1 A. Einstein, Preuss. Akad. Wissen., **pt. 2**, 831 (1915)
2 U. J. Leverrier, Ann. Obs. Paris, **5**, 104, (1859)
3 A. Einstein, Ann. Phys., ser. 4, vol. **49**, 769 (1916)
4 A. Einstein, *Relativity, the Special and General Theory* (Mathuen, London, 1920)
5 R. Moulton, *An Introduction to Celestial Mechanics* (Dover Publications, Inc., NY, 1970)
6 I. Newton, *Mathematical Principles of Natural Philosophy and His System of the World* (Univ. of California Press, Berkeley, CA, 1960)
7 . D. W. Sciama, *The Unity of the Universe* (Faber, London, 1959)
8 A. O'Rahilly, *Electromagnetic Theory* (Dover Publications, Inc., NY, 1965)
9 Hall, 1884
10 F. S. Newcomb, Suppl. Am. Ephe. Naut. Alm., (1897)
11 P. Gerber, Zeits. f. Math. und Phys., 43, 93 (1898)
12 K. Demys, Sci. Ethics, **1**, no.5, 1 (1985)
13 G.B. Brown, Science Progress, no. 181, 15-29 (1958); Retarded Action-At-A-Distance (Cortney Publications, Bedfordshire, England, 1982)
14 R.H. Dicke, Nature, **202**, 432 (1964)
15 . R.H. Dicke and H.M. Goldenberg, Phys. Rev. Letters, **18**, 313, (1967)
16 H. A. Hill and R. T. Stebbins, Astrophys. J., **200**, 471 (1975)
17 H. A. Hill, R. J. Bos, and P. R. Goode, Phys. Rev. Lett., **49**, 1794 (1982)
18 . D. R. Long, Phys. Rev., D, **9**, no. 4, 850 (1974)
19 . D. R. Long, Phys. Rev., D, **10**, no. 6, 1677 (1974)
20 D. R. Long, Nature, **260**, no. 5550, 417 (1976)
21 Correspondent, Nature, **260**, no. 5550, 395 (1976)
22 G.W. Gibbons & B.F. Whiting, Nature, **291**, 636 (1981)
23 S.C. Holding & G.J. Tuck, Nature, **307**, 714 (1984)
24 J. K. Hoskins, R. D. Newman, R. Spero, and J. Schultz, Phys. Rev., D, **32**, 3084 (1985)
25 E. Fischbach, D. Sudarsky, A. Szafer, and C. Talmadge, Phys. Rev. Lett., **56**,no. 1, 3 (1986)
26 F. D. Stacy, G. J. Tuck, G. I. Moore, S. C. Holding, D. B. Goodwin, and R. Zhou, Rev. Mod. Phys., 59, no. 1, 157 (1987)

27 C. Brans and R.H. Dicke, Phys. Rev.,**124**, 925-935 (1961) [Mach's Principle and a Relativistic Theory of Gravitation.]
28 Ref. 26
29 E. Whittaker, *A History of the Theories of Aether and Electricity* (Thomas Nelson & Sons, London, NY, 1953)
30 J. C. Maxwell, *Matter and Motion* (Dover Publishers, Inc., NY, NY)
31 A. Einstein, *Ideas and Opinions* (Crown Publishers, Inc., NY, NY, 1982)
32 A. Einstein, *The Meaning of Relativity* (Princeton University Press, Princeton, NJ, 1974)
33 C.W. Misner, K.S. Thorne and J.A. , *Gravitation* (W.H. Freeman and Co., San Francisco, CA, 1973)
34 C. Møller, *The Theory of Relativity* (Oxford at Clarendon Press, London, 1952)
35 M.R. Hass and D.K. Ross, Astrophys. Space Sci., **32**, 3 (1975)
36 K. Schwarzschild, Preuss. Akad. der Wiss., Sitzber., pp. 189 196 (1916)
37 J.C. Curé, Galilean Electrodynamics, **2**, no. 3, 43-47 (1991) [The Perihelic Rotation of Mercury by Newton's Original Method].
38 Ref. 5, p. 363
39 Ref. 124
40 F. S. Newcomb, Suppl. Am. Ephe. Naut. Alm. (1897)
41 R.H. Dicke, *The Sun's Rotation and Relativity*. Nature, **202**, 432-435 (1964)
42 A.K.T. Assis, *Weber's Electrodynamics* (Kluwer Academic Publishers, Dordrecht, Boston, London, 1994)
43 Ref. 8
44 G.B. Brown, *Retarded Action at a Distance* (Cortney Publications, 95/115 Windmill Road, Luton Bedfordshire, Great Britain, 1982)
45 J.C. Curé, Phys. Lett., **116 B**, no.2.3, 158-160 (1982) [A modified version of the Millikan oil drop experiment to test the probable existence of a new electrodynamic field.]
46 J. Lense and H. Thirring, Phys. ZS., **19**, 156 (1918)
47 Ref. 36
48 L.I. Shiff, Proc. Nat. Acad. Sci. U.S., **46**, 871 (1960)
49 J.C. Curé, Memoir, Escuela de Ingenieria, Universidad de Carabobo, Venezuela (1976) [Sobre la Precesión del Perihelio de Algunos Planetas]
50 A. Einstein, Viertel. für gerichtliche Medizin, ser. 3. vol. **44**. pp. 37-40 (1912) [Gibt es eine Gravitationswirkung die der electrodynamichen Inductionswirkung analog ist?]

51 A. Einstein, *Sidelights on Relativity*.(Dover Publications, Inc.,NY, 1983).
52 Ref. 34
53 J.P. Ostriker, in *Proceedings of the Conference on Experimental Tests of Gravitational Theories* (JPL Technical Memorandum 33-499, California Institute of Technology, 1970), p. 4
54 Ref. 41
55 Ref. 41
56 Ref. 46
57 Taken from *Jupiter* (Editor T. Gehrels, University of Arizona Press, 1976)
58 Ref. 32
59 Ref. 32
60 E. Mach, *The Science of Mechanics* (The Open Court Publishing Co., La Salle, Illinois, 1989)
61 G. Burniston Brown, J. Am. Phys., **28**, 475 (1960)
62 A. Sommerfeld, *Mechanics Vol. I* (Academic Press Inc., Publishers, New York, N. Y., 1952)
63 Ref. 6
64 Ref. 6
65 Ref. 62
66 Ref. 6
67 Ref. 6
68 Ref. 60
69 M. Bunge, Am. J. Phys., **34**, no. 7, 585-596 (1966) [Mach's Critique of Newtonian Mechanics]
70 Ref. 60
71 Ref. 6
72 M. Born, *Einstein's Theory of Relativity* (Dover Publication, Inc., New York, 1965)
73 Ref. 72
74 Ref. 37
75 Ref. 30
76 Ref. 30
77 P. G. Bergmann, *The Riddle of Gravitation* (Charles Scribner's Sons, N. Y., 1968)
78 Ref. 6

79 M. G. Evans, Am. J. Phys., **26**, 619 (1958)
80 W. Heisenberg, *Physics and Beyond* (Harper & Row, Publishers, N. Y., 1972)
81 Ref. 31
82 M. Sach, Phys. Today, 22. 51 (1969)
83 Ref. 73
84 A. Koyré, *Newtonian Studies* (The University of Chicago Press, Chicago, Phoenix Books, 1968)
85 L. de Broglie, *Sur les sentiers de la science* (Publisher A. Michel, Paris, 1960)
86 R. H. Dicke, *The Theoretical Significance of Experimental Relativity* (Gordon and Breach, NY, 1964)
87 Ref. 32
88 A. Einstein, Ann. Der Physik, **55**, 241, (1918) [Prinzipielles zur Allgemeinen Relativitätstheorie]
89 Ref. 60
90 Ref. 60
91 P. Gerber, Zeits. f. Math. und Phys., **43**, 93 (1898)
92 H. Poincaré, *Science and Hypothesis* (Dover Publications, Inc., NY, 1952)
93 Ref. 32
94 Ref. 60
95 Ref. 88
96 M. Speiser, *Mach's Principle* (Master of Science Thesis, Approved by Professor J. Weber, Department of Physics and Astronomy, University of Maryland, 1965)
97 In: Y. Chiu and W.F. Hoffman, *Gravitation and Relativity* (W.A. Benjamin,Inc., New York, 1964)
98 F.A. Pirani, Helv. Phys. Acta, Suppl. IV (1956)
99 R.C. Tolman, *Relativity, Thermodynamics, and Cosmology* (Clarendon Press, Oxford, 1934)
100 W. Pauli, *Theory of Relativity* (Translated by W. Field, Pergamon Press, London, 1958)
101 Ref. 29
102 V.F. Weisskopf, *Lectures in Theoretical Physics, vol. 3* (Interscience Publishers, Inc., New York, 1961) [Mach's Principle and the Existence of Gravity Waves.]
103 Ref. 60
104 Ref. 60
105 Ref. 60

106 O.D. Jefimenko, *Causality, Electromagnetic Induction and Gravitation* (Electret Scientific Company, P.O. Box 4132, Star City, West Virginia 26505, 1992)

107 O. Heaviside, The Electrician **31**, 281 and 359 (1893) [A Gravitational and Electromagnetic Analogy]

108 Ref. 42

109 M.F. Tisserand, Comptes Rendues de l'Academie des Sciences de Paris, vol. **75**, pp. 760-763 (1872) [Sur le mouvement de planètes autour du Soleil, d'après la loi électrodynamique de Weber.]

110 Ref. 8

111 Ref. 42

112 A.K.T Assis, *Relational Mechanics* (Published by Apeiron,4405, rue St-Dominique, Montreal, Quebec H2W 2B2, Canada, 1999)

113 Ref, 112

114 Ref. 32

115 Ref. 32

116 H. Thirring, Phys. ZS., **19**, 33 (1918)

117 P. Frank, *Philosophy of Science* (Prentice Hall Inc., Inglewood Cliffs, NJ, 1957)

118 J.P. Wesley, *Advanced Fundamental Physics* (Benjamin Wesley - Publisher, Weiherdammstrasse 24, 7712 Blumberg, West Germany, 1991.)

119 C. Lanczos, *The Variational Principles of Mechanics*. (Dover Publications, Inc., NY, 1986)

120 Ref. 119

121 Ref. 118

122 Ref. 42

123 Ref. 112

124 C.L. Poor, *Gravitation versus Relativity* (G.P. Putnam's Sons, NY, and London, 1922)

125 Capt. T.J.J. See, The Literary Digest, November 8, 1924 (Funk & Wagnalls Company, NY, London)

126 Parviz Merat, GRG, **5**, no. 3, pp. 757-764 (1974) [Analysis of the Optical Data on the Deflection of Light in the Vicinity of the Solar Limb]

127 E.J. Lerner, *The Big Bang Never Happened* (Vintage Books, A Division of Random House, INC., New York, 1991)

128 J.C. Maxwell, *A Treatise on Electricity and Magnetism*. (Dover Publications Inc., NY, 1954)

129 R. Monti, Physics Essays, **9**, no. 2, p. 238, June (1996) [Theory of Relativity: A Critical Analysis]

129* H.W. Olbers, *Gesammelte Werke* (Berlin, 1894)

130 Ref. 6

131 A.K.T. Assis, Can. J. Phys., **70**, 330-340 (1992) [Deriving gravitation from electromagnetism]

132 Ref. 8

133 Ref. 131

134 J.C. Curé and S, Babb, J. Phys. Chem., **48**, 2045 (1968) [Percus-Yevick Isotherm for the EXP-6 Intermolecular Potential]

135 Ref. 131

136 L.R. Dragone, Hadronic J. Suppl., **5**, 299 (1990)

137 T. Jaakkola, Apeiron, **9** and **10**, 76 (1991)

138 A. Peres, Il Nuovo Cimento, vol. **XXVI**, no. 1, 53-62 (1962) [On Cauchy's Problem in General Relativity - II]

139 U.H. Gerlach, Phys. Review, **177**, no.5, 1929-1941 (1969) [Derivation of the Ten Einstein Field Equations from the Semiclassical Approximation to Quantum Geometrodynamics]

140 D. Bohm and B.J. Hiley, *The undivided universe: An ontological interpretation of quantum theory* (Routledge, 29 West 35th , New York, NY 10001, and Routledge, 11 New Fetter Lane, London EC4P 4EE. England, 1993)

141 Ref. 33

142 Ref. 45

143 Ref. 45

144 A.K.T. Assis, Commun. Theor. Physic. **18**, 475-478 (1992) [Centrifugal Electric Force]

145 H. Thirring, Phys. ZS., **19**, 33 (1918)

146 A.K.T. Assis, J. Phys. Soc. Japan, **62**, no. 5, pp. 1418-1422, May (1993) [Changing the Inertial mass of a Charged Particle.]

147 Ref. 145

148 Ref. 146

149 V.F. Mikhailov, Annales de la Fondation Louis de Broglie, **24**, 161–163 (1999) [The Action of an electrostatic Potential on the Electron Mass.]

CHAPTER 7

EINSTEIN'S THEOLOGICAL BELIEFS AND SCIENTIFIC THEOLOGY

TABLE OF CONTENTS

"Put off thy shoes from thy feet, for the place whereon thou standest is holy ground."

Exodus 3:5

Introduction.

In this chapter, we establish a new relationship among physics, philosophy and theology. To accomplish this task, we face many serious semantic problems with respect to different verbal terms such as mind, spirit, soul, consciousness, ideas, percepts, phenomena, concepts, material complexification, immaterial connections, intelligence, person, Being of all entities, religion, physical reality, psychic realm and some other important concepts. We will see the debunking of many misconceptions, but at the same time, we will see the emergence of an advanced conceptualization of consciousness, one of the most important subjects of this chapter.

In the past decades, many authors published books concerning the concept of *consciousness*. What we should expect from these authors is a reliable definition or, at least, a semi-explanation of what they understand by consciousness. But, in most cases, this is only a wishful thought from their readers. It is incomprehensible that so many people write about consciousness, but have no credible definition of it. Even physicists have entered into this field of consciousness with their books of *quantum mechanics* and their positivistic, pragmatic and agnostic philosophies. They try to formally understand the microcosmic *reality* which they claim does not exist. These attempts are paradoxical or absurd. However, we believe natural philosophers may eventually help in elucidating the *mystery of consciousness* when they replace formal quantum mechanics with an ontological quantum mechanics. This ontological approach might convert the mystery of consciousness into a solvable problem of science, but first, any solution will require an initial transitory *definition* of consciousness. To reach a solution, we must follow a circuitous route which will take us many places, but we feel, in the long run, will give us more solid answers.

I-1. Science and religion have a common foundation.

Before we analyze the transrational knowledge of the *cosmic religious feelings* experienced by Einstein, we should look into the common root from which Science and Religion have evolved. In general, a religious person *knows* that he *believes*. On the other hand, in general, a scientist *believes* that he *knows*. A well-educated religious person is aware that whatever he knows about his religion is based on beliefs, which in turn is founded on dogmas of faith. A well-educated scientist *knows* that he has to *believe* in the truth of scientific principles or axioms of his theoretical knowledge which is, also, founded on faith. If the scientist is an empiricist, he knows a vast amount about experimental facts, but does not necessarily understand the relationships among these many facts. The truths of principles and axioms *cannot* be deduced, cannot be demonstrated logically. Thus, the validity of principles and axioms must be *assumed;* they must be *believed.* The scientist, who has no training in philosophy of science, is not aware that he is a person who *blindly* (irrationally) believes in the truths of principles or axioms, principles and axioms which he uses every day. Principles, axioms, postulates and dogmas cannot logically be further reduced. They are the irrational, or perhaps we can say, they are the transrational foundations of any hypothetical deductive structure. Einstein is one of the few physicists of the 20th century who emphasized this dogmatic beginning of any scientific theory. In his book *Ideas and Opinions* [1, p. 272], in the essay titled *On the Method of Theoretical Physics*, he writes:

> "Apart from that, these latter [fundamental principles] are free inventions of the human intellect, which cannot be justified either by the nature of that intellect or in any other fashion *a priori* . . . These fundamental concepts and postulates . . . form the essential part of a theory, which reason cannot touch."

Einstein seems to be one of the few epistemologists who emphasized the feigned character of the fundamental principles of any theory. He said the conception about the foundations of scientific theories was almost completely absent in the minds of scientists, epistemologists and philosophers of the 19th century. Even in the 20th century, the *fictitious character* of principles of scientific theories remained hidden from the education of scientists. The creation of a scientific theory and the creation of a religious theology requires a set of principles and a set of dogmas, respectively. The common fundamental root on which scientific theories and theologies are based is the necessity to irrationally *believe* in the *truth* of principles and in the *truth* of dogmas. Formally, there is no difference between a scientific principle and a religious dogma. Thus, sciences and religions are created from the same foundations, from *unproven* principles and *unproven* dogmas. Let us hope that all scientists of the twenty-first century will learn and keep this fundamental lesson in mind. Any well-educated scientist in epistemological matters knows that he is a person of unbroken *faith* in the truth of the scientific theory which he practices. One disadvantage is, this unbroken and unquestionable faith *could* transform the scientist or the religious person into a fanatical individual. Faith is nearly always expressed by religious believers, but, unfortunately, is seldom confessed by scientists. For believers, faith is "*the substance of things hoped for, the evidence of things unseen."* Now we can understand that it is not without reason that *religious people know they believe, while scientists believe they know.*

In the previous quotation, Einstein writes that we cannot justify *a priori,* the fundamental principles of a theory. This means the creator and followers of a theory can only hope, in the beginning, that the fundamental principles are *true* and *adequate*. The *a priori* confidence (faith) in the truth contained in the principles or axioms of a scientific theory is reinforced later by *a posteriori* evidence. This evidence is shown by the experimental verification of a particular theoretical conclusion based on the fundamental general principles of the theory. The more experimental verifications a theory accumulates, the greater is our confidence in the fundamental principles of the theory. That is to say, the more verifications we gather, the greater is our *faith* in the truth of the principles.

Again, let us go back to the previous quotation when Einstein said: "*. . . the purely fictitious character of the fundamentals of the scientific theory.*" The following list of synonyms of the word *fictitious* provides a better understanding of the "fundamental principles" of any theory and makes us begin to wonder. The synonyms for fictitious are: fanciful, fictional, imaginary, invented, assumed, fake, false, feigned, fraudulent, apocryphal, disputed, doubtful, unauthenticated and unverified. The logical development in deducing consequences of a theory could be impeccable, but if one experiment disagrees with one of the theoretical conclusions, the theory is *false*, and our "scientific faith" in the fundamental principles collapses, but not instantaneously. We sadly see in the history of science where wrong theories are needlessly prolonged, and cause stagnation of scientific knowledge.

I-2. Theoretical scientific knowledge is relative

In the so-called *relativist* 20th century, very few scientists, especially relativist physicists, knew that all rational or theoretical scientific knowledge is epistemologically *relative knowledge*. By relative knowledge, we mean knowledge which depends on, or is a function of the truth contained in the fundamental principles of a theory. On the other hand, *absolute knowledge* is completely independent from any analytical or deductive mental process. The factual, sensorial, experimental, empirical, phenomenological knowledge is *not* absolute knowledge. Factual knowledge is the most *primordial theoretical or primitive conceptual* knowledge. Empiricists naively believe that empirical knowledge is knowledge about uncontaminated reality; i.e., knowledge not infected by *preconceptual* prejudices. They still do not realize that factual or empirical knowledge is the conceptualization (mental interpretation) of ideas, mental images, noetic illuminations or phenomena (mental lights or images). Any act of acquiring facts about nature is biased, not only by the measuring instrument, but by the ignorance or wisdom of the human mind. The author has never met or heard of an empiricist who has built a measuring instrument and does not know what to measure with it. Consequently, the logical *particular* conclusion of any scientific theory inherits the truth or falsehood of the principles or axioms of the theory; i.e., the truth of particularly deduced theoretical knowledge is *relative* to the truth or falsehood of the principles of the theory. Similarly, the logical *particular* conclusion of any rational theology inherits the truth or falsehood of the dogmas of the corresponding theology; i.e., the truth of particularly deduced theological knowledge is *relative* to the truth or falsehood of the dogmas of the sacred doctrine or theology.

Of course any knowledge depends on one or more factors. Thus, human knowledge, obviously, depends on, is a function of, is relative to the human mind. This is mental relativism. Knowledge also is relative to the environment in which the human mind creates conceptual knowledge. Therefore, we might talk about relative knowledge in respect to social factors, economical factors, political factors and religious factors. We are only interested in the relativity of rational knowledge, i.e., we are interested in the elaboration of conceptual or rational knowledge beginning with nonrational principles or dogmas.

I-3. Elements of Theory of Knowledge.

The history of theories of knowledge shows three gnoseological periods: (1) contemplation, (2) assimilation, and (3) creation. For the ancient Greek world, knowledge was a *contemplation*. For the medieval world, knowledge was an *assimilation*. For the modern world, knowledge is a *creation*.

Knowledge as Contemplation. To contemplate is to see an *image* of a *thing* or *object*. The word *thing*, as we know, translates into Latin as *res* from which the word *reality* is derived. Thus, reality is the world of visible material things or objects. In this book, we use the word thing or object to refer only to a *real thing* or *real object* which is outside the human mind. What is the ontological cause of the images we have in our minds? The ontological causes of the images we have in our minds are the real things which exist in reality, outside our minds. This last assertion is *true* (is logical, is coherent) because of the Principle of Reality which we established in chapter 1. The science we have in the 21st century explains the *electronic connection* between the cause and effect of this *phenomenon.* This scientific explanation of today, between the real thing and the image of that real thing, is in our minds. It would have been considered only pure *metaphysical speculation* in the past, even by the well-known *positivist* Ernst Mach at the beginning of the 20th century. If we do not know any biochemistry, any biophysics, any bioelectronics of the nervous system, then our position is no better than Locke and Hume, particularly Hume, who gave the impression that he did not know that *principles* or *axioms* were not deduced logically from any other propositions. Hume did not know that principles cannot be justified in any manner. He did not know they were free inventions of the mind, as Einstein so eloquently stated and proceeded to solidly demonstrate with the creation of his General Relativity Theory. The scientist or philosopher who still believes that out of one million experiments of particular cases, he can logically deduce or justify a *generalization*, a universal statement, a conceptualization of a principle, knows nothing about Einsteinian Epistemology.

Knowledge as Assimilation. Aristotle saw in the "hungry craving" for knowledge, the "appetitive" capacity of the soul. The will, according to Aristotle, is the appetite which moves us forward to look for the spiritual food so we may assimilate it and absorb it into our minds. Without a doubt, this was a primitive metaphor which was used to lecture to the common people about knowledge.

In the ancient Greek period, as well as in the mediaeval one, the *object* played an important role in the process of knowing. The *being* (*ousia, essentia*) of a thing or object was a mystery in these two periods of philosophy of knowledge. It was Kant who displaced the metaphysics of the object to a metaphysics of the subject. Nicolas Berdyaev, [2, p. 10] in his book *The Beginning and the End,* is extremely clear when he said:

> "Kant strikes a blow at the old metaphysics which were based upon a confusion between the product of thought and reality. It is interesting to note that in Kant the limitations and metaphysical weakness of reason are associated with its cognitive activities. Reason is not active in cognition only. It is active also in the formation of the object-world itself, of the world of phenomena."

Let us analyze part of the last sentence in the above quotation: "*It is active also in the formation . . . of the world of phenomena,*" the formation of the world of *mental lights*. This sentence has two important words: *formation* and *phenomena*. These are both powerful words, but let us take one at a time starting with formation. We will discuss phenomena later. To clarify our understanding, let us first see a list of synonyms for the word "formation": generation, inception, conception and creation. Thus, in the previous quotation, we read the "mind is active also in the *creation* of the world of phenomena." This "creation" of knowledge, obviously, is *engendered,* or *conceived* by the human mind.

Knowledge as Creation. Knowledge is a relationship between a *subject* and an *object*. This relationship or knowledge does not exist simply because there are a subject and an object. This relationship is created by the subject, and therefore, knowledge is a mental creation done by the subject. It is in this respect, as we said above, knowledge is *relative* to the human mind; this mental activity of humans is *knowledge*.

Etymologically speaking, theory of knowledge is alluded to by the word *gnoseology*. It comes from the Greek word γνωσιζ (gnosis=knowledge). If we ask for the nature of this relationship, we immediately jump from *gnoseology* and fall into ontology or metaphysics. We have said the human mind creates or conceives the relationship with the object. Thus, this relationship between the subject and object is an attribute of the subject. This consideration allows us to reduce the concept of *knowledge* to only two entities: the subject who knows; i.e., who actively conceptualizes or conceives the knowledge of the object, and the object that is known. The human mind dresses the image of the object with properties created by the human mind itself. But the human mind does something much more *inconceivable*. The human mind, perhaps in an act of desperation, deceives itself by inventing the *substance* of the object.

Is it possible for the subject to know the object *in-itself*? What do we mean by *in-itself*? The self of an object is its very *nature,* that is, the being of the object. Thus, to know an object in-itself is to know the being, the ultimate substance, the very essence of the object. But this search does not belong to gnoseology. It belongs to metaphysics, to ontology. In this matter, it is very difficult to separate gnoseology from metaphysics. However, we have said that knowledge is an invention, a conception or a creation of the human mind, therefore, the knowledge of the being or ultimate substance of an object can be conceived or invented by the human mind through metaphysical speculations. Of course, this speculative method is definitively *indirect*, and therefore, relative to logical and metaphysical or ontological principles. For centuries we have been *inventing* the *substance* of objects, and for centuries we have been *assigning* this mentally invented "substance" or essence to objects. We have been *hypostatizing the entire universe*; i.e., we, the subjects, for centuries have been *assigning substance* to sensorial mental *images* or *ideas* acquired through our senses. Also we, the subjects, for centuries have been assigning substance to conceptions we have created in our minds. This pretension cannot be rationally accomplished by the human intellect. However, the substance of entities can be apprehended by transrational intuitions. We have been doing this hypostatizing mental process, with the objects of the world, for many centuries, but this does not assure us of the truth we predicate about the objects *in-themselves*. In this respect, the images (ideas) we have in the mind of the external world (the sensorial experience) constitute the raw data with which we conceptualize and hypostatize the external world. The human mind is in "itself" a conceptual jail. We cannot escape from the world of concepts, the mental prison, the noumenal bondage, the human factory of *substances*. We want to escape from this mental illusion in order to reach the object in the external reality and apprehend it as it is. We have to succumb to the conclusion that our minds can never reach this very essence of ideas by rational means. Soon we will see that this task can be accomplished transrationally by metanoumenal intuition.

It is extremely important to avoid metonyms in our discourses; i.e., we have to avoid assigning the meaning of a word to *another* word. Many philosophers, theologians and particularly scientists refer to the word *concept* by using the word *idea*. Some of these people take the words *concept* and *idea* as synonyms. We will expand on this later, but first a very important translation is necessary.

The word "image" translates to Greek as εικων (eikon = icon). On the other hand, the Greek word ειδωζ (eidos = knowing) and the word ιδεα (idea), are derived from ιδειν (idein = to see). Thus, the word *idea* is a process which consists of knowing the form of a mental icon or image of an object. No wonder the ancient Greeks saw the eyes as being the windows of the soul (mind) of human beings. We may even stretch this description of an idea by saying that an idea is the *formal structure of the mental image* of a real object.

Plato used to say that philosophy is due to the sight, and he sustained that to *know* was to *see*. On the other hand, Aristotle referred frequently to images related with the sense of touch. For the Greek philosopher to know, he had to touch and to apprehend. Of course, a philosophy based only on the sensorial icons of touch would be too limited and completely colorless. Helen Keller confessed she never was able to know the idea of "whiteness" when she apprehended the shape of a swan. It is obvious that no branch of Astronomy could have been developed by blind people. Another way to refer to mental icons or ideas is with the word *phenomenon*. This word is composed of two Greek words: φωζ (phos = light) and *noumenon* derived from νουζ (nous=mind), to designate what the mind conceives beyond the mental image (idea), beyond the mental light (phenomenon). Any *concept* or noumenon cannot be perceived as an idea (mental image or phenomenon), because any concept is a rational creation of the human mind. However, there are some trans-concepts which can be apprehended intuitively, without any help from our rational minds. Kant rejected this intuitive possibility arguing that man does not have this mental capacity. For Kant, the sensorial images or mental appearances are called *phenomena*. However, if the entity is a creation of the mind (concept), then we should call it noumenon (concept). It is interesting to note that for Kant, as well as for Plato, *noumena* (concepts) refers to the intelligible world, while *phenomena* (sensorial images = ideas) refer to the sensorial world.

In this chapter, we will use the word *idea* to refer only to *mental sensorial images*. These mental sensorial images are indeed *mental sensorial light* or *phenomenon*. Thus, we take as synonyms the words ideas and phenomena. We also consider as synonyms the words concepts and noumena. We will never introduce the metonym of assigning the meaning of the word *idea* to the word *concept* or to the word *noumenon*. We hope, in the 21st century, scholars will be more precise in their writings.

For centuries, philosophers have taken for granted the capacity of *human reason*. They have done the same with the *intellectual capacity* of human beings. What about the *rational power* in human beings? It also has been taken for granted. Perhaps in the 22nd century a new computer will ask an old computer - "Who created us?"- The old computer will answer - "We were created in the image of man."- This is exactly what we did in the 20th century when we *reified* the *noumenal* activity of John Babbage of the 19th century. Babbage established the main four functions of a computer; (1) Input of data, (2) storage of data in memory, (3) CPU or Central Processing Unit, where the data is treated rationally, and (4) output of information or processed data. Some people criticize some scientists because these non-scientists wrongly believe that the human brain (hardware) and the human mind (software) are compared to computers. These people should ask themselves from where Babbage got his concepts. A recent book by Nick Herbert [3], titled *Elemental Mind or Human Consciousness and the New Physics*, published in 1993, is rich in detailed scientific descriptions of the brain. In chapter 3, Herbert refers to similarities and differences between human brain/mind and computer hardware/software. He also tries to identify the Human Central Processing Unit (HCPU) with physiological material. As a matter of fact he says:

> "In our brain model, the reticular formation plays the part of a computer's CPU ; the sensory/motor cortex, along with basal ganglia and cerebellum, handles input/output routines. Memory in the brain is not segregated into one particular location as in a computer but is distributed in some unknown way among the brain input/output machinery. Since present computers possess (as far as we know) no internal experiences, there is a natural limit to our analogy."

It is obvious that no one teaches little children to *reason*, to learn *intellectual capacity* or to learn *rational power*. For centuries philosophers took these capacities of the human mind for granted, because they knew nothing about the CPU in modern computers. Also philosophers wanted the world to know that man is a magnificent animal, and that all he has in his mind is developed only by sensorial experiences, phenomena and sensorial images. Without *a priori* HCPU, we would still be monkeying around the branches of the tree of knowledge.

Herbert, in the above quotation, says that present computers do not possess inner experiences. The author remembers that in the 1960s the IBM 1620 computer generated modulated electromagnetic waves each time the computer moved zeros from one memory location to another. This internal experience of that particular computer was unknown to human operators. Then one night a human operator detected demodulated electromagnetic waves in his portable radio which were coming from the active computer. This inner experience of the IBM 1620 sounded like organ music. After that event, many computer programmers spent many hours programming, in machine language, pieces of classical and popular music to be interpreted by the noumenal activity of an electronic machine.

It is said that a picture is worth 1,000 words. Perhaps an allegory is worth only 100 words. What follows is an allegory of the human mind. Let us imagine the mind is composed of four chambers in an ascending row. All the chambers are communicated in sequence. Let us name the chambers as follows: (1) chamber of sensorial images (ideas, phenomena), (2) chamber of reason or HCPU, (3) chamber of concepts (noumena), and (4) chamber of transcendental (transrational) inner experiences.

From the chamber of sensorial images, the HCPU gets ideas which are conceptualized (universalized). The particular images or ideas are organized and categorized. It is in this *rational* chamber where the human mind establishes relationships between the ideas of the chamber of images. This relationship is a rational thought, concept, knowledge which is created by the human mind. All these concepts are organized and categorized in the chamber of concepts or rational chamber. This conceptual chamber constitutes the world of noumena. This chamber is the *noumenal* world of concepts which are the rational ethereal products of the mind. It is here where man created the first immaterial model of a computer. It is in this noumenal chamber where the external reality is conceptualize by the human mind. It is in this noumenal chamber where the mind creates ontology by hypostatizing the content of the chamber of ideas, of phenomena. It is in this noumenal chamber where the mind creates the transitory, fictitious, relative, *scientific theoretical knowledge* and invents metaphysical theories. It is in this noumenal chamber where all *technology* comes into mental existence. It is in this noumenal chamber of the human mind where the *application* of science (technology) comes into conceptual existence. This noumenal technology is later reified to become *new ideas* in the minds of users. However, the users of technology have no previous concepts about the *unreal* technology in the minds of engineers until the inner mental conceptuality is converted into *real* technology in the external world of things. Ancient philosophers had around them an extraordinary primitive craftsmanship. They did not degrade themselves by doing manual work which was reserved only for slaves. They developed the Olympic attitude of only thinking metaphysically. This attitude was maintained by all metamathematicians of the 20th century. On the other hand, Eastern philosophers have been much more inclined to passively receive transrational knowledge rather than to create knowledge. They have, more or less, ignored the teachings of 25 centuries ago when Confucius taught that - "*Knowledge is good, but better is its applications.*"- (Philosophers should study science and engineering before they sink into the bottom of the ocean of metaphysics.) The fourth, transrational, or transcendental chamber, brings us to the most unknown chamber of the human mind, especially in the Western world. It is in this chamber where the mind transcends the rational process of acquiring meta-conceptual knowledge. It is in this fourth transrational chamber where an altered state of mind takes place, and allows humans to acquire *absolute* knowledge; i.e., knowledge independent of any noumenal or rational activity of the mind. It is in this transrational chamber where only a few humans have mystical experiences. It is in this mystical chamber where the mind knows and has the *feeling* of knowing when new knowledge has flooded into this mental chamber. You *know* you know, and you do not care in the least why or how you know. You are tranrationally convinced you possess *absolute knowledge*. You know you do not need your eyes to contemplate the object of this nonsensorial knowledge. You know you do not need your rational mind to get this meta-conceptual knowledge. This transrational knowledge has a totally different source. This absolute knowledge is not an *idea:* It is nonrational knowledge, it is transrational knowledge, it is the acquisition of meta-concepts.

When you know this absolute way, you do not need your hands or your eyes to apprehend them. You do not need ideas in your mind in order to assimilate or contemplate them. When you know this absolute way, you do not need your rational tools to conceive or to create these meta-concepts. As a matter of fact, you need to do nothing except passively wait for the great moment when you will receive a crumb of knowledge. At that moment, you are re-connected with the Being of all entities, with the Cosmic Universal Consciousness. After receiving this revelation in your mind, you feel an imperative need to share this transrational (mystical) experience with your fellow men. However, be careful, do not share these mystical pearls in the beginning. Those around you may fail to understand and may attack and try to destroy your integrity. It is in this same tranrational chamber of the human mind where *cosmic religious experiences* takes place.

Noetic is a term derived from the Greek word νουζ (*nous* = mind). It seems that Plato created the term "noetic" to refer to direct knowledge of the truth. This noetic path transcends the rational process of the intellect when creating concepts. When the word "noetic" entered the English language it meant "rational knowledge." Later on, William James assigned to the word "noetic" the pristine Platonic meaning of "direct cognition of absolute truth," meaning mystical or transcendental, spiritual or religious, transrational cognition of absolute truth. However, by the middle of the 20th century, and after the work of Edmund Husserl, the term *noetic* has become synonymous with *noumenal*; i.e., rational, losing its Platonic meaning of transrational cognition of absolute truth. In this book, we will use Husserl's meaning to avoid further confusion.

Later we will see that Einstein, when writing about God, refers to this transrational, transcendental, mystical experience as a *cosmic religious experience*. William James [4], one of the fathers of American Pragmatism, published his encyclopedic book, *The Varieties of Religious Experiences,* at the turn of the 19th century. In the chapter *Mysticism,* he offers a criterion to recognize a mental experience as a mystical one. He warns his readers by writing that: "*The words 'mysticism' and 'mystical' are often used as terms of mere reproach, to throw at any opinion which we regard as vague and vast and sentimental and without a base in either fact or logic.*" This is indeed a reproach. But in this reproach, we see an excellent characteristic of any "mystical experience" caused by the ignorance of the censurer. It is true that any "mystical experience" is "*without a base in either facts or logic.*" It is without a base in either sensorial images or concepts. Three-quarters of a century later the author was reproached by his peers in physics, reproached for talking to them about mystical experiences. At that point, the author introduced the word "transrational" as a synonym for mystical, and only then did the reproaching creatures of the physics department begin to ask interesting questions about transrational experiences. They admitted their status of "ignorant sages," making their first step toward wisdom. To reinforce their interest in "*a priori* synthetic judgments," the author paraphrased Einstein saying - Sometimes God feels an enormous pity for our mental incapacity and reveals little details of Himself.-

James' criterion to recognize an altered state of mind as a transrational or mystical experience, demands the experience to adequately follow four characteristics: (1) Ineffabilty, (2) Noetic quality, (3) Transiency, and (4) Passivity.

(1) **Ineffability**. Every mystical experience is ineffable; i.e., it is beyond verbal description; it is indescribable, inexpressible. Even if the mystical experience conveys trans-concepts, it is more a feeling or vivency (insight) than an intellectual activity. The transrational knowledge may be presented into the mystical experience like very familiar images. These images *are not* what they represent. For example the images may represent a school of fish swimming around you. Until you grab one slippery fish with your transrational hand, you know nothing about the *essence* of the fish. The fish may be metaconcept(s), or even a question like this - Are all electrodynamic forces proportional to the mathematical structure of inertial accelerations? - Other mystical experiences convey the religious absolute knowledge of the *unbroken wholeness of the entire universe*. In this case, you feel enclosed in the womb of the Being of all entities, *before you are, while you are, and after you are*. Trying to express this intuition—this vivency in common words, is next to impossible. Insights are ineffable, and you will only waste your time trying to rationalize these mystical experiences or sharing them with skeptical people. William James says - "*No one can make clear to another who has never had a certain feeling, in what the quality or worth of it consists.*" - Because of this characteristic of ineffability, we referred to Quantum Mechanics, in chapter 2, as Quantum Mysticism. This western super abstract quantum mysticism, nevertheless, allows the modern western quantum mystics to apply it to the unrealistic microcosmos to observe average magnitudes in the reality of laboratories. Quantum reality, in this quantum mysticism, is not an idea, is not a concept, is not even a metaconcept. Quantum reality does not exist for these modern quantum mystics.

(2) **Noetic quality**. The mystical experience is a transrational one. It seems that James prefers to label an altered state of mind "mystical experience" when it conveys more feelings of *re-connection* with the "whole" than conveyance of knowledge. James also seems to prefer the qualification "noetic experience" when the altered state of mind conveys more knowledge than feelings. This noetic knowledge is metaconceptual because it is not *conceived* or *created* by the human mind, but is *received* in a passive way. Writing about mystical states, James writes - "*They are states of insight into the depth of truth unplumbed by the discursive intellect.*" It is here where James establishes, like Plato, that noetic knowledge is metaconceptual, is transrational. These metaconcepts pass into the human subject as an indelible sense of authority along with a feeling of importance, and as yet, an unaccomplished mission.

We have a semantic confusion which we must clarify. After studying William James' thoughts, we should have understood the phrase "noetic knowledge" as knowledge acquired *independently* from a rational discourse and independently from a sensorial experience. However, after James, Edmund Husserl came, as we previously mentioned, introducing the words noetic, noematic, noema and noesis. In Husserl's philosophical writings the word "noetic' means "relative to thoughts," relative to *noesia*, which is the very act of thinking. The word *noema* means the product of thinking, the created concepts beyond the phenomena. Finally, the word *noematic* makes reference to *noema*, the object of the mind. To be more well-defined, because of this Husserlian terminology, we will not use the word *noetic* as used by Plato or by James. From now on, we will refer to metaconcepts, or "knowledge" acquired through "mystical experiences," with the word *transcept* to signify *reception* of knowledge without the intervention of the rational mind. Thus, a transcept is a metaconcept, is a transrational piece of knowledge. Thus, James' "noetic quality" translates to a "transceptual quality." Presently the noun *noetics* means *The Science of Consciousness*. The reader, interested in this new science of noetics, will find more information by just typing the word "noetics" and let any search engine on the Internet uncover detailed material for them.

(3) **Transiency**. Altered or transceptual states of mind are transient, brief, ephemeral. As these mystical experiences are so short, they are not too intense, consequently, they quickly fade away and recollection is difficult if not impossible. For this reason, it is highly recommended to record such transrational experiences as soon as conveniently possible. Also, *never* think about the meaning or content of a transceptual experience, especially if the subject is aware of his altered state of mind. The thinking process automatically disconnects the human mind from the transrational realm, and the mystical experience immediately vanishes.

(4) **Passivity**. The individual will of a subject plays an active part when the human mind delves into the transrational realm; "*. . . the mystic feels*" - says James - "*. . . as if his own will were in abeyance, and indeed sometimes as if he were grasped and held by a superior power.*" This feeling of connection or union with a superior power is the purpose of yoga practices. The meaning of *Yoga* is the path which leads the individual human subject to a union with the divine. This Yoga practice causes the human being to feel the universal love which, in mystical and ontological terms means the *invisible essence of unity*. This unity is *the* connection, or if we prefer, it is the re-connection (religion) with the transrational realm.

At first sight, it is almost incomprehensible that William James [5], the founder of pragmatism, had written such a learned book on *The Varieties of Religious Experiences*. A pragmatist is supposed to be such a functional person, an empiricist, a positivist, a materialist, but seldom a mystical or spiritual person. This impression is not true in the case of William James. In another book by James [6], *Essays in Pragmatism*, he defines *Pragmatism* not as an elaborate philosophical doctrine, but as a simple method. In this respect he writes:

> "The pragmatic method is primarily a method of settling metaphysical disputes that otherwise might be interminable."

We should keep in mind this pragmatic purpose when we discuss metaphysics of the past. However, a serious thinker has trouble in agreeing with William James' concepts of truth. Writing about a concept expressed as a proposition or judgment, James writes - "You can say of it then either that «*it is useful because it is true*» or that «*it is true because it is useful*»"- The converse or latter statement evidently is false. Nevertheless, the important objective of William James was to establish a criterion to determine the truth or falsehood of metaphysical statements. Another criterion, we should have, is to determine the *real* existence of concepts, because it is elementary to affirm that not every concept, which exists in our minds, has a corresponding *existence* in reality. Before we discuss "revealed" or transrational theology, let us first review the three gnoseological paths which are open for us to acquire knowledge.

Ideas or mental images are directly acquired by sensorial means. These mental images are not the result of intellectual processes, of logical deductions. The acquisition of ideas is, therefore, not rational, but *irrational*. Thus, the three gnoseological paths to acquire knowledge are; *irrational, rational*, and *transrational*.

The irrational gnoseological process allows us to *perceive*. To "perceive" derives from the Latin word *percipere*, composed from *per* and *capere* (to take, to apprehend, to grasp). The prefix, in this case, means "the result of" grabbing, seizing. On the other hand, the words *percept* and *perception* have the same relationship as the words *concept* and *conception*, and the words *transcept* and *transception*. Both perception and transception imply *reception*. Percept is knowledge (of external facts) received by means of the senses. Thus, a percept is a mental image, an idea, a mental light or a phenomenon. Concept is not a reception of an entity but a gestation of knowledge created or conceived by the mind. Finally, transcept is knowledge received by means of *intuition* from the transrational realm. Having said this, we can now talk about the three kinds of knowledge; *perceptual, conceptual*, and *transceptual, o*r if we prefer, *sensorial, intellectual*, and *intuitive* knowledge.

In these three types of knowledge, we find that knowledge is a *relationship* between a subject and an object. In respect to sensorial and intuitive knowledge, the mind is a passive receptacle. In the case of the reception of percepts, the *connection* between the subject and the object is established by the senses of the subject. In the case of the reception of transcepts, the *connection* between the subject and the object is established by the intuition of the subject. Jean Paul Sartre says that "*intuition is the presence of the consciousness in the thing.*" Based on Sartre's concept, we should say that "intuition" is the " *identification of the subject's consciousness with the entity*" to be known absolutely.-

Later *consciousness* will be discussed at length, but to establish a firmer base, let us concentrate on the three different words which refer to the gnoseological processes of knowing, and the three different words which refer to the type of knowledge obtained by the corresponding processes.

In Spanish and other romance languages, we have two verbs to express the verb "to know"; in Spanish they are *conocer* and *saber*. In German, we also have two verbs, *kennen* and *wissen*. To know ideas is to "conocer" or "kennen." To know concepts is to "saber" or to "wissen." In our contemporary English language, we have only one verb to refer to two such essentially different noumenal processes. At any rate, in all these languages we have the word *intuition* (German: *Erkenntnis* or *Erfahrung*) to say "to know directly." Bertrand Russell [7], in his delightful little book, *The Problems of Philosophy*, refers to a distinction in the English word "know" when used in a different context. Russell writes:

> "The word 'know' is here used in two different senses. (1) In its first use it is applicable to the sort of knowledge which is opposed to error, the sense in which what we know is *true*, the sense which applies to our beliefs and convictions; i.e., to what are called *judgements*. In this sense of the word we know *that* something is the case. This sort of knowledge may be described as knowledge of *truths*. (2) In the second use of the word 'know' above, the word applies to our knowledge of *things*, which we may call *acquaintance*. This is the sense in which we know sense-data. (The distinction involved is roughly that between *savoir* and *connaître* in French, or between *wissen* and *kennen* in German)."

Here we face two ways to know: one is intelligent, rational; the other one is sensorial, non-rational. In the intelligent way to know, we know concepts. In the sensorial way to know, we know ideas. When we can explain (deduce) what we know (concepts), then we *know*. When we can describe but not explain what we *are acquainted* with (ideas), then we *ken*. The old English verb *to ken* is a wonderful verb to enrich the modern gnoseological terminology expressed in the language of Shakespeare. To ken is to perceive, hence *conocer*, *connaître*, *kennen.* To finish this subsection of the Introduction to this chapter, let us remind ourselves that we humans try to *conceptualize* what we perceive and what we intuit; i.e., we try to rationalize the irrational ideas and the transrational intuitions.

I-4. Theology by revelation and theology by reason.

Theology or the *study of God* is a branch of Theoretical Philosophy or Metaphysics. The etymological meaning of *metaphysics* is *beyond-physics*, beyond sensorial perceptions, beyond the visible causative reality of the percepts, beyond the phenomena. The term *metaphysics* was created after the death of Aristotle. Lovers of words, or philologists and non-philologists as well, have found a fertile garden in metaphysics to plant a variety of words, spelled differently, but having the same meaning. Today, metaphysics is not a garden of meaningful lexical terms, but a terrifying jungle of unreasonable terminology. We should have only one tree in this jungle: the Cartesian tree of knowledge. *Philosophy* is the root of this tree. *Natural Philosophy* is the trunk of this gnoseological tree of Descartes. The different sensorial recordings of data or percepts are represented by the many branches of this gnostic tree. To do this branchial work, philosophy is not necessary! We may keep the name *Natural Sciences* for this wonderful collection of incomprehensible facts or phenomena from which later a *Rational Science* will be created by the human mind, but always keep in mind that metaphysics of the past is the physics of the present. Some say, physics today is the restored, crippled, obsolete, metaphysics of the past. If this is so, and if theology is a department of metaphysics, then theology of the past, too, can be restored to a modern Physics or Natural Philosophy of God. In the mean time, and before we introduce a Scientific Theology, let us do a fast review of two theologies, revealed theology and rational theology.

Revealed Theology. To deny the existence of the transrational subjective acquisition of knowledge is to reveal unforgivable ignorance, but ignorance has never been an excuse. William James in his magisterial collection of documents on transrational experiences, *The Varietes of Religious Experiences* [8], leaves no room for opinions of skeptical ignorant sages. James' documentation of religious experiences is not rational proof, of course, but it is a source from which James himself draws common qualities from the many altered states of mind reported in his book. Before we deny the true existence of a transrational realm and the re-connection with it, we have to rid ourselves of endemic ignorance. From all those privileged human beings who have had what Einstein calls a *cosmic religious experience* or *common-union* with the transrational realm, we can perceive a common essential characteristic. In section 9 of this chapter, we will expand on the meaning of the "transrational realm." In the meantime, let us look at James' findings on the common characteristics in the minds of those who have *received* transcepts (beliefs).

> "1. That the visible world is part of a more spiritual universe from which it draws its chief significance;
> "2. That union or harmonious relationship with that higher universe is our true end;
> "3. That prayer or inner communion with the spirit thereof - be that spirit "God" or "law" - is a process wherein work is really done, and spiritual energy flows in and produces effects, psychological or material, within the phenomenal world."

James adds: "Religion includes also the following psychological characteristics":

> "4. A new zest which adds itself like a gift to life, and takes the form either of the lyrical enchantments or of appeal to earnestness and heroism."
> "5. An assurance of safety and a temper of peace, and, in relation to others, a preponderance of loving affections."

The discovery of these common characteristics, in mystical people, is a stupendous application of James' *pragmatic methodology* to establish a criterion to recognize when a human being has had a transrational experience, when a human being has had a revelation or has received a transcept. A collection of revelations does not constitute a "revealed theology." A set of revelations or transcepts are the basic ingredients to create a "theology." These remarks are analogous with respect to a collection of percepts, or empirical phenomena, or sensorial facts. To have a mountain of perceptual facts does not mean we understand them; on the contrary, it means to understand nothing, because to understand them is to conceptualize them into a rational theory. A collection of transcepts or revelations does not constitute a Theology. A collection of revelations constitute an incomprehensible set of *commands* pertaining to a dogmatical, but intellectually blind, religion. Any theology, to be called the *logos of God,* must be conceptualized, must be theorized like any scientific theory. Any theology must have a *hypothetic-deductive* structure. The basic or fundamental hypotheses (axioms) of any transrational or revealed theology are the trancepts or *a priori* revelations. On the other hand, *Rational Theology or Natural Theology*, is a rational theory of God based on a conceptual creation of the mind and perceptual experiences. The purpose of any theology is to establish the existence of God and His nature and attributes. A mystical person needs no rational theology. Neither does he need a revealed theology. A mystical person needs no belief in the existence of God, because he *knows* God transrationally.

In the western world, the best known *Revealed Theology* is the Christian Theology which is based on the revealed word of God contained in the Sacred Books (Bible). These divine biblical revelations would have remained a set of transrational or divine commands had it not been for Saint Thomas Aquinas (1225-1274) who converted the Christian Religious Doctrine into Revealed Theology. Saint Thomas had, of course, forerunners in his endeavors, but he is the Master of *Scholastic Theology* erected on the philosophical foundations of Aristotelian first principles and first causes. Thomas Aquinas did some outstanding intellectual work in the middle ages, considering the high degree of ignorance of Natural Philosophy at that time. In the present, we respectfully censure the five *demonstrations* of the existence of God adduced by Saint Thomas in his *Summa Theologiae*. These five demonstrations lack the logical rigor of Euclid's demonstrations centuries before. The following quotations and remarks refer to the magnificent conceptual evolution in the Scholastic Philosophy of the Catholic Church which is beginning to honor its name of *Catholic* which means *Universal*. We can see in the Greek word καθολικαζ (katholikós) the term "holikós" which is derived from ολοζ (olos = whole, all). The quotations will be extracted from the Encyclical Letter *Fides et Ratio* [9] (**FR**, Faith and Reason) of Pope John Paul II to the Bishops of the Catholic Church, in 1998. This is an Epistle not only addressed to Catholic Bishops but to any theologian, to any philosopher, to any scientist, to any intelligent man at the entrance of a new millennium. The one regret, we feel, is that this letter was not written at the end of the first millennium, but their conclusions point to the fact that reason had not yet been properly cultivated in those years. Unfortunately, we cannot direct our comments to every aspect of this Encyclical Epistle, but only to those points which have a close relationship between the acquisition of *transrational* knowledge and the creation of *rational* knowledge. This Encyclical Epistle is a wonderful proposition with which to embark the human mind upon a majestic noumenal trek where no mind has ever gone before. This Mind Trek, fueled with *faith* and *reason*, will take human minds to the transrational realm of Universal Consciousness where all *truths* are reduced to *One*.

At the very introduction of **FR,** the Pope calls our attention to contemporaneous philosophers for their part in neglecting their interest in the search for the ultimate truth. He also points out the abandonment of the ontology of subject and object, and instead, preferring the creation of theories of knowledge. This book, *Einstein on Trial*, is an effort to *substantiate* Natural Philosophy as Pope John Paul II would like Philosophy to be ontologized in the 21st century. In this respect, the Pope writes:

> "Therefore, following similar initiatives by my Predecessors, **I** wish to reflect upon this special activity of human reason. **I** judge it necessary to do so because, at the present time in particular, the search for ultimate truth seems often to be neglected . . . Abandoning the investigation of being, modern philosophical research has concentrated instead upon human knowing."

In chapter I of **FR,** the Pope continues his criticism of the excessive dedication to gnoseological matters. However, he uses gnoseological arguments to make his point clear. The Pope writes:

> "The first Vatican Council teaches, then, that the truth is attained by philosophy and the truth of Revelation are neither identical nor mutually exclusive. «There exists a twofold order of knowledge, distinct not only as regards their source, but also as regards their object. With regard to the source, because we know in one by natural reason, in the other by divine faith. With regard to the object, because besides those things which natural reason can attain, there are proposed for our belief mysteries hidden in God which, unless they are divinely revealed, cannot be known». (7) Based upon God's testimony and enjoying the supernatural assistance of grace, faith is of an order other than philosophical knowledge which depends upon sense perception and experience and which advances by the light of the intellect alone. Philosophy and the sciences function within the order of natural reason."

In the **FR** Epistle, the word *faith* is used as *Christian Faith* and also as the *acceptance with great conviction of the truth of a statement which cannot be demonstrated true by rational discourse*. This second meaning corresponds to the definition of faith given in the Bible, Hebrews 11:1. In the above quotation, the Pope is writing about the percepts and concepts as rational elements. He is also writing about transcepts acquired transrationally or divinely. Without a theory of knowledge, it would be extremely difficult to describe the meaning of the word *revelation*. This was the reason we had to present some basic gnoseological elements in our introduction to this chapter. The main problem to solve now is to *actualize* the transceptual and conceptual God who exists in the noumenal realm of human minds. This means to determine if the concept of God also exists in *reality*, independent of any human mind. This is not a task for philosophers, nor is it a task for theologians. This is a responsibility which must be shouldered by natural philosophers or future scientists, because theologians and philosophers have been writing treatises of theology since time immemorial, and still, we are lacking absolute or relative knowledge of a *Real God*. By Real God, we must understand the ultimate *Being* of all visible material entities, and all invisible ethereal entities which exist in the universe. We have taken 7,000 years to develop a very primitive science of the visible material universe. Can we imagine how many more years it will take us to develop a science of the invisible universe? Although scientists have begun to suspect that an invisible Newtonian *Sensorium Dei* may exist in the external world of things.

In 1992, Pope John Paul II, addressed the Pontifical Academy of Sciences for the purpose of informing the Academy about the study and conclusions of the "Galileo case." The title on the front page of the Weekly [10], English Edition of *L'Osservatore Romano*, N. 44, 4 November, 1992, was a forethought of the coming edition of **FR** of 1998: *Faith can never conflict with reason.* On that occasion the Pope said:

> "It is necessary to repeat here what **I** said above. It is a duty for theologians to keep themselves regularly informed of scientific advances in order to examine, if such be necessary, whether or not there are reasons for taking them into account in their reflection or for introducing changes in their teaching."

Unfortunately, to be informed is to ken or to be acquainted with specific knowledge, but it is not to know. However, the Pope's suggestion is a gigantic step toward an implosive neo-renaissance which has already begun in Europe. This neo-renaissance, once it begins to flourish extensively, will bring all the so-called hard sciences within the womb of mother philosophy. A similar but more dramatic suggestion is in order for students working toward their Doctoral of Philosophy degree with a major in science. It is shameful that in the great majority of universities, these candidates do not receive one single *iota* of philosophy and/or theology. For this reason Ortega-y-Gasset called the great majority of scientists *ignorant sages*. In chapter 4 of **FR,** the Pope refers to the creation of Scholastic Theology and Scholastic Philosophy. As we mentioned before, the master of this creation was Saint Thomas Aquinas. The Pope writes:

> "In an age when Christian thinkers were rediscovering the treasures of ancient philosophy, and more particularly of Aristotle, Thomas had the great merit of giving pride of place to the harmony which exists between faith and reason. Both the light of reason and the light of faith come from God, he argued, hence there can be no contradiction between them."

Immediately the Pope, writing about Thomas Aquinas, shares with his readers a new revelation which will be fulfilled in the 21st century. The Pope writes:

> "More radically, Thomas recognized that nature, philosophies proper concern, could contribute to the understanding of divine Revelation."

When the proper concern of philosophy is nature, we are then talking about Natural Philosophy, called physics today. This Papal interpretation of Saint Thomas' thoughts is a confirmation of what we have said above: *This is not a task for philosophers, nor is it a task for theologians. This is a work for natural philosophers or future scientists, because theologians and philosophers have been writing treatises of theology since time immemorial, and still we are lacking absolute or relative knowledge of a Real God.* - We have no doubt that Natural Philosophy "*could contribute to the understanding of divine Revelation.*" In the book, *Theology of Revelation* by Gabriel Moran, F.S.C.[11], the author complains: "*I wish merely to indicate that the large treatises on revelation have astonishingly little to say about revelation itself.*" How can we create a Theology of a transcept we do not know? In this Encyclical Epistle of Pope John Paul II, we find the new rocks with which to build the foundations of a true Universal Church, especially if we read, with an open and critical mind, the biblical message in John 16:25:

> "These things have I spoken unto you in proverbs: but the time cometh, when I shall no more speak unto you in proverbs, but I shall shew you plainly of the Father."

Proverbs, metaphors, allegories and parables are literary forms we use when the audience has little understanding. We strongly *believe* that Natural Philosophy "*could contribute to the understanding of divine Revelation.*"

Under revelation we find Theosophy, the wisdom of God, established by Madame Blavatsky at the end of the 19th century. Madame Blavatsky wrote a series of books titled "*The Secret Doctrine.*" These books, according to Madame Blavatsky, were completely written from transrational experiences.

Rational or Natural Theology is a theology which resorts only to *reason.* We say this but, nonetheless, this is not altogether true. A Rational Theology is a *Theory of God.* As any theory, it requires a set of principles or axioms, a set of definitions (operational definitions if possible), a set of rules to draw logical inferences, a set of many logical conclusions, and a *criterion* to know if the logic-conceptual conclusions have *real* existence in the external visible world of things, or in the invisible immaterial realm.

In respect to the set of principles or axioms, a Rational Theology must accept, under an *act of faith,* the ontological principles. This act is totally equivalent to accepting the truth of religious dogmas or revealed transcepts. Epistemologically speaking, a Rational Theology is no better than a Revealed Theology. In relation to the inferences rules, Rational Theology has to accept, under an *act of faith,* the principles or axioms of Logic. Once more we find, that in order to rationalize, we have to have *faith* in the truths of propositions which we cannot demonstrate rationally. In practical terms; i.e., in terms of William James' "pragmatic methodology," it is entirely futile, irrelevant, superfluous to have any debate about the superiority or inferiority of a Revealed Theology in comparison to a Rational Theology. Both kinds of theologies *must* be founded on faith.

In the last few centuries, we have seen the birth of so-called rational theologies. *Theodicy*, derived from the Greek words *Theo* (God) and *diké* (justice), is the title of one of Leibniz's works (1710), dealing with God's creation of the world in spite of all evils. It is inconceivable that a little insignificant mind of a human being, such as Leibniz, could have the unreasonable conception that he had the authority to write about the reasons God had to create the world. This is a disgraceful attitude for humankind. The interested reader should consult *The Cambridge Dictionary of Philosophy* [12, 1995] for brief descriptions of other rational theologies.

Now we have gathered the basic elements with which to enter a Gnoseological Court to attend our last trial for Einstein. This time it will be on theological matters. We anticipate a wonderful outcome from the fusion of Einstein's General Relativity Theory, not with Pantheism, but with God's Consciousness. Before, however, we must analyze Einstein's four essays on *religion and science*, published in *Ideas and Opinions* [1, p. 36-52], along with many other articles on different subjects written by Einstein over the years.

7.1 Comments on Einstein's essay "Religion and Science" (1930)

The essay, *Religion and Science* was written especially for the *New York Times Magazine, November 9, pp 1-4 (1930).* Two days later the German text was published in the *Berliner Tageblatt.*

In the first part of this essay, Einstein asks what the feelings and needs were that forced human beings to religious thoughts and beliefs. Einstein contemplated that before religious thoughts and religious experiences, a great variety of emotions preceded the appearance of a definite religion. The first religious notions came to the human minds through fear of different things and events. Einstein envisions that in the early stages of human evolution, the primitive mind had not yet created the concepts of cause and effect. However, this primitive human mind had the wonderful capacity to formulate questions, and to look for or invent answers from which to satisfy this almost incomprehensible hunger for explanations. Consequently, Einstein believes that this primitive human mind "*creates illusory beings more or less analogous to itself on whose wills and actions these fearful happenings depend.*" We see in these illusory beings the first step toward an anthropomorphic religion. At this point, Einstein sees the appearance of a *religion of fear.* The wrath of these illusory beings or *anthropomorphic gods* had to be calmed by offering sacrifices. The liturgy, sacred rituals, procedures and protocol of the sacrifices were judiciously transmitted from generation to generation. Therefore, the social group of primitive human beings were able to create a tradition which gave them the assurance that their descendants would continue to follow their *religion of fear*.

To keep the tradition, generation after generation, the primitive human social group created a "priestly cast" to mediate between the people and the invented gods they feared. Einstein thought, in many cases, a leader or ruler, or a privileged class of individuals, combined priestly functions with their civil authority. In this way, the political leaders secured their positions through the cast of the priesthood. In other instances, Einstein thought the political rulers made common cause with the religious cast so they may benefit their own interests. In this respect, Einstein writes:

> "In many cases a leader or ruler or a privileged class whose position rests on other factors combines priestly functions with its secular authority in order to make the latter more secure; or the political rulers and the priestly cast made common cause in their own interest."

Social or moral issues were the reasons for the development of religion, according to Einstein. An undeniable service imposed religion on the individuals of the social group with the intention of controlling the behavior of the so-called rational animals. Up to this point, Einstein writes about the genesis of religion and the administration of religious doctrines. Thus, the *conception* of God, in a primitive religion, was determined by the social behavior of a particular group of human beings. Morality is precisely the set of empirical observations of human behavior in a group of terrestrial rational beings. When this empirical social behavior is *conceptualized*; i.e., is rationalized in a theory to explain the social behavior of human beings, this theory is called *Ethics*. Socrates attempted to create such a theory, but he failed. Centuries later, Spinoza created a theory of Ethics using the propositional formality of Euclid's geometry.

Obviously, in a Revealed Ethics, contained in a Revealed Theology, recourse is made of transcepts or mystical experiences. Naturally any Rational Ethics is a subset of a corresponding Rational Theology. Einstein writes - "*The Jewish scriptures admirably illustrate the development from religion of fear to moral religion, a development continued in the New Testament.*" - Einstein refers to cultural groups who emphasize *moral religion* as civilized societies, and specifically points to people of the Orient.

Now we come to what Einstein calls the *cosmic religious feeling*. Besides the primitive religion in which God is an anthropomorphic conception of man and his moral religion, Einstein points out a third religion based on *cosmic religious experiences*. For centuries, these experiences have been called *mystical experiences*. Einstein expresses the ineffability of this kind of inner experience. He depicts it this way:

> "It is very difficult to elucidate this feeling to anyone who is entirely without it, especially as there is no anthropomorphic conception of God corresponding to it."

About these transrational experiences, Einstein mentions the Psalms of David and some biblical Prophets. He also mentions Buddhism as a path to walk in order to liberate the minds of humans. Einstein writes - "*Individual existence impresses him as a sort of prison and he wants to experience the universe as a single significant whole.*"- These concepts were spoken by Siddhartha Gautama, the Buddha, 25 centuries ago. Susuki [13], a modern Zen Buddhist, writes:

> "Gautama felt as though a prison which had confined him for thousands of lifetimes had broken open. Ignorance had been the jail keeper. Because of ignorance, his mind had been obscured, just like the moon and stars hidden by the storm clouds. Clouded by endless waves of deluded thoughts, the mind had falsely divided reality into subject and object, self and others, existence and non-existence, birth and death, and from these discriminations arose wrong views—the prisons of feelings, craving, grasping, and becoming. The suffering of birth, old age, sickness, and death only made the prison walls thicker. The only thing to do was to seize the jail keeper and see his true face. The jail keeper was ignorance . . . . Once the jail keeper was gone, the jail would disappear and never be rebuilt again."

Einstein's advice about this cosmic religious experience is the following. - "*In my view*" - he writes - "*it is the most important function of art and science to awaken this feeling and keep it alive in those who are receptive to it.*"

Einstein now delves into a delicate and controversial subject. He complains about unjust charges against science for undermining morality. His defense of science is based on the principle of cause and effect. Einstein writes about a man who is thoroughly convinced of the universal validity of the principle of cause and effect:

> "A God who rewards and punishes is inconceivable to him for the simple reason that a man's actions are determined by necessity, external and internal, so that in God's eyes he cannot be responsible, any more than an inanimate object is responsible for the motions it undergoes . . . A man's ethical behavior should be based effectually on sympathy, education, and social ties and needs: no religion basis is necessary."

If human beings would act 100% rationally 24 hours a day, Einstein would be right. Unfortunately, this is not the case. If there is something good in any religion based on fear, it is precisely the conventional moral rules. These rules, which are based on fear, have a strong tendency to control the decent behavior of the members of a social group. Authority has always been based on a variety of punishments for misbehavior so a fear is always there. In this respect, religion and its moral rules is a necessary evil. If *"man's actions are determined by necessity, external and internal,"* as Einstein asserts, then man is not responsible for his acts in the eyes of God or in the eyes of any rational creature. This statement invalidates the "freedom of will," and man becomes an irresponsible robot. In order to preserve any social group of humans, man must be pushed to accept responsibilities which do not belong to him as an individual. These responsibilities are collective obligations of the social group which forces synergism on the irresponsible and compels them to become responsible individuals.

In the last part of this essay, Einstein refers to a long historical irreconcilable encounter between religion and science. Here Einstein foresees a harmonious relationship between *fides* and *ratio*, between *faith* and *reason*, between religion and science. In the history of science, Einstein sees the minds of the great scientists, who have opened significant new paths of noumenal light, and who have been sustained for years in solitary labors by these cosmic religious feelings. This transrational *faith,* in spite of the sometimes ferocious attacks of their peers, is the spiritual force which keeps them in the adventures of their minds. Only a man like Einstein, who was a true natural philosopher, an epistemologist who knew exactly the noetic land he explored, who had these cosmic religious experiences, can talk about transrational intuitions with such convictions. He was a man who was lifted to the shoulders of giants, the same as Newton was in his time, to "see" the invisible being of God. Let us finish these comments while listening to his mind coming from the past, and being transmitted to the future:

> "Only one who has devoted his life to similar ends can have a vivid realization of what has inspired these men and given them the strength to remain true to their purposes in spite of countless failures. It is cosmic religious feelings that gives a man such strength. A contemporary has said, not unjustly, that in this materialistic age of ours the serious scientific workers are the only profoundly religious people."

7.2 Comments on Einstein's essay "The Religious Spirit of Science" (1934).

This is a short writing of two paragraphs written in 1934. In the first paragraph, Einstein differentiates the religious feeling of *"the profounder sort of scientific minds,"* from the religiosity of the naive man. In the rest of this first paragraph, Einstein writes about the psychological conception of God which belongs to the naive man. Like a child, he fears and expects punishment from this celestial divine Father. Einstein describes this naive man as believing he has a *personal relationship* with this spiritual Father. Here we see the first sign of pantheism in Einstein: the rejection of the concept "personal God." We will come back later to this important pantheistic *belief* of Einstein.

In the second paragraph, Einstein shows how the scientist, whose mind is immersed in the ocean of time, creates the concept of a *Supreme Intelligence*. Einstein says "*the scientist is possessed by the sense of universal causation*." This statement of Einstein is in total contradiction with Bertrand Russell's assertion that the principle of cause and effect is as useless as royalty in modern times. But Russell is wrong. In the middle of the paragraph and, unexpectedly, Einstein says "*There is nothing divine about morality; it is a purely human affair*."

Though the author agrees with Einstein in the source of morality, more interesting is to see how Einstein presents the concept of *Supreme Intelligence*; he writes about the scientist:

> "His religious feeling takes the form of a rapturous amazement at the harmony of natural law, which reveals an intelligence of such superiority that compared with it, all the systematic thinking and acting of human beings is an utterly insignificant reflection."

This quotation shows not only the vivency (insight) of those scientists who have the transrational capacity to know the Ultimate Reality, to know the Absolute, to know directly without thinking analytically but also shows the answer to a very unreasonable (synonym of stupid) question. For centuries some sages have asked the favorite question in metaphysics - Why things *are* the way they *are*, instead of *being* different? Some of the so-called modern thinkers claim that *nothingness* is much simpler than some*thing*. Hence, these modern people, like Jean Paul Sartre, write books about *L'Être et le Neant* (Being and Nothingness), instead of studying quantum chemistry. A well-educated person in Natural Philosophy can teach these modern sages that *things* are the way they are because there are *natural laws* which determine the *form* of things. *Things* would be different if different *natural laws* would have been in play.

Thus, in the above quotation, we see Einstein describing the conception of a *Superior Intelligence* through "*a rapturous amazement at the harmony of natural laws*." This mystical experience, of course, is not lived by every scientist. It is only reserved for a few. The conceptual prediction of a new natural law is, in itself, an ineffable experience. This could be a natural law which no one in the world knows about except the one who experiences it. Its existence is only noumenal. The rapturous amazement comes and floods the mind of the scientist when he verifies the real existence of that *"child"* of his mind. When mother Nature confirms that his noetic adventures were not in vain, then his concept will become an idea for the rest of the world. After that, the scientist will spend days as if drunk by the liquor of Bacchus, as Einstein once was. But the liquor was a mystical elixir of knowledge which could kill if too many drops were drunk at one time. Carl G. Jung [14], in *The Secret of the Golden Flower,* analyses the devastating effect on the human mind when it grazes the Universal Wisdom. No one can truly understand these altered states of the human mind unless they, too, have experienced them. A very interesting concept of Einstein, which has to do with his pantheistic position, is his concept of Supreme Intelligence. We will expound later on this point.

7.3 Comments on Einstein's essay " Science and Religion" (1939)

This essay is composed of two Parts. Part I is from an address at Princeton Theological Seminary in 1939. Part II is from a Symposium on Science, Philosophy and Religion in Their Relation to the Democratic Way of Life, New York, 1941.

PART I OF "SCIENCE AND RELIGION," (1941)

At the beginning of this essay, Einstein points to the arrogant position of rationalists of the 18th and 19th centuries. These people claimed there was an irreconcilable clash between knowledge and belief, between reason and faith. They claimed that belief was pure superstition, and that any school should replace belief by thinking and knowledge. In this extreme *ratio*nalistic position, human *reason* became the little god to be worshiped. Let us repeat it once more, *the god of reason has feet of faith.* To unveil this truth is the great contribution of Einstein in the epistemology of the 20th century. In 1933, Einstein [1, p. 272], in a lecture delivered at Oxford, says - "*The view I have just outlined of the purely fictitious character of the fundamentals of scientific theory was by no means the prevailing one in the eighteenth and nineteenth centuries.*"

After these epistemological considerations, Einstein goes into the teleological sources of human existence, particularly into the fountain of human aspiration. He concludes that the rational conception of our existence is incapable of deducing any purpose for this existence. In other words, from the knowledge of efficient causes, we cannot infer teleological causes. Einstein says:

> "But mere thinking cannot give us a sense of the ultimate and fundamental ends . . . They come into being not through demonstration but through revelation, through the medium of powerful personalities. One must not attempt to justify them, but rather to sense their nature simply and clearly."

If we read the above quotation, not knowing that it was written by Einstein, we would probably say it was written by a mystic. However, the same Albert Einstein, as we saw before, wrote about scientific principles, saying that:

> "Apart from that, these latter [fundamental principles] are free inventions of the human intellect, which cannot be justified either by the nature of that intellect or in any other fashion *a priori*."

After these considerations, Einstein goes into different subjects such as divination of a nation, education, politics and fear, international politics, legislation and organization in general. Part I finishes by reminding us that goals, purposes and objectives for the human race will not be fulfilled if behind these teleological causes, we do not have a living spirit.

PART II OF "SCIENCE AND RELIGION," (1941)

In the first paragraph of this Part, Einstein reconsiders the subject matter of his essay on science and religion. Einstein wonders what religion is about. Obviously, he has no doubts as to what science is. He says this about science:

> "To put it boldly, is the attempt at the posterior reconstruction of existence by the process of conceptualization."

In other words, and according to Einstein in this statement, science is the *a posteriori* rationalization of "*the perceptible phenomena of the world.*" However, this *a posteriori* conceptualization of science is in total contradiction to another definition of science given by Albert Einstein in 1933, in which science is an *a priori* activity of the mind. On that occasion, Einstein said: "*Our experience hitherto justifies us believing that nature is the realization of the simplest conceivable mathematical ideas.*" In this case "nature" is "*the perceptible phenomena of the world.*" In this epistemological matter, Einstein behaves as a Buddhist or Taoist poet, making *contradiction* the main character of his thoughts. In 1941, Einstein evidently is referring to theoretical science, i.e, he is making reference to the rational *a posteriori* reconstruction of the perceived reality. About any conception of religion, Einstein remains silent. He never mentions the etymological meaning of the word "religion."

But Einstein, nevertheless, needs a definition of religion in order to compare it with the definition of science, which he gave us above. In the second paragraph, Einstein decides to write about the *aspirations* of a religious person, according to what he understands by a "religious person." Such a person, in Einstein's opinion, has "*. . . liberated himself from the fetters of his selfish desires and is preoccupied with thoughts, feelings, and aspirations to which he clings because of their superpersonal values.*" These concepts of Einstein, also, have a Buddhist's resonance. Indeed, Einstein recognizes that these "superpersonal values" are characterized as religious personalities such as Buddha and Spinoza. Einstein is referring here to what we called transcepts, mystical experiences, religious experiences where truth transcends the personal rational analysis of them. These "superpersonal values" will play the role of dogmas or principles in any theology which may be developed later. In this paragraph Einstein will tell us what religion is. He writes:

> "In this sense religion is the age-old endeavor of mankind to become clearly and completely conscious of these values and goals and constantly to strengthen and extend their effect."

Now Einstein compares his definition of science and his definition of religion, concluding that "*a conflict between them appears impossible*." Evidently both definitions are very abstract, not to say very obscure. To justify the lack of conflict between these Einsteinian definitions of science and religion, Einstein says that "science can only ascertain what *is*, but not what should be." In this respect, Einstein is correct because natural laws are what they are and not what we want them to be. Now if we want to conceptualize or theorize about the social interaction (behavior) of human beings, we have two methods: an empirical one, based on empirical observations of human social behavior called *moral* or *morality* and the other method is a theoretical one. If we want to deduce the social moral behavior (interaction) of human beings, we must create a theory of the actual moral behavior, or what *should be* the moral behavior. Any theory which explains rationally the moral behavior of human beings is, as we said before, called *Ethics*. In the 17th century, Baruch Spinoza developed a theoretical Ethics which required transrational principles or dogmas like any scientific theory. However, the similarities between science and religion are better understood if we consider any religious doctrine as the technology or application of a given theology. Thus, the comparison should be between theoretical science and theology. In this case, there is no epistemological conflict between theology and theoretical science.

In paragraph 3, Einstein shows biblical conflicts when a religious community claims absolute truthfulness of all assertions recorded in the Bible. Of course, if the creation of the biblical universe has nothing to do with the actual universe studied by scientists, then it is impossible to have any conflict between science and biblical religion. However, when a religious community insists that the biblical universe is the same one studied by scientists, then an enormous conflict arises between science and religion. The reason is misinterpretation of the sacred scriptures, or theoretical misinterpretation of the empirical data. In either case, the conflict is rooted in ignorance. Einstein also sees that conflicts between science and religion is created when scientists try to arrive at fundamental religious dogmas by means of the so-called scientific method. If this last statement were true, then a scientific theology would be impossible to develop. Soon we will see that correcting the theological and scientific "fatal error," indicated by Einstein in this paragraph, will allow us to create a scientific theology. These "fatal errors," pointed out by Einstein, correspond to the "biblical misinterpretations" pointed out by Pope John Paul II, in 1998.

In paragraph 4, Einstein stresses the role of *faith* in the minds of the creators of scientific theories. Einstein says, that the world, for them, is a constant rational process which can be understood by human reason. Undoubtedly, this is a creed of faith, a dogma, an axiom whose truth has to be accepted or believed under an *act of faith.* In this respect Einstein writes:

> "I cannot conceive of a genuine scientist without that profound faith. The situation may be expressed by an image: science without religion is lame, religion without science is blind."

Today we may say that science without faith is impossible, because no one would believe the unproven truth of the principles at the very foundations of any scientific theory. On the other hand, "religion without science is blind." To suffer intellectual blindness, i.e, fanaticism, *is not to know* we *do not know*. On the other hand, to suffer intellectual ignorance is to know we do not know. Thus, religion without science is pure fanaticism.

In paragraph 5, Einstein asserts that: "*The idea of God in the religions taught at present is a sublimation of that old concept of the gods*." We should say, "The concept of God," because God is not a sensorial "*image,*" God cannot be an "*idea.*" Anyway, the old concept of the gods was created in man's own image according to Einstein.

In paragraph 6, Einstein comes back to the "irresponsibility" of man's acts. He also criticizes the concept of a "personal God." This personal God is omnipotent, just and omnibenevolent. Einstein sees in the omnipotency of this superior being the essential element of an unwritten essay titled "God is on trial." This should be only a trial of the "anthropomorphic God" of terrestrial beings. The following is Einstein's unconvincing argument: "*That is, if this being is omnipotent, then every occurrence, including every human action, every human thought, and every human feeling and aspiration is also His work.*" This proposition has the correct structure of a mathematical theorem: *If p, then q*. What is missing in this proposition is the proof or demonstration of it. Einstein's unproven assertion is reduced to this: *If God is omnipotent, then God is responsible for man's acts.* If we think this is a true statement, then man has been liberated from any of his responsibilities, and any religious set of commandments are irrelevant in the social life of any highly immoral planet. We must keep in mind two points. One is that Einstein's critical analysis is about any *anthropomorphic God.* The other point is that Einstein never proved his assertion presented above. In paragraphs 7 through 12, Einstein will reject the concept of a *personal God.* He writes:

> "The main source of the present-day conflicts between the spheres of religion and of science is this concept of a personal God."

After this proposition, Einstein goes into a lengthy discourse about what science is and its applications or technology. This whole discourse is established to show that there is no such thing as *personal knowledge* in science. That science offers *objective* knowledge to man, not *subjective* knowledge. But this is not true. Einstein's own Special Relativity Theory is magnificent proof that the subjective interpretation of empirical data created by *one subject* called Albert Einstein, became *objective* when a group of uncritical followers kept on misinterpreting the so-called empirical facts, as their master did in the beginning. It is true that in science the unproven content of principles must be general, universal, not personal or specifically particular, because particular knowledge about nature can be deduced only from universal statements. The concept of God as the Being of all entities is a universal conception that has not a ιοτα of a *personal* entity.

If we reject the concept of a *personal God,* then - what would be the essential function of religion, yoga, TM or satori? - To deny the concept of a *personal God* is not only to deny the noetic existence of God but also the real existence of God. It is to deny the possibility of re-connecting the individual *personal consciousness* with the

Universal Consciousness, of who God is, as we will prove later. The flaw in Einstein's rejection of a personal God is rooted in his lack of knowledge of the concept of consciousness. What we should understand by *personal God* is the interaction between an individual human person with the Universal Consciousness who is God. In section 11 of this chapter, we will expand upon this subject.

Einstein himself, without being aware of it, expressed mathematically the concept of Universal Consciousness. In the last paragraph of his essay Einstein wrote:

> ". . . the path to genuine religiosity does not lie through the fear of life, and the fear of death, and blind faith, but through striving after rational knowledge."

Here we see Einstein again envisioning the future creation of a scientific theology and the creation of an advanced religion.

7.4 Comments on Einstein`s essay "Religion and Science Irreconciliable?" (1948)

Reconcilable means to *be in agreement, to be in accord.* In the first two paragraphs of this essay, Einstein wonders if there could be a reconciliation between science and religion. To investigate this subject, Einstein needs to recognize what he understands by science, and what he understands by religion. He first offers another definition of science:

> "As to science, we may well define it for our purpose as «methodical thinking directed toward finding regulative connections between our sensual experiences.»"

Let us compare this definition of science with the other one offered by Einstein in the previous essay:

> ". . . [it] is the attempt at the posterior reconstruction of existence by the process of conceptualization."

This author believes that the last quotation offers a better definition of science for the purpose of finding a reconciliation between science and religion. This last definition means that terrestrial science is the human effort to rationally explain the "existence" of all that exists, of all that *is*. This explanation has to be *ontological* (metaphysical, theological), because we must identify the *Being* of all *entities*. We must determine the very nature or *common essence* of the *being* of all that exists. In Rational Theology, we find that God is the Being of all entities. God is the Ultimate Substance out of which everything *is*. Can science offer the same kind of explanation? For centuries, science was unable to offer an essential explanation of the *universe*. This ontological incapacity reached its culmination in Natural Philosophy in the 20th century when it degenerated into mathematical physics, as a formal science, in the minds of metamathematicians.

In section 10 of this chapter, we will offer an identical explanation of "existence" as Einstein foresaw it. In the second paragraph, Einstein offers another theological definition of religion:

> "Religion is concerned with man's attitude toward nature at large, with the establishing of ideals for the individual and communal life, and with mutual human relationship."

Let us compare this definition with another one offered by Einstein in the previous essay:

> "In this sense religion is the age-old endeavor of mankind to become clearly and completely conscious of these values and goals and constantly to strengthen and extend their effect."

Summarizing, we may say that "science is the human effort of conceptualizing, rationalizing, theorizing *existence* as sensually experienced by man." On the other hand, "theology is the human striving for conceptualizing, rationalizing, theorizing *existence* as transrationally experienced by man." The great majority of human beings, even today, are not intellectually prepared to understand theological principles, whether *revealed transrationally* or *created noumenally*. If the theological principles or dogmas are not included in a rational theory of God (theology), then the set of dogmas, without any ontological and logical analysis, constitute a Religious Creed. Usually the set of dogmas are abstract, completely incomprehensible for the masses. These dogmas become rules or commandments which are delivered to people in a narrative way through mythical stories or allegories. It is at this point that Einstein sees the source of discrepancies between science and religion. He says: "*It is this mythical, or rather this symbolic, content of the religious traditions which is likely to come into conflict with science*." The conflict is rooted in the *misinterpretation* of the narrative mythological stories which have been taken for the essentials of the Religious Creed. These misinterpretations, over the centuries, have been the main causes of virulent disagreements or irreconcilable attitudes between religious leaders and scientists. In paragraph four of this essay, Einstein compares different religious organizations divested of their myths. Einstein then concludes that: "*they do not seem to me to differ as basically from each other as the proponents of the 'relativistic' or conventional theory wish us to believe.*" All religions have the same two purposes. To teach human creatures to interact with two entities. One is God, and the other is a group of human beings. The divine interaction is through *reconnecting* the human consciousness with the Universal Consciousness. The other purpose of any religion is to teach human beings to act morally correct. Had this planet been populated with more intelligent beings, no religion based on a *fearful* God would have been necessary. At the dawn of the eighth millennium of historical records, human beings are still in their theological infancy. We should never forget, when we criticize any religion, that one of the most outstanding functions they perform is to teach human beings to live with dignity and civility among each other.

In paragraph six, Einstein thinks that the main causes of the lack of fraternal coexistence among human beings are *personal ambition* and *fear of rejection.* In our modern jungles of steel and concrete, we still find the primitive cave instinct of preservation of the individual - a brutal sophisticated daily *competition.* The optimistic messages of religious ideals will perhaps be attained when the individual consciousness of each human being evolves into an integrated global planetary consciousness which will then be aware, not only of one personal individual consciousness, but also of the consciousness of the whole human race.

In the last paragraph of this latest essay, Einstein refers to the dependence of science on the religious attitude. He says:

> "While it is true that scientific results are entirely independent from religious or moral considerations, those individuals to whom we owe the great creative achievements of science were all of them imbued with the truly religious conviction that this universe of ours is something perfect and susceptible to the rational striving for knowledge."

Without the conviction of an *Essential unity (Love) with a Supreme Intelligence,* no science, as we know it today, would exist. Einstein speaks of the inspiration induced by Spinoza's *Amor Dei Intellectualis* in the spirit of those few, who like their

Creator, create enlightenment for the world with an inexhaustible devotion. As Einstein knows, without the *cosmic religious experience* no scientific progress is possible. In these cosmic religious experiences, we are given the transrational knowledge of principles to create scientific theories, and also, we are given the transrational knowledge of dogmas to create theological theories with their corresponding religions. After this noumenal adventure in the mind of Einstein, let us abstract Einstein's theological beliefs.

7.5. Einstein's Theological Beliefs.

After our analysis of Einstein's four essays, let us reassemble the beliefs of Einstein in theological matters.

1 In Einstein's essay of 1930, we can say that for Einstein, God exists. God's existence is not proven by Einstein on a rational terrain, but on a transrational or cosmic religious terrain. Concerning certain attributes of God, Einstein believes that in "God's eyes" man cannot be responsible for his actions.

2 In his essay of 1934, we note he definitively rejects the concept of a personal God. Einstein also believes that morality has no divine source: He believes that theory is a mere invention of humans. Finally, Einstein believes in the existence of a Supreme Intelligence. He draws this conception from the harmony found in natural laws.

3 In his essay of 1941, we see that Einstein believes the purpose of human existence cannot be deduced rationally, but should be received by revelation. Revelation, for Einstein, is the acquisition of knowledge by transrational means; i.e., by a cosmic religious experience. Einstein also believes that faith is at the root of religion and science. He also believes that morality is not divine but a human matter. Einstein rejects emphatically the concept of a personal God, and points to it as one of the main causes of disagreements between science and religion. Finally, he believes that rational knowledge will eventually provide humans with the path to a true religiosity.

4 In his 1948 essay, Einstein noted that the mythical or allegoric religious tradition is another cause for the conflicts between religion and science. In this fourth essay, Einstein, in a very disguised way, identifies nature or the universe with God. Finally, Einstein believes the minds of a few creators of scientific theories are flooded *"with the truly religious conviction that this*

> *universe of ours is something perfect and susceptible to the rational striving for knowledge.*" Once more we note that, Einstein foresees the birth of a scientific theology.

In conclusion, Einstein never was an atheist, but rather a pantheist, and after studying Spinoza's pantheism, we conclude the two thinkers had comparable theological beliefs.

Brief description of Pantheism. The word "pantheism" is derived from the Greek words παν (*pan*=everything, all), and the word θεοζ (*theos* = god). Thus, etymologically, pantheism means "God is everything" or "God is all." Therefore, if God is everything, the entire universe is *God.* The question - Who created the universe? - is tantamount to this absurd question - Who created God? A pantheist will say that *God is a part of His creation*, while a Scholastic theologian will say that *God is apart from His creation.* In *The Cambridge Dictionary of Philosophy* [15], we read:

> "Spinoza is the most distinguished pantheist in Western philosophy. He argued that since substance is completely self-sufficient, and only God is self-sufficient, God is the only substance. In other words, God is everything. Hegel is also, sometimes, considered a pantheist since he identifies God with the totality of being."

The Jesuit priest, Frederick Coplestone [16], in his *A History of Philosophy*, Vol. IV (1959) makes a very interesting interpretation of Spinoza's identification of God with substance. Spinoza was certain that the human mind could only grasp two of God's attributes; thought and extension. Coplestone argues that "*The logically prior state of substance under the attribute of extension is motion-and-rest.*" He then writes:

> "Using the language of a later time one can say, then, that the total amount of energy in the universe is an intrinsic property of the universe and that it remains constant. The physical universe is thus a self-contained system of bodies in motion. This total amount of motion-and-rest, or of energy, is what Spinoza calls the 'infinite and eternal immediate mode' of God or Nature under the attribute of extension."

Here we find a fundamental identification between energy and Spinoza's concept of "infinite and eternal immediate mode of God under the attribute of extension." Pure energy, then, is a mode or modality of God's manifestation. This is a laudable physical interpretation of a philosopher. However, the same philosopher F. Coplestone, S.J. [17], in Vol. III (1953) of his *A History of Philosophy*, declares that:

> "It is, however, fairly obvious that science cannot disprove the validity of faith or of theological beliefs. Physics, for example, has nothing to say about the Trinity or about the existence of God."

As we have seen in this book, if science assuredly disproves the validity of faith, it will prove instantly that science is false. Correspondingly, if science disproves the validity of theologies, whether revealed or rational, it will instantly destroy the hypothetic-deductive structure of all scientific theories. The last part of the previous quotation is a typical assertion of an unfortunate lack of knowledge of Natural Philosophy, and therefore of Physics.

Two other characteristics of Spinoza's pantheistic theology are the strong denials that (1) God is a person, and (2) that God acts according to final or teleological causes. It is clear that Spinoza's conceptions influenced the writings of biblical criticism and literary works in the last two centuries. As we have seen, Einstein undoubtedly was influenced by Spinoza's conceptions. We saw that Einstein recognized Spinoza's *Amor Dei Intellectualis* in the spirit of those few, who like their Creator, generated enlightenment for the world with an inexhaustible devotion. For Spinoza, "*The greatest good of the mind is the knowledge of God , and the greatest virtue of the mind is to know God.*" The ineffable spiritual joy felt by the mind of a seeker when he apprehends the long awaited answer, which is linked to the concept of God as eternal cause, is what Spinoza calls the *"intellectual love of God."* This love is not the mundane concept of love. This love is the mystical essence of unity, unity with the *Supreme Intelligence*. This is when the seeker finds his eyes brimming with tears and wonders why. He, at last, has grazed the essence of God after grasping for answers which he yearned for, but which have eluded him for so long. Let us finish our analysis of Einstein's Theological Beliefs by reminding ourselves of the different instances, in his four essays on science and religion, when he foresaw the future creation of a theology and a religion not based on fear but on an *"intellectual love of God."*

7.6. Science, including Logic, are not completely rational.

This short section is for scientists who overlooked chapter 1 of this book. A collection of experimental facts constitutes a set of *particular empirical knowledge*. The elements of this set are unrelated. To understand this empirical set we must rationalize it; i.e., we must create a hypothetic-deductive structure which puts all elements of a specific set of experimental facts into a relationship with each other. In other words, we have to create a theory to deduce logically each one of the particular pieces of empirical knowledge. For this author, "science proper" is any collection of theoretical knowledge.

In any rational theory, we distinguish the following fundamental components: (1) principles, (2) operational definitions, (3) logic rules of inferences, and (4) many logical conclusions. These logical or rational conclusions must be verified experimentally. The principles of any theory are the universals, or conceptual inventions of the human mind, or transrational meta-concepts received by the human mind. This is a repetition of what we have said before, but we cannot overemphasize, that the content of principles cannot be deduced logically; they are *nonrational*. The truths of principles are accepted under acts of faith in exactly the same manner as the dogmas of any religious doctrine. It is impossible to logically deduce the truths of principles, unless, of course, we create another theory to deduce them. But this new theory also requires the acceptance of another set of principles which are nonrational. Any theory of Logic must be erected on a set of principles which cannot be deduced, and, they too, must be accepted by faith. Thus any scientific theory, or philosophical theory, or theological theory, or theory of Logic must start with nonrational principles. By nonrational principles we mean irrational statements invented by the human mind or transrational statements received by the human mind. If we like, we may say that Science, including Logic, is not completely rational because it is founded on irrational or transrational principles. This is a lesson for scientists and theologians that is long overdue.

7.7. Foundations of Scientific Theology or Cosmotheism.

Theology is the Theory of God. As any theory, theology should be composed of four sets of propositions: (1) principles, (2) definitions, operational or not, (3) logical rules of inference, and (4) a large set of rational conclusions. The principles of Theology either are free creations of the human mind or are directly acquired (revealed) by a transrational experience. Thus, we can distinguish Rational Theology from Revealed Theology. In both cases, we have to believe in the truth expressed by the principles, whether conceptualized by humans, or revealed in transrational experiences. All rational theologies have mainly two purposes: (1) to prove the existence of God, and (2) to determine the nature and attributes of God. Obviously, any revealed theology is in a better position than a rational theory. A revealed theology need not prove the existence of God, because God, Itself, reveals the principles or dogmas of faith so that the receiver may initiate the construction of the revealed theology.

After 23 centuries from the time of Aristotle's death, it is fundamental to reconsider Aristotle's First and Second Philosophies. Aristotle never used the word metaphysics in his writings. It is believed that this term was introduced by Andronicus of Rhodes when he was editing the works of Aristotle around 70 BC in Rome. Andronicus called certain aspects of Aristotle's work *metaphysics* (beyond the writings of nature). These aspects were themes which "followed" the writings of physics. Aristotle's First Philosophy, it is said, corresponds to Andronicus' *metaphysics*. This author disagrees with this terminological identification. For Aristotle, the First Philosophy was the search and establishment of first principles and *first causes*. The Second Philosophy was composed of the empirical sciences, mainly established by observations of natural phenomena. The First Philosophy should be considered as a *methodology* to create theories. Below this First Philosophy, we find substantial principles and formal principles, see Fig. 7.1.

The substantial principles are the *ontological principles* which we presented in chapter 1. The formal principles are the *logical principles*, also presented, in chapter 1. Ontological principles, as well as logical principles, constitute the scaffolding of any theoretical structure. Thus, below The First Philosophy or Philosophical Methodology, we should have the three modes through which humans acquire knowledge. *Gnoseology* is an excellent word to refer to the study of knowledge in general. Now, the three modes to acquire knowledge are *transrational gnoseology, rational gnoseology* , and *nonrational gnoseology*. In Fig. 7.1, we illustrate the new classification of philosophy, which we are proposing.

Transrational gnoseology is the mystical, religious, revealing or intuitive mode through which the human mind receives knowledge. This transrational knowledge, acquired by the mind of a few humans, is absolute in the sense that is not relative to the physical senses nor to any intellectual process. This knowledge comes to the human mind with such conviction or certainty of truthfulness that it remains indelible in the memory of the person.

Rational gnoselogy is the noumenal mode humans use to create or *conceive* rationally, intellectually new concepts. The creation of these new concepts requires the apprehension of nonrational sensorial knowledge, and/or the intuition or insights of transrational knowledge. Rational gnoseology is the creation of knowledge deduced rationally from *hypothetical deductive* mental structures called theories. It is in this realm where humans create their *Theoretical Philosophies.*

We must distinguish Theoretical Philosophy from Theoretical Science. The first is highly speculative, logically impeccable, neatly rational but completely unverified pragmatically in reality. Theoretical Science, on the other hand, is highly speculative, logically impeccable, neatly rational, but with many of its theoretical conclusions verified pragmatically in reality. We have no need to introduce the term *Metaphysics* in our discourse; however, to establish a connection with Andronicus' classification, we propose to *identify* Theoretical Philosophy with Metaphysics. Someday, perhaps, philosophers will learn to connect Metaphysics with reality through Pragmatic Methodology, as shown in Fig. 7.1.

Nonrational gnoseology is the sensorial mode humans undergo to acquire mental images or ideas. Aristotle referred to this gnoseological mode as *Second Philosophy*. Galileo, almost 2,000 years later, called the same gnoseological mode *Experimental Philosophy*. In this new century, we should try our best to convert speculative metaphysics into pragmatic metaphysics. This conversion would reconnect each individual mind of humans with the external world of things or reality.

Under Theoretical Philosophy (Metaphysics), we have different disciplines such as Theology and too many Theoretical Philosophies ending in ***ism***. To create any new theory, we have to adopt the *fundamental* ontological and logical principles to make an intelligible or rational theory. In addition to the fundamental principles of the methodological or first philosophy, we have to create the specific axioms or postulates of the new theory along with the corresponding definitions. The concept of God of any theology, obviously, is an intellectual God who has, undoubtedly, a noumenal existence. This theoretical God has only a metaphysical existence; i.e., this God does not have a pragmatic existence in reality. Therefore, and evidently, this conceptual God is not real, unless Its existence is verified in the world of things (Latin *res* = thing, from which the word *reality* is derived). If this verification is possible, then this conceptual, revealed or metaphysical God would have a *real existence* in the entire universe of material things. In this way, the *metaphysics of God* would become the *physics of God*, the physics of a real God defined in the following theological terms:

> "God is the Being of all entities, material or immaterial, visible or invisible."

To develop a Scientific Theology or *Cosmotheism,* we will add to the ontological and logical principles, of the First or Methodological Philosophy, the axioms or postulates of all physical theories and its experimentally verified theoretical conclusions. Finally, we will *identify* the above definition of God with a theoretic-empirical proposition derived from physics to prove the real existence of God. We will proceed in the next section to demonstrate the physical existence of God, outside the mind of any human being.

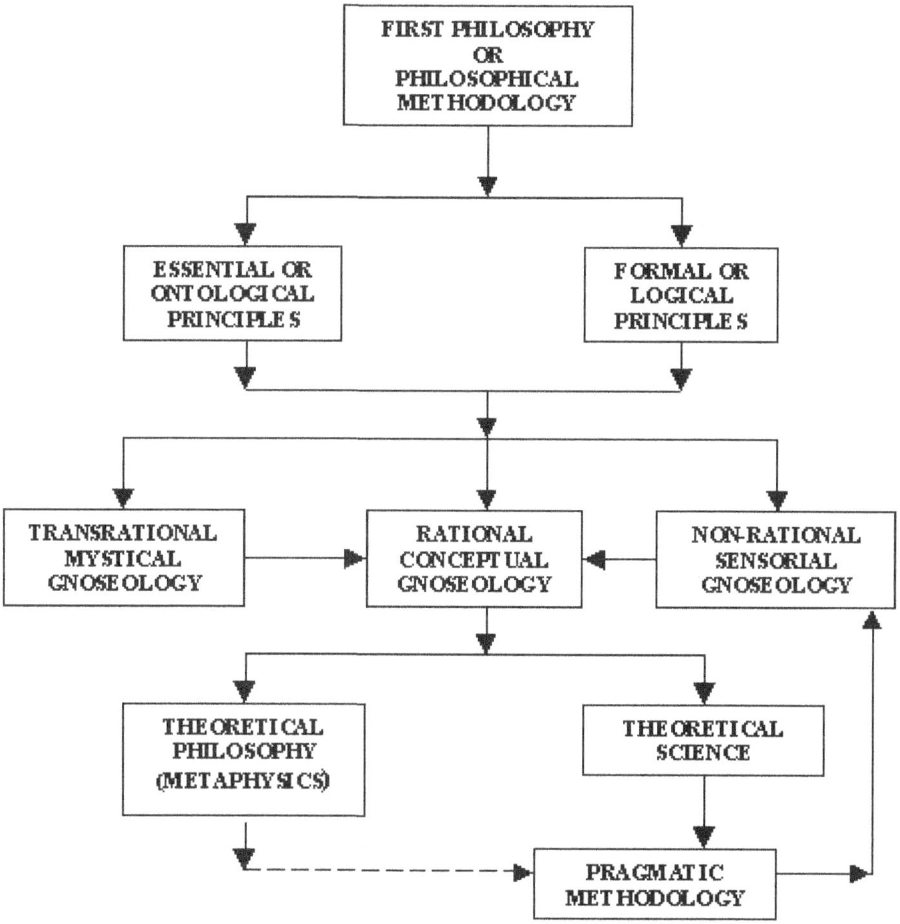

Figure 7.1 Reclassification of Philosophy

7.8. Does God Exist in Reality?

It is evident that in any theology God exists conceptually, theoretically, noetically, noumenally, mentally, intellectually, or rationally. It is also evident that not every entity which has theoretical or noetic existence should, also, have real existence. It is irrelevant from a universal and realistic point of view whether God exists or not in the minds of humans. The most transcendental and important event on the entire planet earth would be to prove that God exists outside the human mind; that God exists in the universe of entities, material and immaterial, and exists independent of the existence of human or extraterrestrial beings.

In chapter 5, we identified the cosmic ether with gravitational potential energy. For centuries philosophers thought the cosmic ether was a medium which pervaded the entire universe, and allowed light to propagate from the stars, planets, moon and sun into the eyes of human beings. Light was thought to be the propagating vibrations of the cosmic ether. In those days, the cosmic medium was named *luminiferous ether*. The nature of this cosmic ether was thought to be mechanical; i.e., a material substance with an incredibly low density and high elastic coefficient. Unfortunately, in the 19th century and later, these mechanical characteristics of the cosmic ether were determined to be very erroneous. Even after light was identified with the propagation of electromagnetic waves, no one attempted another kind of ontological identification of the cosmic ether. In the 20th century, in 1920, Einstein identified the cosmic ether with the components of the metric tensor of his General Relativity Theory. This Einsteinian conception identifies the cosmic ether with gravitational potential instead of gravitational potential energy.

Perhaps the first scientist who identified or interpreted the physical concept of the cosmic ether in theological terms was Isaac Newton. Newton first identified the concept of *absolute space* with the concept of cosmic ether. Now, the most outstanding characteristic of absolute space is its *immobility*. This attribute of immobility is the most feared characteristic of absolute space by relativists. This Newtonian identification allows us to predicate that the cosmic ether is at total absolute rest. This last statement, undoubtedly, is metaphysical. Newton, in his book *Optiks* [19] Book Three, Part I, Query 28, writes about the Æther from different points of view. At the end of Query 28, he writes about the Æther in theological terms, interpreting the cosmic ether as the *Sensorium Dei* (Sense Organ of God):

> "And these things being rightly dispach'd, does it not appear from Phænomena that there is a Being incorporeal, living, intelligent, omnipresent, who in infinite Space, as it were in his Sensory, sees the things themselves intimately, thoroughly perceives them, and comprehends them wholly by the immediate presence to himself."

In this quotation, we not only find theological attributes of a supreme Being, but also the foundations of these attributes on natural phenomena. An important aspect of Newton's speculation is the ontological identification of absolute space with the cosmic ether, and the cosmic ether, or primordial energy field with the *Sensorium Dei*. It is in this *Sense Organ Of God* where we see the manifestation of the Being of all entities.

It has taken approximately 7,000 years for human beings to begin to understand just a little about the visible material universe. Can we imagine how many more millennia it will take for human beings to begin to understand just a little about the probable existence of an immaterial invisible universe? Before any conceptual attempt to understand the immaterial invisible entity, which may be in the vast vacuum of interstellar and intergalactic space, we must first prove the *real existence* of such an entity. We must first prove the *physical existence* of such an entity. The reader should pay close attention to the identification we have made between "real existence" and "physical existence." Of course, the immaterial invisible entity is called *cosmic ether*. This cosmic ether, undoubtedly, has a conceptual or a metaphysical existence. But the important knowledge we want to obtain is the physical existence of the cosmic ether. What would the criterion be for us to establish the real existence of the cosmic ether? The best criterion or proof we can present of the physical existence of the cosmic ether is, precisely, the disapproval of the strongest argument against the real existence of the cosmic ether.

> This strong argument against the real existence of the cosmic ether is the lack in physics of a wave equation for the density of the cosmic ether.

Obviously, a perturbation in a medium generates variations in the density of the medium which propagate away from the perturbing center. It is an elementary statement to say that such a density wave exists if the medium *does* exist. Or we can say it the other way around. A medium *does* exist if a density wave exists. This means that a necessary and sufficient condition for a medium to exist is to show the existence of a density wave of that medium. To express it mathematically, let us verbalize it in the following way. If and only if the medium exists, then does a density wave of the medium exist. As far as the knowledge of this author is concerned, there is no density wave equation of the cosmic ether in the history of physics. In the last two centuries no one knew anything about the ontology of the cosmic ether. Today, many physicists agree that the essence or very nature of the cosmic ether is gravitational potential energy, but according to James C. Maxwell, *Energy is one and the same*. Therefore, we can remove the gravitational characteristic of the energy of the cosmic ether, and simply write that the being or essence of the cosmic ether is *energy*.

Being that light is an electromagnetic phenomenon, we should be capable of explaining what light is and how it propagates in vacuum. Let us see some explanations. Gerald Holton [18], in his book *Thematic Origins of Scientific Thoughts*, offers the quantum mechanical "essence" of light. If we ask the question - What is light? Holton writes:

> "The answer is: the observer, his various pieces and types of equipments, his experiments, his theories and models of interpretations, *and* whatever it may be that fills an otherwise empty room when the light bulb is allowed to keep on burning. All this together is light."

If you dare to criticize this answer, you will be accused of being ignorant of Quantum Mechanics. If you want to get even, you may say that the question demands an ontological answer, because in the question, we find a form of the verb *To Be*. However, the answer shows complete ignorance of ontology. But the answer is good because Quantum Mechanics is a formal theory which has nothing to do with the essence of any of its terms. Another definition of light which is typical in an advanced course of optics, is the following: "*Light is the solution of a linear partial differential equation of the second order, homogeneous or not, with the proper boundary conditions.*" If we agree that light is a wave, the previous "definition" of light is mathematically correct, but we still know nothing about the essence of light. Thus, presently, physicists do not know what light is and how light can propagate in a vacuum. If light is a wave, then vacuum must be an immaterial and invisible medium. This medium is the energetic cosmic ether through which any modification of energy density propagates.

Now, all we have to do is give respectability to these ontological speculations by mathematically deducing a wave equation for the density of energy in empty space using Maxwell's electromagnetic theory. It is interesting to note that Maxwell's propagation of energy was, in the beginning, a metaphysical conclusion because it had no experimental verification. Let us recall that Hertz was the man who transformed a metaphysical theory into a physical theory when he experimentally proved the *real existence* of electromagnetic waves. In chapter 5 of this book, we deduced a partial differential equation for the density of electromagnetic energy. Therefore, the existence of a wave of energy density, necessarily (logically) implies, the existence of a universal medium of energy. Here is the theoretical (metaphysical) proof of the noetic existence of a universal energetic medium.

The first experimental proof of the real existence of the cosmic energetic ether is the almost null result of the famous Michelson-Morley experiment of the 19th century. The rational theoretical explanation of this experiment was also presented in chapter 5 of this book. Another dramatic experimental proof of the real existence of the luminiferous cosmic energetic medium, would be the realization of Einstein's experimental proposal of 1884, when he was a teenager. This was also presented in chapter 5 of this book. Another empirical proof of the existence of the luminiferous cosmic ether, or energetic medium, is the theoretical deduction of Merat's empirical law of starlight deflection by the sun. We presented this mathematic-physical work in chapter 6, based on the primordial energy field theory developed in chapter 5.

Orthodox quantum physicists began to accept, under other experimental phenomena like the Casimir Effect, the real or physical existence of the universal energetic ether. This inexhaustible source of energy is called *Zero Point Energy of Vacuum* today, but it is the same ancient entity called *Æther*. Are there any relationships between energy and mass? Of course there are different relationships between energy and mass. It is interesting to mention that the term mv^2 was called *vis viva* (alive force) by Leibniz. The "vis viva" is twice the kinetic energy $E_K = ½ mv^2$. However, Einstein deduced a more general relationship between mass and energy; $E = mc^2$. He said:

> "It followed from special theory of relativity that mass and energy are both but different manifestations of the same thing."

Here we are in front of an ontological statement which goes beyond the equivalence between mass and energy, expressed by the equation $E = mc^2$. Einstein is telling us that *mass and energy are one and the same entity*. Newton in his book *Optiks* [19], Book Three, Part I, Question 30, wrote:

> "The changing of Bodies into Light, and Light into Bodies, is very conformable to the Course of Nature, which seems delighted with transmutations."

The term "body" in Newton's writings meant "mass," and the term "light" for us, presently, is a synonym of "energy." Let us translate the above Newton quotation into our modern terminology:

> The changing of mass into energy, and energy into mass, is very consistent with the way Nature is, which seems delighted with transmutations.

Of course, we cannot take this conception of Newton's too seriously in terms of an operational point of view; nevertheless, this conception is a forerunner of Einstein's famous equation $E = mc^2$. Another forerunner of the equivalence of mass and energy is H.P. Blavatsky. Let us examine her book, *The Secret Doctrine* [20], published in 1888. If we translate the term *force* by *vis viva* (alive force=energy) in her quotation from W.Q. Judge's *Path* magazine (January 1887, 297), we will be surprised to read the following statement:

> "As declared by an American theosophist «The Monads (of Leibniz) may from one point of view be called *force*, from another *matter*. To occult Science, *force* and *matter* are *only two sides of the same* SUBSTANCE.»"

To say that - Energy (alive force) and matter are only two sides of the same *substance*, is to make a plain ontological statement by *identifying* two seemingly different entities. This kind of statement is very significant for a natural philosopher but completely irrelevant for positivist physicists. The above quotation is found in the recent book (1994) by Sylvia Cranston [21], titled *The Extraordinary Life and Influence of Helena Blavastky, Founder of the Modern Theosophical Movement*. In this book we read:

> "A niece of Einstein reported that a copy of *The Secret Doctrine* was always on his desk.[11]"
> "Another witness, Jack Brown, reports similarly in an article, «I visited Professor Einstein.»[12]"

When a man steps beyond the limit of human knowledge, he realizes that he is the only one in the whole world who possesses that knowledge. He feels overwhelmed by the gnoseological drink, but at the same time, he is not completely sure of the truth of the new knowledge. Then he resorts to all kinds of writings in search of some support for his new noetic discovery. It seems to this author that, in this sense, we should decipher the presence of a copy of *The Secret Doctrine* by Mme. Blavastky, on Einstein's desk. No seeker of the truth should be ashamed of any of his readings. Up to this point, we have demonstrated that in Einstein's mind, mass and energy have been identified noetically, ontologically, metaphysically. In 1932, Cockcroft and Walton demonstrated it experimentally, physically. But what about a relationship between mass and matter? Newton was extremely clear when in Book I, comments of Definition I, he wrote:

> "It is this quantity [of matter] that I mean hereafter everywhere under the name of body or mass. And the same [matter] is known by the weight of each body,"

Please, let us pay very close attention to three *names*: matter, mass, and body, designating one and the *same* entity. For Newton it was more familiar to use the term matter than the term mass. He mentioned the term mass in his *Principia* only a few times. To say that "mass is the quantity of matter in a body" is not only a metaphysical obscurity, as Mach would have said, but a true tautology after the identification proposed by Newton.

Today a great number of physicists agree that mass or matter is condensed energy. Other scientists are not only concerned about the *essence* of elementary particles, but also about the *form* of these elementary particles. What is happening in the world today is an essential noumenal transfiguration of the physicists mind. They are, again, becoming natural philosophers who want to unveil the ontology of the basic "substance" of every entity. Also, they have begun to speculate ontologically and geometrically about the form of these microscopic entities. Bergmann and Wesley [22a], for example, have proposed a very successful geometrical model for the electron in the form of a *toroid*, ring or doughnut. It is possible that elementary particles may evolve just like galaxies. They began with a globular shape of photons, or stars, revolving around an axis. Then due to gravitodynamic forces: pseudo-Coriolis, pseudo-centrifugal, axial and pseudo-Euler, the globular shape evolves into a double saucer-disc shape. Later in this evolution, the elementary particles (weakly condensed energy), or the galaxy, adopt the shape of a spiral vortex with a small opening at the center of the vortex. The opening begins to increase in diameter, and the spiral branches separate from the ring-core, fusing with other spiral branches. Finally, according to Bergman and Wesley, the density of energy increases in the final stable annular (toroid or ring) of the electron or the annular shape of galaxies. If this is the process, then we may, again, repeat the words of Thoth: "*As above so below, for the fulfillment of unity.*" Here we reach the following final paragraphs of this section titled "Does God exist in Reality?" We highly recommend the following books about the genesis and evolution of vortices. One is *Aethro-Kinematics* by Steven Rado [22b]. Two others are *Spiral Grain of the Universe* [23], and *Unified Spiral Field and Matter* [24], both written by Vladimir B. Ginzburg

Scientists have demonstrated theoretically and experimentally that the material universe is condensed energy. They also have demonstrated that the inter-material space is pervaded by a universal medium whose essence is pure subtle, or less condensed, invisible energy. Now, in the dawn of the 21st century, we are able to answer scientifically the following question - What is the being of all entities, material or immaterial, visible or invisible? The scientific answer is:

> **Energy is the Being of all entities, material or immaterial, visible or invisible.**

Let us compare this physical scientific conclusion with the theological definition which we mentioned before:

> **God is the Being of all entities, material or immaterial, visible or invisible.**

In the physical conclusion, we predicate about *Energy* using the same predicate we have in the theological definition of *God.* Therefore ontologically, and logically, the subjects of the physical statement and the subject of the theological statement refer to one and the same entity. Thus the nature or essence of God is *pure energy*, is *pure act*. We will see later that this energetic nature of God is manifested in three seemingly different modes. Here we finish the scientific proof that God not only exists noetically in the minds of men of good faith, but more importantly, God exists outside the minds of any universal creature; i.e., God exists in reality!

7.9. Is God Universal Consciousness?

In this section, we will explore the meaning of the concept *consciousness* with the true intention of generalizing the limited conception we have of it. The limitation consists in applying it only to human consciousness. We will also try to find a relationship among *complex systems* of interacting material particles, and the concept of *interaction* among these particles. We also need to find a probable relationship between consciousness and *complex system*. After this work is finished, we will be able to establish a *pragmatic criterion* with which to recognize, in a system of interactive material particles, the presence of *consciousness*. The concepts of "quantum collective potential," "cosmic collective potential," "Mach's Principle," and "Einstein's matter-energy tensor," all play an unexpected role in relation to Universal Consciousness.

7.9.1 Preliminary Conceptions of Consciousness

Conscience (ME) is knowledge within oneself, or inward knowledge, or consciousness. In Latin *concientia* means "con scientia", signifying *with science*, or if we prefer *with knowledge*. But this knowledge refers to knowledge of oneself. It is in this case where the dualistic Western dichotomy of subject-object becomes *monistic*. It is in this case where the object is *identical* to the subject. It is in this case where in Sartre's terms "*the consciousness of the subject is present in the object*." Better yet, it is in this case where "the consciousness of the subject *is* the consciousness of the object." When this identification is realized, only then do we experience the transrational acquisition of knowledge. Only then do we apprehend the essence of entities, of any and all entities. Consciousness is exactly what we mean by awareness, cognizance, knowingness, sentience. Thus, consciousness and awareness are synonyms. In Latin the word *scio*, which is derived from *sciens-ntis*, means knowing, or with knowledge. Alice A. Bailey [25], in her little book *The Consciousness of the Atom*, writes that *scio* means *that with which we know*. The great majority of people are convinced that the concept associated with the word "consciousness" is only applicable to human beings. This is a very restricted conception which we must generalize in order to apply it to other entities.

From an ontological point of view, an *operational definition* is completely useless. However, this type of definition is very useful from a physical point of view. An operational definition is useful because it permits us to measure a physical attribute of an entity. If the attribute of the entity can be quantified, then the attribute can be expressed mathematically. Once the attribute has been expressed mathematically, we can then introduce it into principles or natural laws which have been expressed in the language of Pythagoras. The concept we have in mind to define operationally is the concept of *consciousness*. Our task appears to be extremely difficult, if not impossible. While searching the Web (World Wide Web: WWW) for an operational definition of consciousness, the results were discouraging: *zero entries found*. Some *expert* professors "on line" answer your questions about an operational definition of consciousness by telling you there is no such definition, and there can never be one. Many other "on line" professors never acknowledge your electronic mail. Perhaps they have no reason to respond because they have no answers. In what follows, we will, nevertheless, attempt to establish an *operational definition of consciousness.*

After consulting and studying many books on the theme *consciousness*, the author has concluded that, in the present, the majority of authors are trying to *create* the *Science of Consciousness*. The great majority of the works deal with *human consciousness*. Now, each of these works is biased by the profession of the researcher. This "specialty" in writing of consciousness is natural, but undesirable, because a research project on the neurology or brain-basis of consciousness is not understood by an author writing about consciousness from a quantum mechanical point of view. On top of these two unconnected points of view, we must remember the work refers only to a *particular* field of human consciousness. But human consciousness can be approached, as it has been done, from many other points of view. For example, the religious aspects of consciousness, the biological aspects of consciousness, the physical-chemical aspects of consciousness, the psychological aspects of consciousness, the ethical aspects of consciousness, the social aspects of consciousness, and even the mathematical aspects of consciousness after mathematics has been ontologized. Someone is claiming he has already done this latter one.

The future of consciousness studies is difficult to predict. Perhaps all *consciousness* investigators are enlightened, and they are in the process of creating the proto-science of consciousness, the pseudo-science of consciousness. Thomas Metzing [26] is quoted in the *Journal of Consciousness Studies*, vol. **4**, No. 5-6, pp. 385-8 (1997), referring to the Elsinore meeting, saying that:

> "... consciousness studies [were] in a chaotic, pre-paradigm state - somewhat akin to nuclear physics at the beginning of the century."

It seems that Metzing is quite benevolent with his analysis of the studies of consciousness. This proto-science is closer to medieval physics. If we want to compare the consciousness studies with physics, we must try hard to define consciousness operationally, and free it from the restricted field of human consciousness. One should expect some kind of quantitative aspects of consciousness in the quantum mechanical studies of consciousness. But here we have a serious problem. Rick Grush [27], in 1999, wrote an article in the Web on *quantum consciousness : theories of.* Among the different approaches to draw some connections between quantum mechanics and consciousness, Grush writes:

> "Consciousness appears to be an extremely mysterious phenomenon . . . Quantum mechanics also seems to be very mysterious . . . So perhaps they are the same mystery. Nobody phrases it that way, of course, but this seems to be a line of intuition that motivates people."

Of course Grush is very serious when he says that "Quantum Mechanics seems to be very mysterious." In chapter 2, we quoted Murray Gell-Man [28], a well-known quantum physicist, saying:

> "Quantum mechanics, that mysterious, confusing discipline which none of us really understands but which we know how to use."

On the other hand, the book *Mind, Matter, and Quantum Mechanics*, written by Henry P. Stapp [29], makes 42 references to the word consciousness in the Index. In the Glossary of this book, we find a definition of consciousness:

> "*Consciousness* That luminescent presence of coming-into-beingness that constitutes our inner world of experience. It is present during our wakeful states, and during our dreams, but is extinguished during dreamless sleep, and in the state of unconsciousness induced by a severe blow to the head."

In essence, Stapp is telling us that *consciousness is our inner world of experience*. At any rate this definition refers to human consciousness. On page 176 of the book of Stapp, we have an extraordinary comment and reference to *The Principles of Psychology* by William James. Stapp wonders if maybe the main thesis of James' epoch-making treatise was:

> ". . . that each conscious thought is essentially a complex whole: each thought has components, which can be examined by subsequent analysis, but, *as given*, is a unified whole that cannot be reduced to a collection of parts without destroying the essence."

This conception of *conscious thought* of William James anticipates the conception of *consciousness* of Teilhard de Chardin as we will see shortly. Stapp's book is an excellent book for those interested in getting acquainted with a good bibliography of different studies of consciousness using quantum mechanics and other viewpoints. Another extraordinary source of articles online is found on the Website http//:ling.ucsc.edu/~chalmers/

mind.html. Here is an excellent collection of 569 online papers plus other sources of bibliographies which refer to about 2000 offline papers. This is a wonderful source which is at the fingertips of the interested reader. The collection was prepared by Professor David Chalmers.

7.9.2 System of Complex Interactive Material Entities and Complexification

Our main purpose in section 7 is to see if we can find a way to *generalize* the concept of consciousness beyond what is known about human consciousness. We also hope to find some initial steps in the direction of an *operational definition* of consciousness. This generalization of human consciousness already exists. It was created by Pierre Teilhard de Chardin, a French Jesuit priest, paleontologist, geologist and a scholar in science, philosophy and theology. From now on we will refer to this scholarly priest by the name *Teilhard*. In the year of his death, 1955, his book *Le Phénomene Humain*, was finally published. The English translation, *The Phenomenon of Man* [30], by Bernard Wall with an introduction by Julian Huxley, was published in 1959. In January,1999, when this author began to write this chapter, he recalled a seminar he attended, in1957, on *El Grupo Zoologico Humano* [31] (The Human Zoological Group) also written by Teilhard. After more than three decades, the word *complexification* was still ringing in the back of this author's consciousness. Finally, the word *interaction* emerged to create the trilogy consciousness-interaction-complexification. In what follows, we will try to establish a temporary relationship between consciousness with interaction and complexification. We should be aware that we may have to change this relationship. Functionally expressed, this relationship is:

$$C = f(i,c) \tag{7.1}$$

where C stands for Consciousness, i represents interaction, and c stands for complexification. The concept of *interaction* immediately induces us to think in terms of dynamic actions. These dynamic actions are exerted at least between two material neutral or electrical particles. On the other hand, the concept of *complexification* forces us to think about a system of many entities. This system could be composed of many electrons, protons, and neutrons which constitute an atom. If the system corresponds to an iron atom, then this atomic system is a complicated one, showing a definite degree of complexification. We can think of another system more complicated if we imagine an enormous collection of atoms of hydrogen, oxygen, carbon and nitrogen. The electrons and protons, of each atom in this enormous collection, interact with all the electrons and protons of the other atoms. This collection of atoms constitutes a molecule. Now, this molecular system is much more complicated than an atomic system, showing, this time, a much higher degree of complexification. Obviously, if by an act of magic, all the interaction between the atoms would disappear in the molecule, then the whole structure of the molecule would collapse into a chaotic system of atoms which are in total disorganized motion. The high degree of complexification in the molecule is possible only because of the interactions among *all* the atoms in the molecule. Thus, we must distinguish between two types of complexification: one is structural or organized complexification caused by the *whole* system of interacting entities on one and each of the individual entities of the system. The other type of complexity in the system is chaotic, or disorganized complexification caused by the total absence of interaction among the entities of the system. An interesting point to observe is that the concept of interaction implies an *action* acting on one entity, and a *reaction* acting on the entity which caused the *action*. This explanatory note leads us to other consequences.

Imagine a system of interacting particles with a very high degree of complexification. The total action of the *whole* system, minus one particle, will *act* on the particle left *alone,* and conversely the particle left alone will *react* on the rest of the system. This means that the particle left alone will *react* on each one of the particles of the rest of the system. If the number of material particles is N, physicists can calculate the intensity of the *action* of the (N-1) particles on the particle which was left alone. Physicists can do something more interesting. They can calculate the potential gravitational energy caused by the other (N-1) particles in the immediate neighborhood of the particle which was left alone. Is this all the potential gravitational energy found in the immediate neighborhood of the particle which was left alone? Most certainly not. The particle, which was left alone from the system, is immersed in *the potential energy field* caused by the interaction of every one of the particles of the collection of (N-1) with each other and with (N-2) particles of the system. From a mathematical point of view, this system is very complicated, showing a very high degree of organized complexification caused by the interaction between *all* members in the collection of particles. Using Bohm's terminology, we may say that Teilhard's concept of complexification of a system of particles shows an *unbroken wholeness of the entire system of particles*. The phrase, "the particle which was left alone" only has a didactic value, because, ontologically speaking, that "lonely" particle can *never* be alone. This is so because of the Principle of Inseparability. But these last physical considerations make the Principle of Inseparability more intelligible. We also see that *organized complexification* depends on the global *interaction* of the system of material particles. This dependence will have a consequence in eq. (7.1), as we will see later. Another point we have to observe in this paragraph is that this type of complexification is not static. On the contrary, it is a very dynamic complexification, constantly changing at microscopic levels. This permanent change in the atomic or molecular complex structure allows the system to capture, in their *energetic field*, other atoms or other molecules, developing another molecular structure with a much higher degree of complexification. This molecular development is called by some, "structural mutation;" other people call it "structural evolution." All these processes of complexification, or development, or mutation, or evolution were considered by Father Pierre Teilhard de Chardin in his writings which were much advanced for his time. Even today, after almost eighty years of their conception, Teilhard's work is still scientifically ignored. Before we establish a relationship between consciousness and interaction, let us ponder on Teilhard's unusual work *The Phenomenon of Man*, and, perhaps, we may find some clues for an incipient general operational definition of consciousness.

7.9.3 Consciousness According to Pierre Teilhard de Chardin (1881-1955).

Teilhard, in his fundamental book, *The Phenomenon of Man*, takes the reader down a long but fascinating path of the material evolution of *things*, culminating with the material evolution of man. Then Teilhard, in an incredible act of intuition and conceptualization, takes the reader through an overwhelming path of synthesis. In essence, Teilhard establishes that matter and spirit are two manifestations of the same entity. In this way, Teilhard annihilates, once and for all, Descartes' dilemma between body and soul, between matter and spirit. We will soon come back to this transcendental synthesis. Teilhard's definition of consciousness is ubiquitous in his book. In the Index of this book, we find 38 references to the word *consciousness*. They are distributed throughout his book. Julian Huxley [32], in his Introduction to the *The Phenomenon of Man*, provides an excellent summary of Teilhard's concept of consciousness. Huxley, writing about Teilhard, says:

> "Thus he states that full consciousness (as seen in man) is to be defined as 'the specific effect of organized complexity' . . . we must infer the presence of potential mind in all material systems, by backward extrapolation from the human phase to the biological . . . to the inorganic."

But Teilhard went beyond material systems as a necessary condition for the appearance of consciousness. He searched for a connection between consciousness and *energy*, bringing the problem of consciousness to the field of physics. The problem was that the physics of his time was not ontologically prepared to help him. Consequently, Teilhard's work has been ignored for obvious reasons. He has considerably expanded Darwin's theory of evolution to the point we can say today that: *man descends from elementary charged material particles.* But Teilhard has created *another* theory of evolution. A theory of evolution of the immaterial and invisible realm of the spirit. Spirit, mind, soul, consciousness are all synonyms of *energy*. Both Einstein and Teilhard apprehended the essence of matter. Einstein has shown that the essence of matter is energy, while Teilhard has shown that the essence of spirit is energy. Through the writings of Teilhard, we see that *consciousness* **can be** only because of matter and the interaction of material particles. Through the writings of Einstein, we see that *matter* **can be** only because of energy. Teilhard's theory of evolution of spirit requires that another condition be satisfied. This condition is based on a geometrical configuration of material entities belonging to a material system. Consciousness manifests itself in a material system when a *specific effect of organized complexity* is observed in the system. Thus, the other condition for the manifestation of consciousness, according to Teilhard, is the presence in the material system of an *organized complexification*, as we described before in section 7.9.2 of this chapter. As far as this author's knowledge is concerned, the only book which has developed Teilhard's conceptions into a rational structure was recently published, in 1993, by Joseph P. Provenzano [33]. In his book, *The Philosophy of Conscious Energy,* we see the efforts of Provenzano in fusing physics, philosophy and theology. In what follows, we will continue the thoughts of Teilhard and Provenzano.

Huxley's interpretation of Teilhard's conception of consciousness contains the germ of an operational definition of this elusive concept. We need to unveil what should be the meaning of *specific effect* of the consciousness of a material system of particles. This task is better analyzed in the field of physical phenomena rather than studying elementary systems of material particles. Now, let us try to establish a preliminary *pragmatic definition* of consciousness using a dynamical concept of physics.

> Consciousness is the *capacity* of a system, constituted by interacting material entities, to act significantly on any and each one of its material components.

Therefore, consciousness is a *synergistic* system in which the system as a *whole* has the capacity to act effectively on each one of the individual material entities. This definition of consciousness offers a physical or practical criterion to test all metaphysical speculations about consciousness. The material entities could be elementary neutral or electrically charged particles, or celestial bodies.

When people say the *whole* is greater than the sum of the parts, they simply do not know what they are saying. What they should say is the *whole* system of *interacting* entities acquires *attributes* which are not intrinsic to the individual entities. The new attributes only appear when the individual entities unite to create a completely new entity. This new entity - *the whole* - has properties totally different from the properties of the

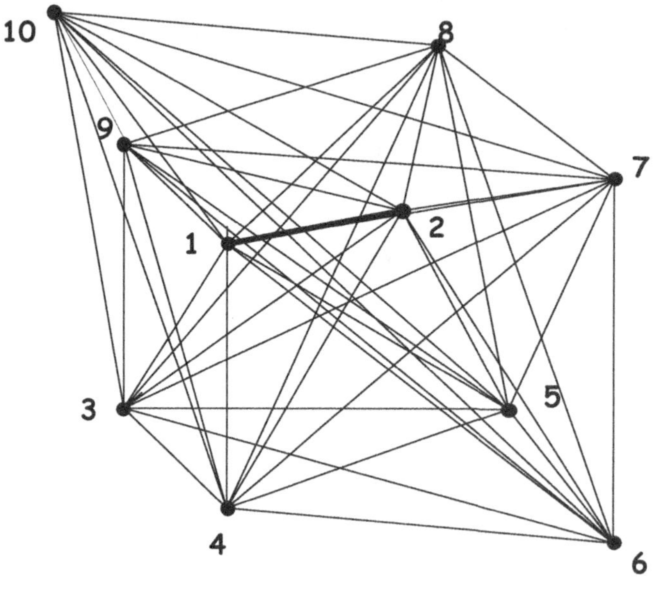

Figure 7.2 Interaction of material entities 1 and 2 with the rest of the system.

individual material entities which constitute the synergistic system. If we consider another system of material entities, which do not interact among themselves, then we have a disorganized, complex and chaotic system. On the other hand, in any system of interacting material entities, the more complex it becomes, the more capacity it has to significantly act on each individual entity of the system, showing at the same time a certain degree of organization, a certain degree of order. This organization is caused by the interaction of all its elements among themselves. Fig. 7.2 illustrates the interaction of material entities 1 and 2, represented by a thick black line. All the other thin lines represent the action of the rest of the system on entities 1 and 2, and the interaction of each one of the entities of the system with all the other entities. Fig. 7.2 is a graphical representation of the Principle of Inseparability which establishes that no material entity of an interactive system can be separated, can be isolated, from the action of the *whole* system.

It is obvious that the interaction between the material entities 1 and 2 of Fig. 7.2, is affected by the action of the material entities 3 through 10 acting on 1 and 2. However - is the interaction between 1 and 2 affected by the interaction of any other pair in the system, like the interaction between 7 and 9, for example? - Our first answer is to say that the interaction of 1 and 2 has nothing to do with the interaction of 7 and 9. From a dynamical point of view, it is hard to see how the forces of interaction of the material entities 7 and 9 can affect 1 and 2. However, from an energetic point of view the answer is in the affirmative. The interaction between the material entities 7 and 9 creates a *potential energy field* which extends to infinity. It is in this potential energy field where the material entities 1 and 2 are immersed. Therefore, the interaction of 1 and 2 takes place in a space where the local potential energy is modified by the contribution of the potential energy created between the material entities 7 and 9. Thus, being that the force is the negative gradient of the potential energy, the interaction between 1 and 2 is definitively affected by the interaction of 7 and 9 and all the other pairs of the system. Now we see that a system of interactive material entities offers a geometrical complexification and a dynamical complexification. In other words, the complexification of the *form* of a material system is *caused* by the complexification of the *interactions* as the system *grows* in its numbers of material entities.

The concept *complexity* (*complexification*) is in the minds of many researchers in the field of consciousness. They assume that consciousness depends on the complexity of the system which constitutes consciousness. Paul Davies, in a conversation with Stephen Jones at the Conference Tucson II , and published in the WWW (Web, 10/5/99), says:

> "I believe that only in systems sufficiently complex will consciousness emerge and flourish. Now whether complexity alone is enough is what I am not sure about."

Paul Davies is not only a physicist, a quantum physicist, but he is also, perhaps, the most prolific writer on the subject of consciousness associated with theology and science. Paul Davies is a man of the *Neo-Renaissance.* This Neo-Renaissance is presently emerging at the beginning of this new century. In 1995, Paul Davies was awarded the Templeton Prize, equivalent to the Nobel Prize, but dedicated to research to link religion and science; to bring together *Fides et Ratio.* If the interested reader wants to meet the mind of this Natural Philosopher, he needs to type in the words "Paul Davies and consciousness." Any search engine of the world-wide-web will find his work. In the above quotation, Davies wonders if consciousness emerges only in systems sufficiently complex. In his conversation with Stephen Jones, Davies adds the following statement:

> "Now, I think complexity, specifically *organised complexity*, is the key to consciousness."

It is unfortunate that after almost 70 years, we read the same conclusions which were reached by Teilhard de Chardin in respect to consciousness and complexification. In the opinion of this author, consciousness does not depend on the complexity of a system. Consciousness is not a function of how complex the system is, but it depends on matter and on the interaction of the material entities of the system. The organized complexity or complexification of a synergistic system of interacting material entities is not the cause of consciousness. For consciousness to exist, to manifest itself, two conditions must be present: *matter* and *interaction* between the material components of the system. Let us not forget that this *matter* can be neutral or electrically charged. The transitory functional relationship given by eq. (7.1) should be changed to the following one:

$$C = f(m,F) \tag{7.2}$$

where m is the mass of the material entities of the synergistic system, and F is the force of interaction between all material entities with all the material entities of the same system. This global interaction of the parts of the *whole* system generates the *collective potential energy field* (CPEF). The ordered complexity of the material system is caused by the interaction of the parts of the system. Organized complexity is possible only in a material system in which the parts or material entities *interact* with each other. Thus, complexity is a property or "accident" of consciousness. We should be mentally prepared to reason and, finally, conclude the existence of a *spectrum of consciousness. The Spectrum of Consciousness* is the title of a fascinating book written by Ken Wilber [34], published in

1982. Chapter 2 of this book, titled *Two Modes of Knowing* is simply magnificent, and constitutes a basic lesson to understand why *formal quantum mechanics* has nothing to say about the ontology of consciousness. However, now we can change eq. (7.2), to read:

$$C = f(m, CPEF) \qquad (7.3)$$

where CPEF is the *collective potential energy field* of the synergistic consciousness. Our *pragmatic definition* of consciousness can be rephrased as follows:

Consciousness is the *collective potential energy field* (CPEF) of an ensemble of interacting material particles, which causes *specific effects* on each and every one of the material particles of the ensemble. This definition of consciousness is not a definition by analogy. It is a definition by identification. It is the identification of consciousness with the *collective potential energy field. Thus, consciousness is the invisible immaterial collective potential energy field of an ensemble of material neutral or electric particles.*

David Bohm, the creator of the *mathematical* quantum hidden potential, wrote a book titled *Wholeness and the Implicate Order* [35]. Going forward with an incredible intuitive and analogical endeavor, he tried to discover the connection between consciousness and the *implicate order* which was manifested as *an explicate order.* The long path followed by Bohm was a consequence of his formal (mathematical) knowledge of his own *quantum hidden potential*. Bohm never knew the ontological (Newtonian) origin of our collective potential energy field. In Bohm's book, chapter 7, titled *The enfolding-unfolding universe and consciousness*, he contrasts mechanistic order in physics with implicate order. "Implicate," from Latin, means *to enfold, to enclose, to enwrap, to fold inward.* After Bohm proposed the new notion of implicate order, he wrote:

> "In terms of the implicate order one may say that everything is enfolded in everything. This contrasts with the *explicate order* now dominant in physics in which things are *unfolded* in the sense that each thing lies only in its own particular region of space (and time) and outside the regions belonging to other things."

From ancient times, the universe has been dichotomized into two aspects. One aspect was perceived by the human senses as discontinuous, material, visible, divided into parts. The other aspect was conceived by the human mind as continuous, immaterial (ethereal), invisible to the senses, wholly or undivided. Western sages of all ages have seen order in the discontinuous or discrete aspect of the universe. Obviously, it is much easier to study the parts of what we see than to study the *whole* we do not see. Let us learn what the task of science in mechanistic terms of parts is, and also in terms of an unbroken wholeness of the entire invisible implicate order of the universe. In this respect, Bohm writes:

> "The task of science is then to start from such parts and to derive all wholes through abstraction, explaining them as the results of interaction of the parts. On the contrary, when one works in terms of the implicate order, one begins with the undivided wholeness of the universe, and the task of science is to derive the parts through abstraction from the whole."

These thoughts of Bohm are exactly the same as expressed by J.C. Maxwell [36], in 1873, in the preface of his *Treatise*. We quoted Maxwell in chapter 3 of this book, but it is worthwhile to read these thoughts about electrodynamic fields and electrodynamic forces between electric particles.

> "I found that in general the results of the two methods coincided, so that the same phenomena were accounted for, and the same laws of action deduced by both methods, but that Faraday's methods [English electromagnetics] resembled those in which we begin with the whole and arrive at the parts by analysis, while the ordinary mathematical methods [German electrodynamics] were founded on the principle of beginning with the parts and building up the whole by synthesis."

Newton's theory of dynamics is founded on the concept of the explicate order, while Einstein's general relativity theory is founded on the concept of the implicate order. As we mentioned in this book before, a physical theory is incomplete if it does not explain both the continuous and discontinuous aspects of the universe. Our concepts are different from Bohm's concepts, though sometimes they coincide inordinately well with each other. For this reason, and to avoid confusion, we will introduce the concepts of *invisible order* and *visible order*. As we saw above, in the pragmatic definition of consciousness, we identified consciousness with the collective potential energy field of an ensemble of particles.

> Now, we identify any collective potential energy density distribution, at any level, with an invisible order, and the corresponding material geometrical configuration of particles with its visible order.

The above statement is equivalent to the following one:

[Tensor of energy-momentum content of space] = [Tensor of geometry of space]
or
[Tensor of geometry of space] = [Tensor of energy-momentum content of space]

These last equations represent Einstein's field equations. Thus, the *invisible* distribution of the density of energy in the local space, caused by distant moving matter and flowing energy (Poynting vector as Wesley has proposed, see reference 37), determines the geometrical distribution of the *visible* material entities in space. In 1993, Bohm and Hiley [37] published the book *The Undivided Universe: An Ontological Interpretation of Quantum Theory.* On page 389 they wrote - *"Although the implicate order is a theory of the whole, it is in no sense a 'theory of everything'."* - Bohm never expressed his *theory of his unbroken wholeness* in mathematical terms. Einstein did it in his GRT, but he did not realize the transcendental meaning of his energy-momentum tensor. Einstein's energy-momentum tensor represents the density of the collective cosmic potential energy field of the entire universe, and therefore, represents Universal Consciousness.

Bohm and Hiley were able to foresee that: *". . . even an electron has at least a rudimentary mental pole, represented mathematically by the quantum potential."* A "mental pole" is a synonym of "rudimentary polar consciousness." This representation of a "mental pole" is not an identification with the quantum potential. Furthermore, the assertion that an electron *"has at least a rudimentary mental pole"* is not based on any ontological or logical necessity. Bohm and Hiley mentioned, on page 381, that many physicists have suggested that consciousness (human) is closely related to quantum mechanics. This conception indicates that in order to understand the quantum formalism, it is necessary to correlate it, somehow, with the concept of consciousness. But where is the formalism of consciousness? Bohm and Hiley said:

> "Throughout this book it has been our position that the quantum theory itself can be understood without bringing in consciousness and that as far as research in physics is concerned, at least in the present general period, this is probably the best approach."

Then both authors wonder whether the intuition of consciousness and quantum mechanics are in some way related. That constituted a good starting point. Bohm and Hiley wrote in this respect:

> "Our proposal in this regard is that the basic relationship of quantum theory and consciousness is that they have the implicate order in common."

Of all the physicists that have tried to relate consciousness with quantum mechanics, David Bohm was the best equipped. Bohm has been the most prolific author writing about the *wholeness* of the quantum potential. Unfortunately, he wrote from formal and logical points of view. Let us make, in his name, the last connection which he never made:

> Our proposal in this regard is that the basic relationship of quantum theory and consciousness is that they have *the collective quantum potential energy field in common.*

In the spectrum of consciousness or spectrum of collective potential energy field, we should expect an enormous variety *of complexifications*. The spectrum should range from extremely low consciousness, with a very low material structural complexity, to extremely high consciousness, with an unduly high structural complexity. Speaking of *elemental or rudimentary consciousness,* we understand the capacity of a system which has a few interactive material entities which are capable of acting significantly on at least one of the material entities of the system. To rationally proceed, we have at our disposal one pragmatic definition of consciousness and physical laws. The pragmatic definition, which we referred to, allows us to establish the existence of consciousness by dynamical means or by energetic means. In what follows, we will see how a rudimentary consciousness is *enfolded* into a more complex consciousness, which in turn is enfolded into another incredibly higher complex consciousness. This leaves the less complex consciousness with a certain degree of independence.

7.9.4 Does the nucleus of an atom have an elemental consciousness?

The nuclei of atoms have a *collective quantum nuclear potential energy field* (CQNPEF), and therefore, the nucleus of an atom has a *nuclear consciousness*. This is an impeccable theoretical or metaphysical conclusion. But how do we prove the real or pragmatic existence of this nuclear consciousness? In chapter 4, we explained the inner structure of the neutron. This structure corresponds to Eddington's model of the neutron which was treated with the new Newtonian relativistic electrodynamics. The nucleus of a complex atom consists of a set of *miniature hydrogen atoms* (neutrons) which are in constant interaction with themselves and with ionized neutrons (protons). In an unpredictable random manner, the CQNPEF, through a tunneling effect, ejects one electron from a neutron. This beta decay (ejection of electrons from radioactive atomic nuclei) is the real, pragmatic, experimental evidence of the physical existence of the CQNPEF, and therefore, the pragmatic existence of nuclear consciousness. This is a very rudimentary nuclear consciousness. This physical phenomenon shows the *specific action* of the *whole* nucleus on one of its parts.

7.9.5 Consciousness of atoms, molecules and cells.

Atoms are composed of a nucleus surrounded by a set of electrons, and also, surrounded by an ensemble of other atoms. This system of electrical particles creates an invisible *whole* or implicate order of electrostatic and electrokinetic potential energy field. This energy field will be well known in the near future as the *collective atomic quantum potential energy field.* Hence, if this collective potential energy field does exist, then *atomic consciousness* also exists by necessity. This is a theoretical conclusion; i.e., the atomic consciousness has a noetic, mental, rational, intellectual existence, but it seems that it has no real or pragmatic existence. Being that consciousness is the collective atomic quantum potential energy field, we then have thousands of books reporting the experimental verifications of a huge number of theoretical conclusions in quantum physics, quantum chemistry and quantum biochemistry. These are the pragmatic proofs that atomic consciousness, indeed, exists in reality; i.e., outside the human consciousness. The atomic consciousness engulfs, enfolds the nuclear consciousness, but it does not affect it in normal conditions. However, under special circumstances, particularly in biological matter, it may affect it.

When an ensemble of atoms get together, a molecule is created, exhibiting attributes, properties, qualities which none of the individual atoms have. These new properties belong to the *whole* of the atoms. These new properties, of this collection of atoms or molecules, are an integrated cooperative quality of a particular set of atoms. The explicate order, or organization of the atoms in the molecule, is guided by a more complex collective molecular potential energy field created by the ensemble of atoms. Thus, here we have the physico-chemical process which explains the gestation of the molecular consciousness, which in turn enfolds the atomic consciousness and the nuclear consciousness.

Now to continue, molecules do gather together in highly complex structures creating macromolecules which in turn get together to actualize complex entities called cells. We find ionized atoms and ionized molecules both on the inside, and also, on the outside walls of the cells. The external walls of the cells are made up of bilipid layers, across which, we have transmembrane proteins. The external walls of the cells are covered by electrical negative residues of Sialic or Nitrogene-Acetyl-Neuraminic-Acid (NANA), along with electrical negative residues of Ribo-Nuclei-Acid (RNA). The protons of these negative residues are trapped by the permanent electric dipole moments of water molecules. Thus, in the near vicinity of cells, we have a semi-crystalline structure of hydrated protons which insulate them electrically. For more details on this subject, see the author's essay *Cancer: An Electrical Phenomenon* [38, 1991]. The most incredible electrical phenomenon exhibited by cells is the magnitude of the intra-membrane electric field intensity. This intensity or force per unit electric charge, in the interior of the bilipid membrane of the cell, is equal to 10,000,000 volts per meter. This is a huge electric field which may induce nuclear transmutations in atoms, as C.L. Kervran [39], in his fabulous book, *Biological Transmutations,* has shown pragmatic evidence of bio-transmutations through his experiments. This highly complex electric-ionic structure of cells creates a *collective monocellular quantum potential energy field*, and therefore, a *unicellular consciousness*, or monocellular mind, or unicellular "intelligence." The so-called "cell communication" must be done through the collective unicellular quantum potential energy field: electrostatic and electrokinetic potential energy fields. We may also say that this cell communication must be done through their individual collective implicate order. This collective unicellular consciousness enfolds the consciousness of ensembles of molecules, inside of which we have ensembles of atoms. Accordingly, inside the ensembles of atoms we have ensembles of nuclear particles.

7.9.6 Consciousness of biological organs, animals and plants.

Nature offers the pragmatic or experimental phenomenon of huge colonies, dominions, ensembles of cells called organs. This dominion of cells is extremely complicated, but still highly organized. In Teilhard's and Bohm's terms, an ensemble of cells shows an enormous complexification, but also shows a marvelous implicate order controlled by the ionic *collective polycellular quantum potential energy field*, and therefore, controlled by the polycellular consciousness. Can science help human beings in making decisions on moral or ethical issues? Perhaps, but it seems that all depends on the human capacity of interpretation. Cell differentiation in a human embryo happens after the embryo reaches a total of a few billions cells. This event transpires somewhere between nine and twelve weeks after the fertilization of the egg. It is interesting to observe that at the very moment the sperm penetrates the ovum, the external membrane of the egg becomes unduly negative, charged by the negative residue of sialic acid. This electrical phenomenon serves two purposes. One is to instantly repel, electrostatically, the negatively charged sperms. The second purpose is to change the transmembrane electric potential of the egg, triggering the process of meiosis or division of the fertilized ovum. Initially, the collective polycellular quantum potential energy field or polycellular consciousness is an elemental consciousness which manifests itself through the proliferation of the embryonic cells. This elemental consciousness proceeds to evolve as the number of undifferentiated cells continues to develop more and more embryonic cells. This embryonic consciousness had not yet reached the highly evolved human consciousness. It is possible that the incipient human consciousness appears when the *differentiation process* is initiated by the totality of embryonic cells (the order of a few billions). This suggestion is debatable, but it offers an opportunity to experimentalists to make a thoughtful study on moral decisions based on scientific knowledge.

To study animal and plant consciousness, it is best to refer the reader to two extraordinary books. One is *The Secret Life of Your Cells* [40,1989] by Robert B. Stone. The other book is *The Secret Life of Plants* [41, 1974] by Peter Tompkins and Christopher Bird.

7.9.7 Consciousness of human beings.

In the past few decades, the world has been flooded with books on "human consciousness," so it would be irrelevant for us to write anything more about this theme. As we read these books, however, one important concept is coming through quite clearly. It is telling us that it is imperative to turn back the pages and slowly re-evaluate all that has been written on human consciousness.

7.9.8 Consciousness of the planetary system.

In Fig. 7.1, we have ten material entities which we may assume represent the sun and the planets of the solar system. The sun is represented by material entity 1 and planet Mercury is represented by material entity 2. If we epistemologically neglect the gravitational *action* of the rest of the planets on planet Mercury, then the elliptical orbit of planet Mercury will remain fixed in stellar space. In accepting this assumption, we have ontologically violated the Principle of Inseparability. We are also left with a crippled system of material entities with no interaction of their parts among themselves. Under these circumstances, we will not observe any significant action of the *whole* planetary system upon planet Mercury. However, when we accept the action of the whole planetary system upon planet Mercury, we can conclude, theoretically, as Isaac Newton did in 1687, that the elliptical orbit of planet Mercury has to rotate or precess in stellar space. Nevertheless, and most importantly from a pragmatic point of view, this perihelic rotation of planet Mercury was established by an astronomical observational method. In 1859, Leverrier detected an excess of perihelic rotation on planet Mercury. This observational and theoretical discovery of Leverrier was made after the calculation of the total influence of the rest of the planets on Mercury. All the mathematical and physical arguments are displayed in section 1 of chapter 6. The *whole* planetary system acts significantly, not only on planet Mercury, but also on Venus, Earth and Mars, showing by astronomical observations that these planets have a direct precessional motion of their perihelia. Thus, according to our pragmatic definition of consciousness and practical astronomical observations, we rationally have to conclude that the solar system has an elemental *collective planetary gravitational potential energy*, and consequently, a *rudimentary consciousness*.

7.9.9 Consciousness of the Milky Way.

Our own galaxy, the Milky Way, has around 100 billion stars. This enormous number of stars constitutes a highly complex ensemble of suns in constant gravitational interaction. From a physical point of view, we can express mathematically the ontological existence of a *collective galactic gravitational potential energy field*, and consequently, a *galactic consciousness*. This galactic consciousness or mind enfolds all other previous consciousness which we have considered. This *collective galactic gravitational potential energy field* is not constrained to the interior of the galaxy. This energy field also extends outside the galaxy. The energy density, associated to this collective energy field, decreases as we recede from the galaxy. Thus, any galaxy must be surrounded by a gigantic envelope of energy. This verbal description of the *collective galactic gravitational potential energy field* gives the reader the impression that it is a wonderful metaphysical fairy tale or allegory. But, is there pragmatic or realistic evidence of its existence? The astrophysical evidence comes from the observation of *galactic lenses*. The envelope of invisible energy, which surrounds each galaxy, functions as a lens which refracts the light coming from stars belonging to the more distant galaxies. Thus, if a radio telescope points in the direction of a radio-star, an optical telescope pointing in the same direction does not show any visible source of light. This is so because electromagnetic waves are refracted more or less by a lens, depending on the wavelength associated with the electromagnetic wave. Thus, the real or pragmatic existence of galactic lenses assures us of the real existence, or pragmatic evidence, of the invisible collective galactic potential energy field, and consequently of the real existence of the consciousness of our own Milky Way.

7.9.10 Consciousness of all galaxies: Universal Consciousness.

Even though we do not know the total number of galaxies in the universe, we can conceive the existence of a *collective cosmic galactic gravitational potential energy field in the entire universe*. Therefore, we must ontologically and logically conclude that a *Universal Consciousness* does exist. This Universal Consciousness enfolds all other consciousness which we have discussed. This seems to be another metaphysical fairy tale, or parable, until we bring realistic or pragmatic evidence of the actual existence of the Universal Consciousness. In chapter 5, we introduced the concept of the Primordial Energy field and proved the existence of a wave of energy density. We deduced a non-homogeneous D'Alambert's equation for the energy density of electromagnetic energy fields. In chapter 6, we explained Merat's empirical law by the energy stellar lens which surrounds our sun. We also showed that the red shift of light sources, belonging to distant galaxies, is caused by an amplitude decay of the electromagnetic wave during the interaction with the zero-point energy of vacuum (energetic ether) in intergalactic space. Finally, let us mention an energy densitometer. This energy densitometer is a luminiferous speedometer. Thus, in the near future, measuring the speed of light in different regions of the solar system and stellar systems, and using eq. (5.10), which associates the speed of light with the energy density of the region in space, we can determine the density of energy in different regions of space. Thus, we have different pragmatic evidence that "the obsolete vacuum of space" is filled with the Universal Consciousness. If the reader wants to call this Universal Consciousness *God* or *Supreme Intelligence*, or *Ultimate Truth,* or *Being of all entities* or any other name, it is his privilege to do so. But his God will have only a noetic existence like the mental existence of a mermaid. The most transcendental event for the human race is to know that God exists not only in their minds, but also, that God exists outside their human minds. That God exists in the real world of things. That God exists in the entire universe. That God is the universe. That God is Universal Consciousness.

Here we come to an extraordinary ancient knowledge of the Tradition. God manifests in three modes. God is the invisible source from which anything and everything **be-comes**, and **is** that it **is**. This mode of manifestation is the Primordial Cosmic Energy Field. In anthropomorphic terms, God is the *womb* in which everything is born, lives and evolves, and returns to. However, this anthropomorphic allegory is very ancient. The second mode of the manifestation of God is the material universe, is the condensed energy of every corporeal entity in the entire universe. In anthropomorphic terms, every material body that is born from and in the *womb* of God, is a child of God; hence the entire material universe constitute the children: the sons and daughters of God. The third mode of the manifestation of God is the Universal Consciousness, is the invisible cosmic implicate order of the collective cosmic potential energy. It is the **Supreme Intelligence** of the *Unbroken Wholeness of the Entire Universe*. This wholeness represents a tri-manifestation of one and the same **Supreme Spirit**. In this conception there is no mystery. Our cosmotheistic conclusions are indeed necessary (logical) inferences, based on our scientific theology, however, our scientific conclusions are by no means new. We have only ratified ancient knowledge of the Holy Trinity. H.P. Blavatsky [42] in *The Secret Doctrine* writes:

> "Ether, the celestial virgin, the spiritual mother of every existing form and being, from whose bosom, as soon is 'incubated' by the Divine Spirit, are called into existence Matter and Life, Force and Action."

On the other hand, Newton never accepted the doctrine of the Holy Trinity. Richard S. Westfall in his excellent book *Never at Rest* [43, p. 309-334], describes Newton's arguments about his tenacious position against the Christian doctrine of the Holy Trinity. In the opinion of this author, Newton was wrong in refusing this transcendental theological knowledge. The reason for Newton's rejection of the Holy Trinity is that his theory of dynamics had to wait more than 300 years to be understood and to be developed. Today, Newton would have reasoned the proposition of the Holy Trinity, and he would have accepted it.

Another point of discrepancy between Pantheism and Christianism is the rejection by Spinoza and Einstein of the belief that God is a person. If Spinoza and Einstein had known the pragmatic definition of consciousness, as we have presented in this chapter, and the definition of *person as an entity conscious of itself* which is a generalization of the mundane anthropomorphic concept "person," then Spinoza and Einstein would have reasoned and reached the conclusion that God is Universal Person. Thus, our scientific theology or cosmotheism has allowed us to conclude that God exists in reality, that God is Universal Consciousness, and that God is Universal Person. These conclusions are logical (necessary) consequences of our theological principles, our pragmatic definition of consciousness and definition of "person." If someone rejects these definitions, he will be left without a scientific theology. As a final remark, we should mention that if we decide to present this new cosmotheism to uneducated people, we must resort to an allegorical description of the doctrine of the Holy Trinity, for example. If we do so, we can present to them the *feminine mode* of the manifestation of God, instead of the lengthy scientific rational approach which will not be understood. In this allegorical narration of the Holy Trinity, we may choose to depict the Universal Consciousness as the Father, because it would be impossible to explain to uneducated people the ontology, physics and mathematics of the collective cosmic gravitodynamic potential energy field, or the collective energy-momentum tensor of geometrodynamics of Einstein's GRT. Finally, we will end the allegory by telling these good but uneducated people that we are, as well as the entire material universe, the sons and daughters of God. This is what we read in the Bible: Psalm 82:6, and John 10:34. All our conclusions in this chapter, though obtained from the new physics presented in this book, belong to *Ancient Wisdom.* Recently we wrote that God is the *womb* in which everything is born, lives, evolves and returns to. In the Bible, Acts 17:28, we read: "*For in him we live and move and have our being . . . For we are also his offspring.*" Much earlier in the past we read in Hindu Rishi's: "*The Universe lives in, proceeds from, and will return to Brahman.* "

Soon the day will come when our grandchildren will not read allegories or metaphors about God, and they will know many attributes of the Being of all entities. At that moment, planet earth will know an Advanced Religion. An excellent book is *Ancient Wisdom - Modern Insight* [44, 1985], written by Shirley Nicholson. This book uses the science of the 20th century to interpret the Ancient Wisdom. As the physics of the 20th century is incomplete in many respects, Nicholson's book necessarily inherits the defects of the physics of the last century. However, the reader can remedy these defects with the physics of the 21st century. Eventually all of us will discover - *there is nothing new under the sun* - On June 17, 1916, when Einstein wrote a letter to Lorentz, admitting that:

$$g_{\mu\nu} = \text{Æther}$$

Einstein was telling Lorentz that the Æther was represented by the components of the metric tensor $g_{\mu\nu}$, which in turn was caused by the energy-momentum tensor $T_{\mu\nu}$ of the entire universe. At the time, Einstein did not realize he was writing both as a physicist and as a future theologian. As a physicist he was referring to $T_{\mu\nu}$ as the density of the *collective cosmic gravitational potential energy field of the universe*. As a future theologian, Einstein was predicting that $T_{\mu\nu}$, some day, was going to be another way to express the Universal Consciousness, to express the human noetic concept of God as an idea of God perceived from the world of things, perceived from the external reality. Today, Einstein may have written the previous equation as follows:

$$T_{\mu\nu} \equiv \text{God}$$

This last expression is not an equation, but an *identity*. This identity establishes that *the whole visible material universe and the whole invisible immaterial universe* ***is God***. Or if we prefer to say, **God is** *the whole visible material universe and the whole invisible immaterial universe.*

7.10 Advanced Religion.

The Knowledge of God. We learned before that the etymological meaning of the word religion is "to tie back," "to relink," "to reconnect." The same meaning belongs to the word yoga. Thus, the purpose of any religion is to teach human beings to reconnect their individual consciousness with The Universal Consciousness, or God. The word reconnect forces us to ask - is it possible that some time ago the human race was connected to God? According to Professor Julian Jaynes, this was exactly the case in the last half of the second millennium B.C. In his book, The Origin of Consciousness in the Breakdown of the Bicameral Mind [45, 1976], he defends this shocking thesis. However appalling you may find his theory, it brings semantic consistency to the word religion.

In front of the author, next to the keyboard of his computer, there are three very interesting books. In retrograde time order these books are *How to Know God* (B1) by Deepak Chopra [46, 2000], *The Mind of God* (B2) by Paul Davies [47, 1993], and *The Secret of the Golden Flower; A Chinese Book of Life* (B3), translated and explained by Richard Wilhelm [48, 1931 and 1962]. The latter has a foreword and commentary by C.G. Jung. Also included here is part of the Chinese meditation text *The Book of Consciousness and Life*, with a foreword by Salome Wilhelm. For fast reference in what follows, we will refer to these books as B1, B2, and B3.

Let us *assume* that B1, indeed, shows a neat process to humanly know God. Next, let us *assume* that following this process we come to know the Mind of God as B2 asserts. If we finally *assume* that B3 expresses the truth in every one of its propositions, then we have a serious problem. According to Jung in B3:

> "Whenever the narrowly delimited, but intensely clear, individual consciousness meets the immense expansion of the collective unconscious, there is danger because the latter has a definitely disintegrating effect on consciousness."

If through any method the individual consciousness of a human being grazes the Universal Wisdom, according to Chinese Taoism, the Universal Consciousness annihilates the individual consciousness, and the personality of the human vanishes from a mindless body that wanders aimlessly in a mental hospital. Now, if Jung's western interpretation is correct, then Paul Davies would be in serious difficulties, and Deepak Chopra would be teaching, in good faith, western human beings to lose their minds in the bottomless realm of Universal Wisdom. The conclusion about these three books is that the first two are strongly founded on the physics of the 20th century. This foundation is no ontological guarantee of anything. Jung's comments on B3 are extremely interesting, but they do not help to establish Noetics or the Science of Consciousness in spite of all the empirical studies on the biology, chemistry, physics, mathematics and electronics of the human brain. Thus, if we still do not understand human consciousness - how are we going to believe that a human being could be intellectually capable of writing about the attributes of God, the Universal Consciousness? This is not only a wishful thought, an irreverent act, but an unbelievable arrogant enterprise. The Universal Consciousness, God, is the Ultimate Mystery. With our insignificant consciousness (intellect, mind, spirit, soul), we should be more than happy to prove the existence and the nature of Universal Consciousness *outside the human mind*. Indeed, we should be satisfied to prove the existence and essence of Universal Consciousness in reality, in the world of material things until human consciousness evolves to higher levels of comprehension. Then we will understand that human consciousness is a concept, a noumenon, created by the Universal Consciousness, and reified by the same Universal Consciousness. Maybe Jung is not completely correct and Davies and Chopra are correct in Spinoza's opinion, when he said: "*The greatest good of the mind is the knowledge of God , and the greatest virtue of the mind is to know God.*"

Evolution of Human Intelligence and Advanced Religion.

An Advanced Religion will be established on planet earth when the intelligence of the human race has evolved. This is exactly the conception of Father Pierre Teilhard de Chardin: evolution of the soul, of the spirit, of the consciousness of all creatures and things on planet earth. When this phenomenon happens, human beings will begin to connect their consciousness with the consciousness of other animal species and with the consciousness of plants. After these accomplishments, human beings will attempt to *reconnect* their minds with "The Mind of God," but still running the risk of being annihilated by an ineffable flood of knowledge. If this future race of thoughtful intelligent human beings survive this gnoseological evolution, then an Advanced Religion will be founded on planet earth. A religion not based on fear, but on a wonderful intellectual love for God. A religion based on mystical love which is the essence of unity, a re-union with the Universal Consciousness. How far into the future will this momentous event finally take place? Most probably it has already started to happen.

Let us assume that it is true that human beings use a maximum of 10% of their mental capacity. What is happening to the 90% of the human brain which is not being used? By now we know that "to think" is to set up a very complex electrical process in the brain. This is transmitted to organs and muscles by means of the electric action potential. The electro-neural connections of the human brain, with the rest of the body, constitutes an extremely complex electrical-network. This has been recently discovered by Dr Björn Nordenström [49, 1983] of Stockholm, Sweden, who has shown experimentally the existence of an electrical circulatory system in human beings and animals. Nordenström told this author that "*Life is an Electromagnetic Phenomenon.*" Life, consciousness and the entire universe are electromagnetic processes. We highly recommend Nordenström's book, plus two extraordinary books written by Dr. Robert Becker which are titled *The Body Electric* [50, 1985], and *Cross Currents* [51, 1992]. In the opinion of the author of *Einstein on Trial,* 90% of the human brain is not used because in the embryonic process, the local electrodynamic environment is not intense enough to establish the many electro-neural connections. Because of this natural electrical condition, 90% of the human brain was left unconnected. The active brain of humans has no access to this disconnected 90% of the human brain. However, there is hope for our grandchildren.

In chapters 3 and chapter 4, we learned that all the electrodynamic force-terms are divided by the square of the speed of light; i.e., by c^2. Thus, if the speed of light decreases, then all the electrodynamic force-terms will increase. The biological effect of this "enlightening" phenomenon "may" increase the electro-neural connections in the brain of human embryos. Here we have a beautiful example of another metaphysical allegory, unless we bring pragmatic evidence that the speed of light has begun to decrease. We know that the solar system, according to *astrologers,* entered the Aquarius Zodiac Sign in May, 1948. What is the meaning of this esoteric statement of *Astrology* from the point of view of Astrophysics? If the solar system has entered a new region (Aquarius region) in our galaxy in which the interstellar density of energy is increasing, then the speed of light must be decreasing. The increase in energy density must be caused by a closer proximity of our sun to other stars in our galaxy. But where is the pragmatic evidence that the speed of light is decreasing?

In 1987, Norman Trevor and Barry Setterfield [52], from Australia, published a report showing that the speed of light has been decreasing. For an update the reader should consult the article, *Is the Velocity of Light Constant in Time?* by Alan Montgomery and Lambert Dolphin [53, April 15, 2000]. If new measurements of the speed of light corroborate these initial calculations, then our metaphysical metaphor has high probabilities of becoming physical, experimental, pragmatic evidence. Hence, in the present, in the following years, and in the following centuries, we should observe human babies and young children revealing more intelligence than their parents. When human intelligence evolves enough, then and only then can an Advanced Religion descend on planet earth in the hands of the envisioned "indigo children," who will know who they will worship.

Conclusions.

The conclusions of this chapter will be divided into three groups: Introduction, Einstein's Theological Beliefs, and Scientific Theology. From each group, we will emphasize only the most important topics.

CONCLUSIONS ABOUT THE INTRODUCTION

I-1. Science and Religion have a Common Foundation.
I-2. Theoretical Scientific Knowledge is Relative.
I-3. Elements of Theory of Knowledge.
I-4. Theology by Revelation and Theology by Reason.

One concept that should be overemphasized, again and again, and put into the minds of the young generation of the 21st century, is that any rational theory, philosophical or scientific, must be built on irrational principles, axioms, postulates or dogmas. These fundamental blocks of any theory cannot be demonstrated, cannot be deduced logically from anything. Principles are revealed transrationally, or generalized from sensorial images that are nonrational. Thus, any scientific theory is as dogmatic as any religion. Both start with dogmas of faith. There must be unbroken faith that the scientific principles or the religious dogmas are true.

In Greek language, the word *theory* means play, theatrical play. Thus a scientific theory is a representation on the stage of a noumenal theater narrating the behavior of natural things and natural events. The same pertains to any philosophical theory. The main difference between these theories is the criterion used to pragmatically verify their theoretical conclusions. Anyway, both types of theories provide new knowledge which is relative to the foundations of the theories: the principles, axioms, postulates or dogmas. The advocates of any theory, "in decaying process," irrationally prolongs its agony.

The apprehension of knowledge by the human mind is accomplished holistically or fractionally. Holistic knowledge is acquired tranrationally, intuitively, mystically or by religious revelation. This holistic knowledge is synthetic, and corresponds to Kant's *a priori* synthetic judgments. Holistic knowledge is a universal judgment known as principle. On the other hand, fractional knowledge is sensorial, nonrational or pseudo factual which constitutes ideas or phenomena which we perceive from nature. Later we create generalizations of this fractional or particular knowledge. These generalizations are the conceptual judgments which are universal, but are created by the human mind. Thus, by generalizations (induction), we create principles from the raw sensorial particular data of nature. In the 21st century, there will be an epistemological necessity to teach students of science that great human minds of the centuries acquired their transcendental knowledge by transrational means, by intellectual and religious revelations. Time has run out for the arrogant cast of physicists from the 20th century to learn to say, *we know we have **to believe** in the foundations of science in order to create scientific theories.*

Any theology or theory of God is a rationalization of revealed dogmas or principles created by the human mind. Thus, we have Theology by Revelation and Rational or Natural Theology. The practical applications of theologies should be called *religion.* We may say that religion is the technology of theology, as engineering or technology is the application of science. Any theology should teach the procedures to reconnect (*religare*, from which the word *religion* is derived) the individual human consciousness with the Universal Consciousness. Also, any theology should teach moral rules of social behavior among human beings. Thus, any religion has this incredible mission of teaching human beings to be rational (intelligent) creatures, to become *homo sapiens.*

CONCLUSIONS ABOUT EINSTEIN'S THEOLOGICAL BELIEFS

In this group of five sections of chapter 7, we commented on four essays written by Einstein on Science and Religion. From these essays, we analyzed Einstein's beliefs on theology.

In Einstein's essay of 1930, we can see that Einstein believed in the existence of God. In this essay, Einstein does not rationally prove the existence of God, rather, he resorts to a transrational or mystical way to know that God exists. Einstein called this transrational way *cosmic religious experience*. In this essay, Einstein expresses his belief that man is not completely responsible for his actions.

In his essay of 1934, Einstein reiterates his belief that morality has no divine source. We must remember that Ethics is the theory of morality; i.e., the rationalization of the actual social-moral behavior of humans. We think that Einstein, after he read the *Ethics* written by Spinoza in Euclidean logical style, realized that Ethics, like any other theory, is a mere invention of humans, and therefore, morality only has a human source. Obviously, terrestrial authorities (politicians) cannot let the *average* human being know of this type of theological speculation. Instead, they prefer to present an "angry god" who eternally punishes the poor devils who disobey "God's divine moral laws."

Einstein's God is not anthropomorphic but the *Supreme Intelligence*, as he calls It. Einstein drew this attribute of God from the harmony he found in the natural laws which govern the entire rational universe. Evidently, the belief in a "rational universe" is another scientific and philosophical dogma of faith. Without this faith, science is impossible.

In his essay of 1941, we saw that Einstein believed that the purpose of human existence cannot be deduced rationally, but should be received by revelation. Revelation, for Einstein, is the acquisition of knowledge by means of a cosmic religious experience, i.e., transrationally. Einstein also believed that *faith was at the root of religion and science*. Let us emphasize that the *italics* we used in the previous sentence is to enhance our total agreement with Einstein about the foundations of science and religion. The foundations of both are faith as we showed in the Introduction of chapter 7. In this essay,

Einstein again reiterates his belief that morality is not divine but a human matter. Einstein rejects, emphatically, the concept of a personal God, and points to it as one of the main causes of disagreement between science and religion. On this point, the author is in complete disagreement with Einstein. By "personal God" we may interpret the phrase in two ways: (1) that humans may have "a personal God" to communicate with, and (2) that God is Person.

With respect to interpretation (1), if humans beings cannot have a personal God to communicate or reconnect (*religare*) with, then all theologies and religions are mere fairy tales or metaphors. However, being that God is Universal Consciousness, it implies *Universal Interaction* of the *Unbroken Wholeness of the Entire Universe* with one or all entities of the universe. This interaction implies the *action* of the Whole on one individual entity of the Universal System, and the reaction of the one individual entity with the Whole or Universal Consciousness. This author is convinced that on this point Einstein is wrong.

In respect to interpretation (2), this author feels that again Einstein is mistaken. In both interpretations, Einstein is not right because he lacked knowledge about the concepts of consciousness and person. This was proven in the latter part of chapter 7 when we defined *person* as an entity conscious of itself. Finally, Einstein believes that rational knowledge will eventually provide humans with a path to true religiosity. Here is where we see Einstein's vision of a scientific theology or cosmotheism. The interested reader should consult chapter 2 of the recent book *Einstein and Religion* [55] by Max Jammer. Here we may analyze the many theological proofs about the existence of a "personal God." These are not scientific proofs.

In his 1948 essay, Einstein noted that the mythical or allegoric religious tradition is another cause for the conflict between religion and science. In this fourth essay, Einstein, in a rather surreptitious manner, identifies nature or the universe with God. Thus, for Einstein, Spinoza or any pantheist, *God is the universe*. Finally, Einstein believes that the minds of a few creators of scientific theories is flooded *"with the truly religious conviction that this universe of ours is something perfect and susceptible to the rational striving for knowledge."* Previously, we mentioned that the belief in a rational universe is another dogma of faith in science and philosophy. Now we read that the rationality of the universe is absolute knowledge, which comes through a cosmic religious experience, through a transrational altered state of the human mind. We can also reach this absolute knowledge in a relative way. If the universe is God, and God is, according to Einstein, *The Supreme Intelligence*, then the universe, by logical necessity, must be intelligent, logical, rational. Once more this author reads, between Einstein's written lines, his message of a future scientific theology and an advanced religion. In conclusion, Einstein never was an atheist, but rather a pantheist, and after studying Spinoza's pantheism, we have to conclude that both of these two thinkers have similar theological beliefs.

In these sections, we find Einstein guilty of nothing. However, more than one religious fanatic believer has accused Einstein of being an atheist. We must emphasize, Einstein was not an atheist but a pantheist, a man who sees God in all (pan = all, in Greek). Some other fanatic believers have said that a pantheist has never had the time to become an atheist. If Einstein had had the time to think about the concepts of *consciousness* and *person*, he would have created a scientific theology, and would have been the first cosmotheist in the past seven millennia. We may refer to these seven millennia as the long dark nights of the human soul.

CONCLUSIONS ABOUT SCIENTIFIC THEOLOGY

7.6. Science, Including Logic, are not Completely Rational.
7.7. Foundations of Scientific Theology or Cosmotheism.
7.8. Does God Exist in Reality?
7.9. Is God Universal Consciousness?
7.10. Advanced Religion.

We initiate this set of sections insisting, once more, that any scientific or philosophical theory (including logic), must be founded upon unproven principles. This is a lesson no scientist or philosopher (including theologians) should ever forget. Any rational theory has to be raised on irrational or transrational principles. Theology cannot escape this appalling truth. In chapter 1, we concluded that the purpose of any scientific theory is to explain the seemingly unexplained reality, starting with unexplainable principles. But reality is the world of visible material things, and therefore, "seeing is believing" in the existence of the *res*, of the things, which we perceive. However, the concept of God is just a **concept**, which has only a mental existence. The first problem to be solved by any Rational Theology is the existence of God in reality, outside the conceptual world of the human mind. This is the most important problem for a cosmic race who believes it is intelligent. For obvious reasons, this problem does not exist for any Revelational Theology. In the western world, we have had four different "rational" methods to prove the actual existence of God: ontological, cosmological, teleological and moral. In the opinion of this author, any of these methods offers a pitiful spectacle of **verbiage**. All these so-called rational arguments about the existence of God have remained in the realm of metaphysics, without any possibility to substantiate them with pragmatic evidence. For these reasons, we decided to pursue Einstein's vision of developing a fifth path to prove the real existence of God. We called this path Scientific Theology or Cosmotheism. We did this hoping to bring pragmatic evidence that God **is** The Unbroken Wholeness of the Entire Universe, independent of the human conceptual existence of God. That God **is** The Being of all entities, visible material entities and invisible immaterial entities.

In the process of establishing the foundations of our Scientific Theology, we found the taxonomic necessity of reclassifying Andronicus' ancient gnoselogical scheme. In this new classification, two aspects are very important. One is that Aristole's First Philosophy constitutes a methodology to build the scaffolding of philosophical theories and scientific theories. The second aspect is the recognition by different scientists that not only can religious knowledge be acquired transrationally, but scientific knowledge can also be acquired through mystical experiences. Thus, we presented a scientific theology based on substantial or ontological principles, formal or logical principles, and scientific principles from which we used many scientific theoretical conclusions verified experimentally, empirically or through pragmatic evidence.

In chapter 5, we scientifically proved that the universe **is energy**. In other words, the Being of all entities is energy: visible concentrated energy and invisible subtle energy. Thus, if the purpose of any rational theology is to determine the nature, the essence, the substance (being) of God, then this scientific theology establishes that God **is cosmic energy**. But this scientific conclusion, presented in this book of the third millennium of the new era, is nothing new. Carlo Suarès, in his magisterial book [xx] *The Sepher Yetsira*, one of the fundamental textbooks of the Kabbala (*Qabala)*, wrote in its Introduction: "*I begin our study by saying that the Qabala is a science and that The Sepher Yatsira is a precise and accurate treatise on the structure of* ***cosmic energy****, written in a hidden code.*" (The boldface letters are ours.) In the next page, Suarès, writing about the accomplishments of 20th century science, says: "*is contending with the mystery of mysteries, the phenomenon of consciousness, about which it admits it understands nothing.*" This is exactly the same conclusion this author reached after one solid year dedicated exclusively to finding something comprehensible about consciousness.

Another purpose of rational theology is to determine the attributes or properties of God. In this respect, this author transcepted a relational conception between God, consciousness and dynamical interaction between material entities. Solid epistemological help came from Pierre Teilhard de Chardin, S.J., through his masterpiece, *The Phenomenon of Man*, which was in relation to the concept of *consciousness*. The fact that this concept is pervasive throughout his book, it is difficult to apprehend the essence of this mystery of mysteries, according to Suarès. However,

Julian Huxley who wrote the English introduction to Teilhard's book clarified, almost completely, Teilhard's conception of consciousness. The inclusion of the concept *interaction* in Teilhard-Huxley's definition of consciousness was an easy task for this author. In this way, we liberated the concept *consciousness* from seven thousands years of bondage under the rigid human consciousness. Chapter 2, of course, was fundamental in the identification of the quantum collective potential energy field. Finally, chapter 6 is the other supporting column of the cosmic collective potential. Thus, we scientifically concluded the existence of *Universal Consciousness*, identified with *God*, identified with *Supreme Intelligence*.

Another purpose of rational theology is to determine the *activities* of God. In the opinion of this author, this is an outrageous and presumptuous objective for any human rational theology. Evident examples are the western theologians of the middle age's who lost themselves in a jungle of meaningless sentences. Perhaps, when the average intelligence or consciousness of humans evolves to higher levels, according to Teilhard de Chardin, our children's children will be able to establish an Advanced Religion which will allow them to reconnect their evolved human consciousness or spirit with the Universal Consciousness, without running the risk of losing their minds.

In this last chapter, we found Einstein not guilty of his theological beliefs. On the contrary, we acknowledge Einstein's vision and motivation for this author to develop the foundations of an incipient scientific theology, or cosmotheism.

(Note of the editor: See VEREDICT after the References)

References.

1 A. Einstein, *Ideas and Opinions* (Crown Publishers, Inc., NY, NY, 1982)

2 Nicolas Berdyaev, The *Beginning and the End* (Harper & Brothers Publishers, NY, 1957)

3 Nick Herbert, *Elemental Mind or Human Consciousness and the New Physics* (A Dutton Book, a Division of Penguin Books USA Inc., NY, 1993)

4 William James, *The Varieties of Religious Experiences* (Collier Books, NY, Collier MacMillan Publisher, London, 1979)

5 Ref. 4

6 William James, *Essays in Pragmatism* (Hafner Press, a Division of MacMillan Publishing Co., Inc., NY, London, 1948)

7 Bertrand Russell, *The Problems of Philosophy* (Oxford University Press, London, Oxford, NY, 1973)

8 Ref. 4

9 Pope John Paul II, *Fides et Ratio* (Encyclical Letter to the Bishops of the Catholic Church, 1998)
10 Pope John Paul II, *Galileo Case* (Weekly English Edition of *L'Osservatore Romano*, N. 44, 4 November, 1992)
11 Gabriel Moran, F.S.C., *Theology of Revelation* (Herder and Herder, NY, 1966)
12 *The Cambridge Dictionary of Philosophy* (General Editor Robert Audi, Cambridge University Press, Cambridge, U.K., 1995, and NY, 1996)
13 D.T. Susuki, *The Essentials of Zen-Buddhism: Selected from the Writings of Daisetz Teitaro Susuki* (Greenwood Press Publisher, Westport, Connecticut, 1973)
14 Carl G. Jung, in *The Secret of the Golden Flower: A Chinese Book of Life* (Translated and explained by Richard Wilhelm, A Harvest Book, A Helen Kurt Wolff Book, Harcourt Brace & Company, San Diego, NY, London, 1962)
15 Ref. 12
16 Frederick Coplestone, S.J., *A History of Philosophy, Vol. IV* (The Newman Press, Westminster, Maryland,1959)
17 Frederick Coplestone, S.J., *A History of Philosophy, Vol. III* (The Newman Press, Westminster, Maryland,1953)
18 Gerald Holton, *Thematic Origins of Scientific Thoughts* (Harvard University Press, Cambridge, Massachusetts, and London, England, 1988)
19 I. Newton, *Optiks* (Dover Publications, Inc., NY, 1979)
20 H.P. Blavatsky, *The Secret Doctrine* (Adyar: Theosophical Publishing House, 1971)
21 Sylvia Cranston, *The Extraordinary Life and Influence of Helena Blavatsky, Founder of the Modern Theosophical Movement* (A Jeremy P. Tarcher/Putnam Book, published by G.P. Putnam's Sons, NY, 1994)
22a D.L. Bergmann and J.P. Wesley, Galilean Electrodynamics, **1**, 63 (1990) [Spinning Charged Ring Model of Electron]
22b Steven Rado, *Aethro-Kinematics* (http://www.aethro-kinematics.com/)
23 Vladimir B. Ginzburg, *Spiral Grain of the Universe* (Amazon.com)
24 Vladimir B. Ginzburg, *Unified Spiral Field and Matter* (Amazon.com)
25 Alice A. Bailey, *The Consciousness of the Atom* (Lucis Publishing Company, NY, and Lucis Press Ltd., London, 1981)
26 Thomas Metzing, quoted in the *Journal of Consciousness Studies*, vol. 4, No.5-6, p. 385 (1997).
27 Rick Grush, *Quantum consciousness, theory of*, (http:// www.artsci.wustl.edu/ ~philos/MindDict/ quantum.html, 1999)

28 M. Gell-Man, *The Nature of Matter, Wolfson College Lecture, 1980* (Clarendon Press, Oxford, 1981)
29 Henry P. Stapp. *Mind, Matter, and Quantum Mechanics* (Springer-Verlag, Berlin, NY, London, Paris, Budapest, Tokyo)
30 Pierre Teilhard de Chardin, *The Phenomenon of Man* (Harper & Row , Publishers Inc. 1959)
31 Pierre Teilhard de Chardin, *El Grupo Zoológico Humano* (Taurus Ediciones S.A. Madrid, España 1957)
32 Ref. 31
33 Joseph P. Provenzano, *The Philosophy of Conscious Energy* (Winston-Derek Publishers, Inc., Nashville, Tennessee, 1993)
34 Ken Wilber, *The Spectrum of Consciousness* (The Theosophical Publishing House, Wheaton, Ill, U.S.A., Madras, India/London, England)
35 David Bohm, *Wholeness and the Implicate Order* (Ark Paperbacks, London, Boston, Melbourne and Henley, 1980)
36 J.C. Maxwell, *A Treatise on Electricity and Magnetism* (Dover Publications Inc., NY, 1954)
37 D. Bohm and B.J. Hiley, *The Undivided Universe: An Ontological Interpretation of Quantum Theory* (Routledge, 11 New Fetter Lane, London EC4P 4EE, and 29 West 35th Street, New York, NY, 1993)
38 J.C. Curé, Health Consciousness, vol. XII, October 1991, Part 1 of 3, p. 83 [Cancer: An Electrical Phenomenon]
39 C. Louis Kervran, *Biological Transmutations* (Beekman Publishers, Inc., Woodstock, N.Y., 1980. 0riginally published in French in 1966 by Le Courrier de Livre)
40 Robert B. Stone, *The Secret Life of Your Cells* (Witford Press, a Division of Shiffer Publishing Ltd., Atglen, PA, USA.,1989. E-mail: shifferbk@aol.com)
41 Peter Tompkins and Christopher Bird, *The Secret Life of Plants* (Harper & Row, Publishers, NY, London, Sydney, 1974)
42 Ref. 20
43 Richard S. Westfall, *Never at Rest: A Biography of Isaac Newton* (Cambridge University Press, Cambridge, New Rochelle, Melbourne, Sydney, 1987)
44 Shirley Nicholson, *Ancient Wisdom - Modern Insight* (The Theosophical Publishing House, Wheaton, Ill, U.S.A., Madras, India/London, England, 1985)
45 Julian Jaynes, *The Origin of Consciousness in the Breakdown of the Bicameral Mind* (Houghton Mifflin Company, Boston, 1976)
46 Deepak Chopra, *How to Know God* (Harmony Books, NY, 2000)

47 Paul Davies, *The Mind of God: The Scientific Basis for a Rational World* (A Touchstone Book, Published by Simon & Shuster, NY, London, Sydney, Tokyo, 1993)
48 Ref. 14
49 Björn Nordenström, *Biologically Closed Electric Circuits* (Nordic Medical Publications, Stockholm, 1983)
50 Robert O. Becker, *The Body Electric* (William Morrow and Company, Inc., NY, 1985)
51 Robert O. Becker, *Cross Currents* (Jeremy P. Tarcher, Inc., LA, CA, 1990)
52 Norman Trevor and Barry Setterfield, Stanford Research Institute-International Invited Report, August 1987 [Title of the Report: *The Atomic Constants, Light, and Time from Flinders University of South Australia 1987*]
53 Alan Montgomery and Lambert Dolphin, *On the constancy of the Speed of Light*, **Website** http:/ldolphin.org/ (Published by Lambert Dolphin, April 15, 2000) [Is the Velocity of Light Constant in Time?]
54 Carlo Suarès, *The Sepher Yetsira* (Shambhala Publications, Inc., Boulder, CO & London, 1976)
55 Max Jammer, *Einstein and Religion: Physics and Theology* (Princenton University Press, Princeton, NJ, 1999)

VERDICT

The final verdict of the book *Einstein on Trial* is that Einstein is not guilty of any of his conceptions about the universe. The sufficient reason we have to explain this verdict is that no creator can be guilty of his creations. If we have to uncover guilt, in the 20th century, for the ontological crimes committed in the temple of Natural Philosophy, they are the followers of SRT. Einstein was an ontological victim of the philosophical ignorance of the 19th century physicists. SRT solved transitorily an urgent problem at the beginning of the 20th century. The behavior of all advocates of SRT is a typical example, in the history of science and philosophy of this planet, of a natural inclination of terrestrial beings to submit their mental capacities to the bondage of an intellectual master. This submission endures until another intellectual master appears to slave the minds of future followers. Some day, we hope this bluish planet will be inhabited by true thinkers. However, we did find Einstein guilty in all counts relating to his false accusations against Newton's *Principia*. Perhaps, we can understand this second verdict if we concentrate on Einstein's uncritical acceptance of Mach's attacks against Newton's *Principia*. Finally, we found Einstein not guilty in his accusation against Quantum Mechanics. Einstein was always right in saying that quantum theory is *incomplete* because it cannot explain a simple atomic system. The ontological Newtonian quantum collective potential energy field has exonerated Einstein from the unfair and unscientific exile he was forced to take from the quantum mechanics establishment. Let us finish this verdict with some thoughts this author expressed in San Francisco, California, USA, on the occasion of a gathering of dissident physicists. We constituted an heterogeneous segregated group at a meeting of the American Society for the Advancement of Science, in 1994. This author said:

> *"Let us never forget that we are here today because of Professor Albert Einstein. Let us admit that without Einstein our academic and scientific existence would have been extremely boring. Let us thank Einstein for his creative endeavors."*

Einstein was a devoted admirer of Baruch Spinoza (1632 - 1677). Any person who has thoroughly studied *The Elements* of Euclid (3rd century, BC), learns Logic through the theorems of geometrical figures. Classical Euclidean Geometry is perhaps the oldest branch of mathematics that transfigures the chaotic mind of youngsters into rational minds. But Geometry, the way French people teach the formal science of Euclid, forces the minds of youngsters to develop *their intuitive minds* next to the *analytical minds* of young students. No wonder Einstein said that if Euclidean geometry "*failed to kindle your youthful enthusiasm, then you were not born to be a scientific thinker.*" We can imagine only the overwhelming spiritual rejoicing of Einstein's mind when he read and studied the most rational *Ethics*, written by Spinoza, in the formal, intelligent and logical mode of Euclidean geometry. Only those who numinously enjoyed studying geometry when they were young, can feel the ineffable experience of studying the *Ethics* of Spinoza. Einstein's admiration and veneration for Spinoza, metaphysician, epistemologist, psychologist, moral philosopher, philosopher of religion and political theorist of the 17th century, can explain the reason Einstein had to write, in1920, a poem dedicated to this Jewish scholar. The poem is "Zu Spinozas Ethik." To close this verdict, this author will quote the first two lines of this poem, extracted from *Einstein and Religion* written by Max Jammer [55], and dedicated to Einstein's spirit:

"How much do I love this noble man

More than I could tell with words . . . "

EPILOGUE

This book is basically a collection of epistemological trials, which is a compilation of material written over a period of a little more than twenty-five years. The essential purpose of this collection, which consists of the thoughts of many of our forerunners, was to extract the essence of their thoughts and rethink them anew. This was done in order to unite the best aspects of all into a harmonious oneness. In the end, the author realized the epistemological synthesis was possible only after an ontological one. The **Being of all entities** had been identified with **energy**, as it was done in the ancient wisdom of the Kabbala, long before Parmenides introduced the concept of the *common substance* of all entities. After thousands of years we identified again the Being of all entities with energy. This identification was done in chapter 5 of this book. Once again we can say *nothing is new under the sun.*

The *Principle of Inseparability,* which we presented in chapter 1, and used it again in chapter 5 after translating it into the powerful language of *Mathematics,* taught us that every visible nugget of condensed energy is connected to all the other nuggets of condensed energy in the universe. We learned in chapters 2 and 5 that this invisible link is also energy, *subtle energy.* However, without the appearance of the ontological collective quantum potential energy field, which we unveiled in chapter 2, it is rationally impossible to understand the identification of the Universal Consciousness with its invisible immaterial being. This *union* or *common-union* (communion) of the *Whole that Is,* is the true and real unity of the entire universe. Now, if we ask a mystic what love is, the mystic will answer: *Love is the invisible essence of unity.* Finally, combining physics and mysticism; i.e., blending western rationalism and eastern tranrationalism, we offer a scientific understanding of spiritual and cosmic religious experiences: the Unbroken Wholeness of the Entire Universe, the Being of All Entities, the Universal Consciousness, God, is Energy, is the Invisible Essence of Unity, is Cosmic Inseparability, is Universal Common Union, is Cosmic Love.

Alphabetical Index

A

B

C

D

E

H

N

P

Q

R

S

Made in the USA
Middletown, DE
28 July 2025

11317989R00245